Quantum Dot Heterostructures

Dieter Bimberg
Marius Grundmann
Nikolai N. Ledentsov
Institute of Solid State Physics,
Technische Universität Berlin,
Germany

JOHN WILEY & SONS

Chichester • New York • Weinheim • Brisbane • Singapore • Toronto

Copyright © 1999 John Wiley & Sons Ltd,
Baffins Lane, Chichester,
West Sussex PO 19 1UD, England

National 01243 779777
International (+44) 1243 779777
e-mail (for orders and customer service enquiries): cs-books@wiley.co.uk
Visit our Home Page on http://www.wiley.co.uk
or http://www.wiley.com

All Rights Reserved. No part of this publication may be reproduced, stored in a retrieval system, or transmitted, in any form or by any means, electronic, mechanical, photocopying, recording, scanning or otherwise, except under the terms of the 'Copyright Designs and Patents Act 1988 or under the terms of a licence issued by the Copyright Licensing Agency, 90 Tottenham Court Road, London W1P 9HE, UK, without the permission in writing of the Publisher

Other Wiley Editorial Offices

John Wiley & Sons, Inc., 605 Third Avenue,
New York, NY 10158-0012, USA

WILEY-VCH Verlag GmbH, Pappellallee 3,
D-69469 Weinheim, Germany

Jacaranda Wiley Ltd, 33 Park Road Milton,
Queensland 4064, Australia

John Wiley & Sons (Asia) Pte Ltd, Clementi Loop #02-01,
Jim Xing Distripark, Singapore 129809

John Wiley & Sons (Canada) Ltd, 22 Worcester Road,
Rexdale, Ontario M9W 1L1, Canada

Library of Congress Cataloging-in-Publication Data

Bimberg, Dieter.
    Quantum dot heterostructures / Dieter Bimberg, Marius Grundmann,
    Nikolai N. Ledentsov.
        p.   cm.
    Includes bibliographical references and index.
    ISBN 0-471-97388-2 (alk. paper)
    1. Quantum dots. 2. Heterostructures. I. Grundmann, Marius.
II. Ledentsov, Nikolai N. III. Title.
TK7874.88.B55   1988 1999
621.3815′2—dc21
                                                          98-7734
                                                            CIP

British Library Cataloguing-in-Publication Data

A catalogue record for this book is available from the British Library

ISBN 0 471 97388 2

Typeset in 10pt Times by Techset Composition Ltd., Salisbury, Wiltshire, England
Printed and bound in Great Britain by Biddles Ltd, Guildford, Surrey
This book is printed on acid-free paper responsible manufactured from sustainable forestry, in which at least two trees are planted for each one used for paper production.

Contents

Preface ix

1 Introduction 1
  1.1 HISTORICAL DEVELOPMENT 1
    1.1.1 From atoms to solids 1
    1.1.2 From solids to 2D heterostructures 1
    1.1.3 From 2D heterostructures to quantum dots 3
  1.2 BASIC REQUIREMENTS FOR QDS IN ROOM TEMPERATURE DEVICES 6
    1.2.1 Size 6
    1.2.2 Uniformity 7
    1.2.3 Material quality 8

2 Fabrication Techniques for Quantum Dots 9
  2.1 QUANTUM DOTS FABRICATED BY LITHOGRAPHIC TECHNIQUES 9
    2.1.1 Overview 9
    2.1.2 Free-standing quantum dots 12
    2.1.3 Selective intermixing based on ion implantation 15
    2.1.4 Selective intermixing based on laser annealing 16
    2.1.5 Strain-induced lateral confinement 16
    2.1.6 Quantum dots grown on patterned substrates 17
    2.1.7 Cleaved edge overgrowth 19
  2.2 QUANTUM DOTS FORMED BY INTERFACE FLUCTUATIONS 19
  2.3 SELF-ORGANIZED QUANTUM DOTS 19

3 Self-Organization Concepts on Crystal Surfaces 22
  3.1 INTRODUCTION 22
  3.2 SPONTANEOUS FACETING OF CRYSTAL SURFACES 24
    3.2.1 The problem of equilibrium crystal shape 24
    3.2.2 The concept of intrinsic surface stress of a crystal surface 26
    3.2.3 Force monopoles at crystal edges 28
    3.2.4 Spontaneous formation of periodically faceted surfaces 29
  3.3 PERIODIC ARRAYS OF MACROSCOPIC STEP BUNCHES 32
  3.4 HETEROEPITAXIAL GROWTH ON CORRUGATED SUBSTRATES 33
  3.5 ORDERED ARRAYS OF PLANAR SURFACE DOMAINS 35
  3.6 ORDERED ARRAYS OF THREE-DIMENSIONAL COHERENTLY STRAINED ISLANDS 37
    3.6.1 Heteroepitaxial growth in lattice-mismatched systems 37
    3.6.2 Energetics of a dilute array of 3D islands 41

- 3.6.3 Ordering of 3D islands in shape 43
- 3.6.4 Ordering of 3D islands in size versus Ostwald ripening 44
- 3.6.5 Lateral ordering of 3D islands 48
- 3.6.6 Phase diagram of a two-dimensional array of 3D islands 49
- 3.6.7 Ordering-to-ripening phase transition for 3D islands 51
- 3.6.8 Kinetic theories of ordering 51
- 3.6.9 Equilibrium ordering versus kinetic-controlled ordering of 3D islands 52
- 3.7 VERTICALLY CORRELATED GROWTH OF NANOSTRUCTURES 53
  - 3.7.1 Vertically correlated growth of nanostructures due to a modulated strain field 53
  - 3.7.2 Energetics of Stranski–Krastanow growth for subsequent InAs and GaAs deposition cycles 54
  - 3.7.3 Strain energy of vertically coupled structures 56

4 Growth and Structural Characterization of Self-Organized Quantum Dots 59
- 4.1 INTRODUCTION 59
- 4.2 MBE OF INGAAS/GAAS QUANTUM DOTS 61
  - 4.2.1 Deposition below the 2D–3D transition 61
  - 4.2.2 2D–3D transition 63
  - 4.2.3 Quantum dot structure 64
  - 4.2.4 Hierarchy of self-organization mechanisms 69
  - 4.2.5 Influence of deposition conditions 71
- 4.3 MOCVD GROWTH OF INGAAS/GAAS QUANTUM DOTS 76
  - 4.3.1 2D–3D transition 76
  - 4.3.2 Quantum dots 78
  - 4.3.3 High index substrates 79
- 4.4 OTHER MATERIAL SYSTEMS 80
  - 4.4.1 InP on GaInP/GaAs 80
  - 4.4.2 Sb compounds 82
  - 4.4.3 Ge on Si 82
  - 4.4.4 III–V on Si 84
  - 4.4.5 II–VI compounds 84
  - 4.4.6 Group III nitrides 85
- 4.5 VERTICAL STACKING OF QUANTUM DOTS 86
- 4.6 ARTIFICIAL ALIGNMENT OF QUANTUM DOTS 92

5 Modeling of Ideal and Real Quantum Dots 95
- 5.1 STRAIN DISTRIBUTION 95
  - 5.1.1 Stress–strain relations 96
  - 5.1.2 Strain in dots 96
  - 5.1.3 Valence force field model 107
  - 5.1.4 Impact on band structure 111
  - 5.1.5 Impact on phonon spectrum 113
  - 5.1.6 Piezoelectric effects 114

- 5.2 QUANTUM CONFINEMENT 115
  - 5.2.1 Particle in a harmonic potential 115
  - 5.2.2 Particle in a sphere 116
  - 5.2.3 Particle in a cone 118
  - 5.2.4 Particle in a pyramid 119
  - 5.2.5 Particle in a lens 122
  - 5.2.6 Stress-induced quantum dots 122
  - 5.2.7 Twofold cleaved edge overgrowth 124
  - 5.2.8 Eight-band k·p theory and pseudopotential methods 125
- 5.3 COULOMB INTERACTION 128
  - 5.3.1 Excitons 128
  - 5.3.2 Type-II excitons 143
  - 5.3.3 Electronically coupled quantum dots 145
  - 5.3.4 Coulomb blockade 146
- 5.4 OPTICAL TRANSITIONS 148
  - 5.4.1 Single quantum dots 148
  - 5.4.2 Quantum dot ensembles 150
  - 5.4.3 Inter-sublevel transitions 154
- 5.5 POPULATION OF LEVELS 155
  - 5.5.1 Thermal versus nonthermal distribution 155
  - 5.5.2 Population statistics 158
  - 5.5.3 Auger effect 165
- 5.6 STATIC EXTERNAL FIELDS 165
  - 5.6.1 Electric fields 167
  - 5.6.2 Magnetic fields 170
- 5.7 PHONONS 174
  - 5.7.1 Phonon spectrum 174
  - 5.7.2 Electron–phonon interaction 174
- 5.8 QUANTUM DOT LASER 177
  - 5.8.1 Basic properties of quantum well lasers 178
  - 5.8.2 The ideal quantum dot laser 181
  - 5.8.3 The real quantum dot laser 182

6 Electronic and Optical Properties 198
- 6.1 ETCHED STRUCTURES 198
- 6.2 LOCAL INTERMIXING OF QUANTUM WELLS 201
- 6.3 STRESSORS 202
- 6.4 SELECTIVE GROWTH 208
- 6.5 EXCITONS LOCALIZED IN QUANTUM WELL THICKNESS FLUCTUATIONS 212
- 6.6 TWOFOLD CLEAVED EDGE OVERGROWTH 220
- 6.7 SELF-ORGANIZED TYPE-I QUANTUM DOTS 222
  - 6.7.1 Luminescence 223
  - 6.7.2 Absorption 239
  - 6.7.3 Electroreflectance 241

        6.7.4   Fourier spectroscopy of inter-sublevel transitions   243
        6.7.5   Carrier dynamics   246
        6.7.6   Vertically stacked quantum dots   252
        6.7.7   Annealing of quantum dots   255
   6.8 SELF-ORGANIZED TYPE-II QUANTUM DOTS   260
        6.8.1   CW properties   260
        6.8.2   Time-resolved experiments   262

7 **Electrical Properties**   265
   7.1 CV SPECTROSCOPY   265
   7.2 DLTS   269
   7.3 VERTICAL TUNNELING   270
   7.4 LATERAL TRANSPORT   274

8 **Photonic Devices**   277
   8.1 PHOTO-CURRENT DEVICES   277
   8.2 QUANTUM DOT LASER   279
        8.2.1   History   279
        8.2.2   Static laser properties   281
        8.2.3   Dynamic laser properties   291
        8.2.4   VCSEL   295
        8.2.5   Inter-sublevel IR laser   299

References   303

Index   325

Preface

Quantum dots, coherent inclusions in a semiconductor matrix with truly zero-dimensional electronic properties, present the utmost challenge and point of culmination of semiconductor physics. Their properties resemble those of atoms in an electromagnetic cage, rendering possible fascinating novel devices.

It was at the beginning of the 1990s that a modified Stranski–Krastanow growth mechanism driven by self-organization phenomena at the surface of strongly strained heterostructures was realized for the fabrication of such dots. This process presents a sound way to fabricate easily and fast large densities of quantum dots. A rapidly increasing number of leading laboratories around the world embarked on the investigation and modeling of the growth, the physical properties, and device applications of the numerous possible material combinations.

Many fundamental facts and phenomena are now at least qualitatively understood, but no comprehensive survey exists to guide newcomers to the field. This book tries to fill this gap. It focuses on phenomena and principles. With regard to collecting *all* existing experimental material it is as incomplete as such a work in a rapidly progressing field necessarily must be.

In Chapter 1 a brief account of the history of quantum dots is given and basic requirements on their properties for making them useful in devices operating at room temperature are formulated. Chapter 2 surveys various alternative techniques used in the past decade to fabricate quantum dots. The chapter ends by introducing the concept of self-organized growth. The following chapter extends this subject and presents a broad review of thermodynamically driven self-organization phenomena at surfaces of crystals.

Results on growth for a number of different quantum dot structures and on their structural characterization are presented in Chapter 4. Knowledge of the geometric structure and chemical composition of dots is a prerequisite for numerically modeling the electronic and optical properties of real dots. Such modeling is presented in Chapter 5, together with general theoretical considerations on carrier capture, relaxation and properties of quantum dot lasers.

Experimental results on electronic and optical properties are summarized in Chapter 6, followed by a rather brief Chapter 7 on electrical properties. The final chapter, Chapter 8, presents results on quantum dot based photonic devices, mainly quantum dot lasers.

ACKNOWLEDGMENTS

The work of the TU Berlin and Ioffe Institute, St Petersburg, teams, which presents one backbone of this book, would not have been possible without the generous support from the Deutsche Forschungsgemeinschaft and the Russian Foundation for Basic Research. The cooperation and exchange of scientists of the teams were particularly supported by the Volkswagen Stiftung, the governments of Germany and Russia in the framework of their general science cooperation agreement, Alexander von Humboldt Foundation and INTAS. We could not have done our research or written this book without their help and we are very grateful to the respective administrations and many anonymous approving project referees.

Many individuals lent us their personal advice. We are particularly grateful to Zh.I. Alferov, A. Madhukar and M.S. Skolnick. Others contributed by their work as members of our teams to this book. Particular thanks to V. Shchukin for his contribution to Chapter 3, to J. Böhrer, J. Christen, F. Hatami, F. Heinrichsdorff, R. Heitz, A. Hoffmann, C.M.A. Kapteyn, N. Kirstaedter, A. Krost, M.-H. Mao, O.G. Schmidt, O. Stier, V. Türck, and M. Veit of TU Berlin, to N.A. Bert, P.N. Brounkov, A.Yu. Egorov, N.Yu. Gordeev, S.I. Ivanov, I.V. Kochnev, V.I. Kopchatov, P.S. Kop'ev, A.R. Kovsh, I.L. Krestnikov, A.V. Lunev, M.V. Maximov, B.Ya. Mel'tser, A.V. Sakharov, Yu.M. Shernyakov, I.P. Soshnikov, A.A. Suvurova, A.F. Tsatsul'nikov, V.M. Ustinov, B.V. Volovik, S.V. Zaitsev, and A.E.Zhukov of the Ioffe Institute, to G.E. Cirlin, A.O. Golubok, G.M. Guryanov, N.I. Komyak, V.N. Petrov of the Institute for Analytical Instrumentation. St Petersburg, and to U. Gösele, J. Heydenreich, A.O. Kosogov, S.S. Ruvimov, and P. Werner of the Max-Planck Institut für Mikrostrukturphysik, Halle/Saale.

D.B.
M.G.
N.N.L.

Berlin, March 1998

1 Introduction

1.1 HISTORICAL DEVELOPMENT

1.1.1 FROM ATOMS TO SOLIDS

By introducing the concept of an electronic 'bandstructure' for an ideal crystalline solid Bloch (1928) (see the textbooks of Kittel, 1963, and Ibach and Lüth, 1991) presented in the late 1920s a revolution in the world of physics dominated by research on atoms. In atoms the energies of bound electrons are discrete and precisely defined within the limit of Heisenberg's uncertainty relation. In solids the electron energy is a multivalued function of momentum resulting in energy bands, continuous densities of states, and gaps. The wavefunctions became completely delocalized in real space. Central in Bloch's theory is an *infinite* extension of the regular array of lattice points in all three dimensions of space. The restriction of the theory to infinite bodies (Fig. 1.1) was regarded as being meaningless from any practical point of view, except next to surfaces. Still small crystallites of a few micrometers in size are very large compared to next-neighbor distances. No observable deviations from the predictions for an infinitely extended crystal were expected, even for small objects.

1.1.2 FROM SOLIDS TO 2D HETEROSTRUCTURES

If the carrier motion in a solid is limited in a layer of a thickness of the order of the carrier de Broglie wavelength (or mean free pass, if this number is smaller), one will observe effects of size quantization. The de Broglie wavelength λ depends on the effective mass m_{eff} of the carrier and on temperature T:

$$\lambda = \frac{h}{p} = \frac{h}{\sqrt{(3m_{eff}kT)}} = \frac{1.22\ nm}{\sqrt{(E_{kin}/[eV])}} \quad (1.1)$$

The mass of charged carriers entering Eq. (1.1) is not the free electron mass but the effective mass of the electron (or hole) in the crystal. As this mass can be much smaller than the free electron mass, size quantization effects can be already pronounced at a thickness ten to a hundred times larger than the lattice constant.

The idea of using ultra-thin layers for studies of size quantization effects was already popular by the late 1950s and early 1960s (see the review by Lutskii, 1970). The main object of research at this time were thin films of semimetals (e.g. Bi) on mica substrates obtained by vacuum deposition. However, thin films of metals and semiconductors were also studied. For these films of Bi (Lutskii and Kulik, 1968) and InSb (Filatov and Karpovich, 1968) the effect of the increase of the effective

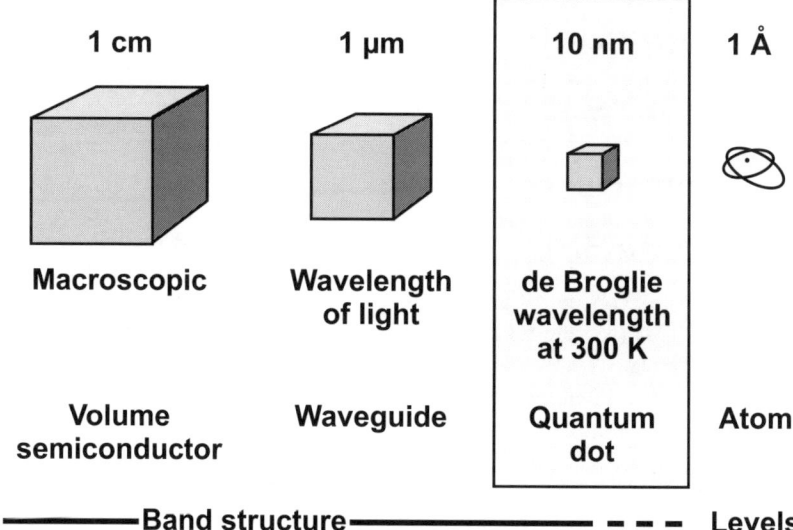

Fig. 1.1 Schematic comparison of typical dimensions of bulk material, waveguides for visible light, quantum dots, and atoms

bandgap with decreasing film thickness was demonstrated. In 1962 Keldysh (1962) considered theoretically the motion of electrons in a crystal with superimposed periodic potential having a period much larger than the lattice constant. In this work the appearance of minizones and negative differential resistance has been predicted. The use of intense sound waves was proposed to modulate the potential. In 1963 Davies and Hosack (1963) proposed to use five-layer 'metal–dielectric–metal–dielectric–metal' double-barrier structures with an ultra-thin intermediate metal layer and dielectric barriers in order to observe the effect of resonant tunneling of electrons. This effect occurs when the Fermi energy crosses the size quantization level in the ultra-thin metal layer upon variation of an external electric field. Independently in 1963 the idea of using resonant tunneling of electrons in double-barrier structures for obtaining negative differential resistance had been proposed by Iogansen (1963) for semiconductor structures with dielectric barriers. Tunneling effects in structures with large numbers of quantum wells were also considered (Iogansen, 1964).

Although this mostly theoretical work found large recognition, experimental studies were limited by insufficient technology at this time. To realize quantum size effects in an ultra-thin layer one needs to have mirror-like reflection of carriers from the surface and/or interface and a large carrier mean free pass parallel to the interface. If the boundaries are rough and the material in their vicinity is damaged, the possibilities of observing size quantization effects are very limited.

With the advent of novel epitaxial deposition techniques like molecular beam epitaxy and somewhat later metal organic chemical vapor deposition in the late

INTRODUCTION

1960s the counter-revolution started. Suddenly it was possible to insert coherent layers, a few lattice constants thick, of a semiconductor of lower bandgap in a matrix with a larger bandgap, restricting carrier movement to only two dimensions. In 1969–70 Esaki and Tsu proposed the use of multilayer periodic semiconductor heterostructures (superlattices) for the creation of artificial materials with controlled width of minizones for carrier transport along the superlattice axis (Esaki and Tsu, 1970). They also proposed formation of delocalized states and negative differential resistance for high-speed electrons reaching the minizone boundary in *nipi* superlattices and quantum wells. Another important effect regarding device applications is *sequential tunneling* in superlattices that was considered first by Kazarinov and Suris in 1970 (Kazarinov and Suris, 1971, 1972). One of the first of these superlattices was grown, also in 1970, by Alferov in the $GaP_{0.3}As_{0.7}/GaAs$ material system (Alferov et al., 1971). A clear demonstration of size quantization effects became possible only after lattice-matched GaAs/AlGaAs structures with planar interfaces were realized by molecular beam epitaxy. In 1974 Chang *et al.* (1974) observed the effect of resonant tunneling, proving the application of quantum mechanics to describe transport phenomena in ultra-thin semiconductor heterostructures. The possibility of creating unipolar long-wavelength laser using radiative transitions between electron size quantization subbands was also considered at that time (Kazarinov and Suris, 1971). Such lasers, so-called 'cascade' lasers, have been realized one quarter of a century later (Faist *et al.*, 1994a, 1994b). The extended possibilities for bandgap engineering by using strained quantum wells were described in Osbourn (1982).

Particularly impressive results have arisen from optical studies of quantum wells and superlattices. In 1974 Dingle *et al.* (1974) directly observed the step-like character of the absorption spectrum related to the two-dimensional character of the density of states in quantum wells (Fig. 1.2). A decrease in the GaAs layer thickness resulted in a shift of the steps toward higher photon energies. Optical studies also underlined the tremendously increased role of excitonic effects (Christen and Bimberg, 1990; Christen *et al.*, 1990) and demonstrated zero-dimensional properties of excitons localized in potential fluctuations of quantum wells. Much increased exciton binding energies, oscillator strengths and temperature stability of excitons were demonstrated. Excitonic effects were found to be of supreme importance for some device applications, such as optical modulators or optical bistable elements (Miller *et al.*, 1984; Miller, 1990).

1.1.3 FROM 2D HETEROSTRUCTURES TO QUANTUM DOTS

By the end of 1980s the main properties of quantum wells and superlattices were rather well understood and the interest of researchers shifted toward structures with further reduced dimensionality—to quantum wires (Kapon *et al.*, 1989) and quantum dots (QDs). Complete reduction of the remaining 'infinite' extension of a quantum well in two dimensions to atomic values (Fig. 1.1) leads to carrier localization in all three dimensions and breakdown of the classical band structure

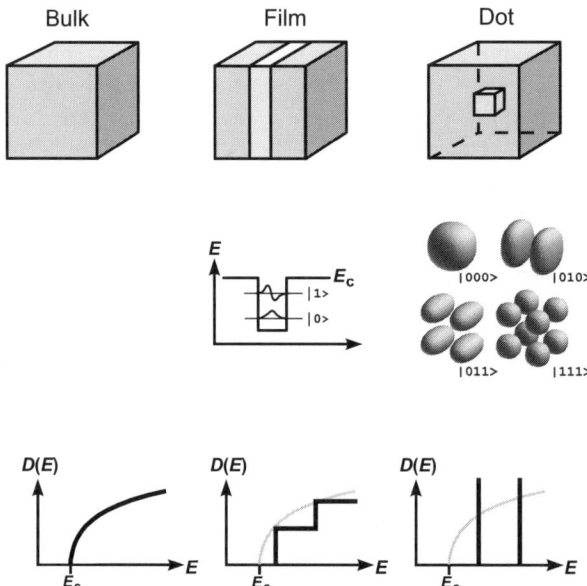

Fig. 1.2 Nature of electronic states in bulk material, quantum wells, and quantum dots. Top row: schematic morphology, center row: quantized electronic states, bottom row: density of electronic states

model of a continuous dispersion of energy as a function of momentum. The resulting energy level structure of quantum dots is discrete (Fig. 1.2), like in atomic physics, and the physical properties of quantum dots in many, although not all, respects resemble those of an atom in a cage. A profound size-dependent change of all macroscopic material properties as compared to the bulk occurs. A qualitative measure when size quantization effects are important and energy separation of sublevels is sufficiently large is again the de Broglie wavelength (Eq. 1.1), now restricting geometry in all three dimensions. A typical size of such a dot is 10 nm and it may still contain 10^4 or more atoms.

The study of single quantum dots and ensembles of quantum dots presents a new chapter in fundamental physics. Moreover, many applications of QDs for novel devices have been predicted, such as single electron transistors or quantum dot lasers. Such devices have exciting novel properties enabling new concepts and systems for many different fields of our society, ranging from information technology to environmental protection.

The road to develop quantum dots useful for devices has been full of detours and obstacles. The first realization of quantum dots were nano-size semiconductor inclusions (e.g. CdSe) in glass (Rocksby, 1932), which have been commercially available as color filters for decades (Fig. 1.3a). Quantum confinement effects in such a system were confirmed experimentally by Ekimov and Onushenko (1984). The usefulness of such color glasses for switches was investigated at a very early

INTRODUCTION

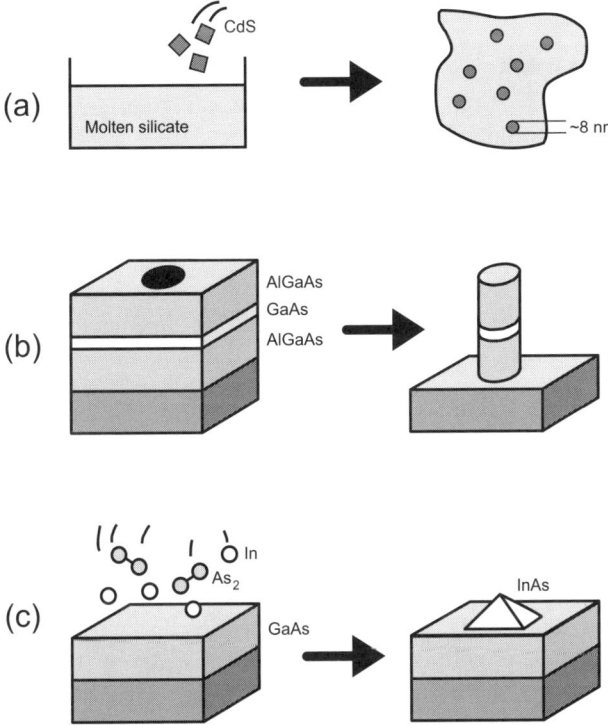

Fig. 1.3 Schematic representation of different approaches to fabrication of nanostructures: (a) microcrystallites in glass, (b) artificial patterning of thin film structures, (c) self-organized growth of nanostructures

stage. The electrically isolating matrix, however, prohibits electric injection and use in electronic and optoelectronic devices. Many alternative approaches have been developed, one of them being artificial patterning of thin layer structures into three-dimensional regions (Fig. 1.3b).

In the last few years nanostructures have been successfully realized using self-organization effects, e.g. occurring during growth of strained heterostructures (Fig. 1.3c). These effects are also called self-ordering or self-assembly. Both thermodynamic and kinetic ordering mechanisms together can create unique three-dimensional patterns of islands within a matrix for many different material systems. The word 'self-organization' in other fields of science, in particular biology, is used for systems away from equilibrium. In contrast, QD formation is in many cases an equilibrium process.

The self-organization effects discussed in this book are a general phenomenon of strained heterosystems. The fabrication process of such quantum dots is compatible with present optoelectronic device technology. A beautiful and well-understood model system for which many examples of results will be presented is InGaAs/AlGaAs. This material combination appears at the moment to be the most promising candidate for immediate device applications. Exciting reports on self-organized quantum dots fabricated in other III–V systems like InP/InGaP, Sb-containing systems such as (In,Ga)Sb/GaAs, structures based on group III nitrides like (In,Ga,Al)N, Ge/Si, SiO_2/Si, and II–VI heterostructures emerge rapidly. In order to concentrate on the most important physics, unfortunately not all of this work can be covered here in detail.

1.2 BASIC REQUIREMENTS FOR QDs IN ROOM TEMPERATURE DEVICES

Quantum dots should fulfill the following requirements in order to make them useful for devices at room temperature:

(a) Sufficiently deep *localizing potential* and small QD *size* is a prerequisite for observation and utilization of zero-dimensional confinement effects.
(b) QD ensembles should show high uniformity and a high volume filling factor.
(c) The material should be coherent without defects like dislocations.

1.2.1 SIZE

The lower size limit of a QD is given by the condition that at least one energy level of an electron or hole or both is present. The critical diameter (for a spherical QD) D_{min} depends strongly on the band offset of the corresponding bands in the material system used. An electron level exists in a spherical quantum dot if the confinement potential, defined by the conduction band offset (in a type-I heterostructure) ΔE_c (see Eq. 5.41), exceeds the value:

$$D_{min} = \frac{\pi \hbar}{\sqrt{(2m_e^* \Delta E_c)}} \tag{1.2}$$

where m_e^* is the effective electron mass. Assuming a conduction band offset value of ~ 0.3 eV for GaAs/$Al_{0.4}Ga_{0.6}$As heterostructures, the diameter of the quantum dot should be larger than 4 nm. This is the lower limit of the QD size. In quantum dots of this or slightly larger size the separation between the electron level and the barrier energy is very small, and at finite temperatures thermal evaporation of carriers from QDs will result in their depletion. For the InAs/AlGaAs system the conduction band offset is much larger while the electron effective mass is smaller, so the product of $\Delta E_c m_e^*$ is comparable and the critical size is also about 3–5 nm depending on the nonparabolicity effects in the InAs conduction band.

There also exists a limit for the maximum size of a QD. A thermal population of higher-lying energy levels is undesirable for particular devices like, for example, lasers. The condition to limit thermal population of higher-lying levels to $5\% \approx e^{-3}$ can be written as

$$kT \leqslant \frac{1}{3}(E_2^{QD} - E_1^{QD}) \tag{1.3}$$

INTRODUCTION

Here E_1^{QD} and E_2^{QD} are the energies of the first and the second levels in the QD, respectively. This equation establishes an upper limit at room temperature for the size of GaAs/AlGaAs QDs of ~12 nm, and of ~20 nm for InAs/AlGaAs QDs if electron levels are considered. This maximum size is, of course, a function of operating temperature.

Consideration of hole quantization in QDs leads to size limits quite different from those derived for electrons above. The maximum dot size to achieve sufficient separation between hole sublevels at room temperature according to Eq. (1.3) should be 5–6 nm for both material systems discussed above. This means that quantum dots with sufficient sublevel separation for both electrons *and* holes (at room temperature) are difficult to realize in most III–V material systems like GaAs/AlGaAs and InAs/AlGaAs because of the large electron/hole mass ratio. For electrical applications, where confinement has to be realized only for one sort of carrier, the problem of different size requirements for simultaneous efficient hole and electron confinement does not exist. In group III nitride and II–VI materials, having more similar electron and hole masses, strong localization of both carrier types seems to be possible.

1.2.2 UNIFORMITY

If a single device is based on more than one quantum dot, the issue of uniformity arises. Even if an individual device is based on one single quantum dot, uniformity is important in order to attain similar characteristics for different devices. In principle, all structural parameters of a QD such as size, shape, and chemical composition are subject to random fluctuations, even in the presence of ordering mechanisms. In most cases a dense array of equisized and equishaped quantum dots is desired.

The main impact of size fluctuations is a variation in the energy position of electronic levels. Such variation is typically Gaussian (Section 5.4.2). For a device that relies on the integrated gain in a narrow energy range, such as a quantum dot laser (Sections 5.4 and 8.2), the inhomogeneous energetic broadening should be as small as possible. This means that for a given *average* size of the QDs one needs to ensure the smallest possible size and shape dispersion values. If wavelength multiplexing is attempted, a certain width of the distribution of energetic levels is desired (e.g. see Section 8.1). In order to realize a reasonable width of a QD ensemble luminescence spectrum, comparable to that of a quantum well (QW) at room temperature (20–30 meV) for a GaAs QD size of about 10 nm, one needs to have a size dispersion lower than 1 nm.

If device operation is based on many quantum dots, typically a minimum number of dots is necessary to obtain a given performance, e.g. a QD laser needs a certain minimum number of QDs to overcome the losses. Those dots have to be packed into a given volume, making it a prerequisite to achieve a minimum packing density or volume filling factor.

1.2.3 MATERIAL QUALITY

The density of defects in a QD material and its interface to the surrounding matrix should be as low as possible, on the level of *in situ* grown quantum wells and interfaces used in state-of-the-art devices. QD fabrication using self-organized growth seems to be predestined to achieve this goal since all interfaces are formed *in situ* during crystal growth.

2 Fabrication Techniques for Quantum Dots

Since the 1980s considerable efforts have been devoted to the realization of semiconductor heterostructures that provide carrier confinement in all three directions and behave as electronic quantum dots. Initially, mainly techniques like lithographic patterning and etching of quantum well structures (Section 2.1.2) have been employed. Technologically related techniques include selective intermixing of quantum wells (Sections 2.1.3 and 2.1.4) and the use of stressors (Section 2.1.5). Substrate encoded epitaxy, which allows the fabrication of small nanostructures from a much larger template, is discussed in Section 2.1.6.

Interface fluctuations in quantum wells can also cause three-dimensional confinement, similar to quantum dots, when 2D excitons become completely localized in space at low temperatures (Section 2.2). Self-organized formation of quantum dots without any artificial patterning is the main focus of this book, which is historically the most recent technique and discussed in Section 2.3.

2.1 QUANTUM DOTS FABRICATED BY LITHOGRAPHIC TECHNIQUES

2.1.1 OVERVIEW

By the end of the 1980s the fabrication of QDs by patterning of quantum wells was considered to be the most straightforward way of QD fabrication. Patterning has several advantages and still attracts much attention because:

(a) QDs of almost arbitrary lateral shape, size and arrangement can be realized depending on the resolution of the particular lithographic technique used.
(b) A variety of processing techniques, like dry and wet etching, that are continuously improved are at the researcher's disposal.
(c) It is generally compatible with modern VLSI (very large scale integrated) semiconductor technology.

Lithographic techniques comprise (Forchel *et al.*, 1988; Beaumount, 1991; Sotomayor Torres *et al.*, 1994):

(a) optical lithography and holography,
(b) X-ray lithography,
(c) electron and focused ion beam lithography (EBL, FIBL),
(d) scanning tunneling microscopy (STM).

Optical lithography, e.g. based on excimer lasers, presently provides resolution below 0.2 μm. Improvements of resolution below the 100 nm range have been predicted (Brunner, 1997). Current developments concentrate on UV optics and resists with steep photosensitivity curve and exposure close to the photosensitivity threshold. Overdeveloping a resist can also reduce the size of a confining potential. Subsequent photoresist deposition steps, where the mask is shifted laterally, also reduce the characteristic dimensions of the resulting pattern. Resolution of optical lithography based methods will nevertheless not be sufficient in the foreseeable future to fabricate structures of lateral dimensions of 20 nm or less. A maximum size of ≈ 20 nm is required such that carriers are still sufficiently confined at room temperature in typical semiconductors. Optical lithography, however, can be applied for fabrication of structures investigated only at low temperatures where confinement is easily reached and for fabrication of patterned substrates that can be used subsequently, e.g. for 'substrate encoded growth' of quantum dots.

X-ray lithography (Warren *et al.*, 1986; IBM, 1993) has the advantage of much shorter wavelengths and can be used for nanostructure fabrication. X-ray lithography represents a contact printing process, since high-resolution X-ray lenses are presently not available. An additional process is required to fabricate the mask. A big advantage of X-ray lithography is in its potential to mass produce nanoscale structures.

For direct lateral patterning the most developed approaches are electron beam lithography (Howard *et al.*, 1985; Beaumount, 1991), focused ion beam lithography (Komuro *et al.*, 1983), as discussed below, and contact imprinting (Krauss, 1997). Scanning tunneling microscope techniques present novel approaches for creating nanoscopic patterns and structures (Snow *et al.*, 1993; Snow and Campbell, 1994). The lack of parallelism is yet a problem remaining to be solved.

2.1.1.1 Electron beam lithography

The electron beam is usually emitted from a high-brightness cathode (e.g. LaB_6) or a field emission gun (cold emission) and focused on the substrate by a multilens system. To define a focus in the nanometer range a short working distance between the final lens and the substrate is required. Resolution in the 10–20 nm range has been already demonstrated at the beginning of the 1980s (Craighead *et al.*, 1983; Stern *et al.*, 1984).

Computer control is employed to scan the beam over the surface and to define the image. To minimize the working distance and to obtain large scanning fields the deflection system is usually placed in the final lens. Typical deflection fields for high resolution lithography machines are about 100 μm × 100 μm when addressed in reasonable time. For beam currents in the 10 pA range a focus diameter of 8 nm can be maintained throughout this field at an accelerating voltage of 50 kV. The address grid resolution in a high-resolution field is 2.5 nm or even smaller. In order to minimize distortion effects which occur in the electron beam exposure over large fields all modern systems are equipped with appropriate software correction algorithms. Laser-controlled interferometers can stitch adjacent fields with an accuracy of a few nanometers. The final resolution of the process of around 10 nm is limited by the resists due to finite length of the organic molecules and the grain size.

A novel way to improved resolution is 'mask' formation without resists. Ultrathin layers of a material having high-temperature stability (e.g. 1 ML thick AlAs film on GaAs) is sufficient to suppress the thermal evaporation of the underlying material. Local formation of such 'masks' can be realized by using electron-stimulated adsorption or desorption processes. Similar protective layers can also be used for selective gas etching of structures and for electron-beam-stimulated epitaxy. These techniques, however, are not sufficiently developed at present and there is still a long way to go before the full potential of electron beam lithography can be realized in practice.

Periodic nanostructures can also be fabricated using an electron beam interference technique (Fujita *et al.*, 1995), which is particularly advantageous for mass-scale fabrication of quantum wires and dots.

2.1.1.2 Focused ion beam lithography

Focused ion beam lithography has a number of properties advantageous for the fabrication of nanostructures, although the present minimum beam diameter of ≈ 30 nm is somewhat larger than that of an electron beam. FIBL can be used for:

(a) maskless etching (ion beam sputtering or stimulated chemical etching),
(b) maskless implantation of dopants,
(c) deposition of metallic structures,
(d) patterning of resists with strongly reduced proximity effect

Ion beam lithography is conceptually very similar to electron beam lithography. Its success is based on the realization of bright liquid metal ion sources. Such a source typically consists of a tungsten needle emerged in a reservoir with liquid metal and wetted by it. Due to capillary effects a very uniform and sharp metal tip is formed. Metal sources are either elemental (e.g. Ga) or alloy (e.g. Au/Be/Si) materials. The electric field applied to the tip extracts an intense beam of ions which is focused by a series of lenses. The extracted particles have a large mass/charge ratio, requiring magnetic fields for focusing that are larger than those generated by conventional magnetic lenses. Instead, electrostatic lenses are used. The minimum diameter of the beam of ≈ 30 nm is limited by the ion energy spread and by chromatic aberration of the lenses. A focused ion beam system usually includes a mass separator to select single charged and multiple charged ions, or to switch between different elements in alloy sources. If the lateral resolution further improves, focused ion beams might become a basis for fabrication of quantum dot arrays, e.g. by focused implantation of In ions in GaAs/AlGaAs quantum wells.

Applications of such structures for devices depend also on further improvements of their crystallographic properties. The point defect concentration caused by the implantation process has to be reduced by high-temperature annealing. Otherwise a giant increase in the diffusion coefficient of constituent or doping atoms might result.

2.1.2 FREE-STANDING QUANTUM DOTS

Free-standing quantum dots have now been fabricated for a decade from quantum wells and were studied by many groups (Scherer and Craighead, 1986; Forchel *et al.*, 1988; Beaumount, 1991; Sotomayer Torres *et al.*, 1994). In Fig. 2.1 a regular array of columns is shown that has been etched from an AlGaAs/GaAs superlattice. The process sequences for the definition of such a quantum dot pattern by either positive or negative resists are given in Fig. 2a and b, respectively. Four major steps are needed in both cases. In the case of a positive process, the sample is first coated by two thin layers of PMMA of different molecular weights (with a typical thickness of 50 nm). The bottom layer has a lower molecular weight and therefore a higher sensitivity than the top layer. After exposure of the resist, a mixture of methylisobutylketone and isopropylalcohol is used for development. Narrow gaps are created in the resist. Due to the higher sensitivity of the bottom PMMA layer the wall profiles have a negative slope (undercut). The surface of the future quantum dot is open and the rest of the surface is covered by the resist after this procedure. Now a thin layer of a metal (e.g. Cr or Ni) or an insulator (e.g. silicon nitride) is evaporated on the surface, and with a suitable solvent the unexposed resist and the metal layer on top of it is removed (lift-off). The metal deposited directly on the sample surface remains and defines the quantum dot for a subsequent dry etching process. Dry etching results in erosion of the sample surface *and* of the mask. To minimize damage the mask thickness is chosen in such a way that the mask is not completely removed at the end of the etch procedure. Almost arbitrary shapes can be created, as

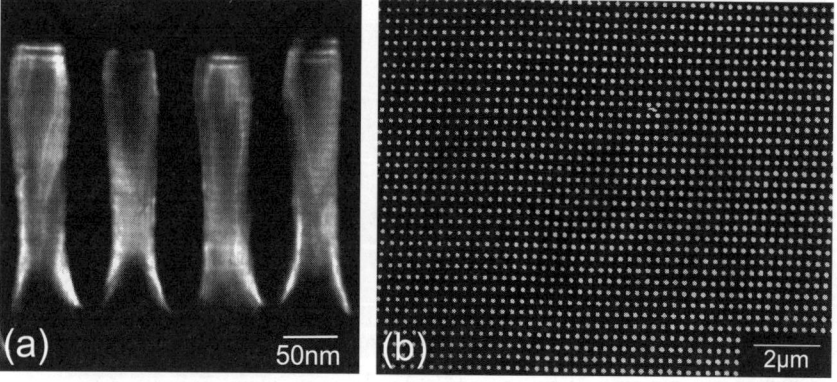

Fig. 2.1 (a) Cross-sectional and (b) plan-view SEM (scanning tunneling microscopy) image of dry etched AlGaAs/GaAs columns. (After Scherer and Craighead, 1986)

FABRICATION TECHNIQUES FOR QUANTUM DOTS

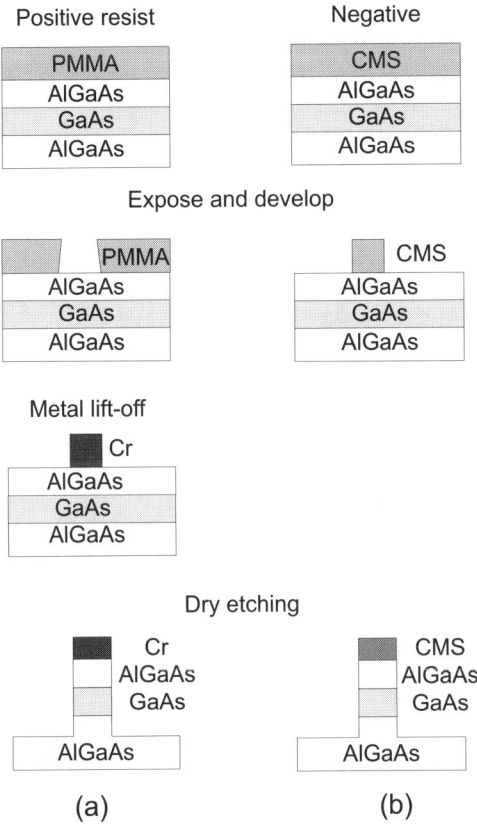

Fig. 2.2 Sequence of the definition procedures for fabrication of a free-standing quantum dot using (a) positive or (b) negative resists. (After Forchel *et al.*, 1988)

shown in Fig. 2.3 (Forchel *et al.*, 1996b), where electron beam lithography and wet etching have been used for fabrication.

There are three major effects affecting the properties of free-standing etched quantum dots: the depleted region, the dead layer and residual strain.

A thick (of the order of tenths to hundreds of nanometers) depleted layer is manifested in luminescence and optical reflectance studies of MBE (molecular beam epitaxy)-grown GaAs layers (Schultheis and Tu, 1985). Fermi level pinning by surface states results in formation of electric fields in the structure, in ionization of neutral shallow impurities and in charge transfer to surface states. For sufficiently large dots a significant electric field can be formed between the mesa 'core' and the side-wall regions. This field can also result in partial lateral separation of electrons and holes and can affect radiative lifetimes, peak energies and luminescence polarization.

Dry etching usually results in the formation of a highly damaged or even amorphized layer at the surface. For nanostructures in the InGaAs/InP material

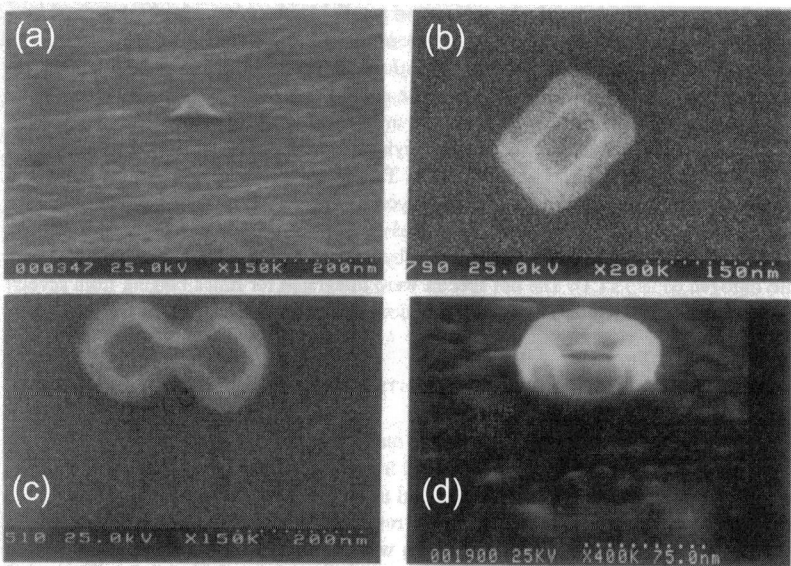

Fig. 2.3 SEM images of quantum dot structures of different shapes defined by electron beam lithography and wet etching: (a) cone-shaped InGaAs quantum dot (diameter ∼ 50 nm), (b) rectangular cross section (central area: $83 \times 47\,\text{nm}^2$), (c) coupled dot structure (molecule, ∼ 80 nm), and (d) ring structure (outer diameter ∼ 75 nm, inner diameter ∼ 30 nm). (After Forchel *et al.*, 1996b)

system subjected to Ar/O$_2$ reactive ion beam etching (RIE) the width of this layer was estimated to be about 19 nm independent of the lattice temperature. In such a layer nonradiative recombination is predominant and the effective size of the binding potential of the electronic quantum dot is reduced. The concept of the 'dead' layer was introduced to explain the lateral size dependence of luminescence efficiency of etched quantum dots (Forchel *et al.*, 1988). The effective thickness of this layer seems to depend strongly on etching parameters. Reactive ion etching of quantum dots with BCl$_3$ was also claimed to result in formation of the dead layer, but of much smaller thickness (Clausen *et al.*, 1989). Using SiCl$_4$ (Kohl *et al.*, 1989) reduces the luminescence efficiency of 70 nm GaAs/AlGaAs quantum wires at low temperatures compared to unpatterned material by a factor of 30. Particularly strong degradation of properties appears for etched II–VI nanostructures (Sotomayor Torres *et al.*, 1992). Significant recovery was observed, however, after annealing of the structures.

Reactive ion etching is also known to result in a loss of stoichiometry at the surface (Cho and Arthur, 1975), with a tendency of depletion of the more volatile element (As in the case of GaAs). This effect can result in a significant deviation from stochiometry and in a pronounced strain in etched structures. One possibility for compensation is annealing in a group V gas atmosphere. A compressive uniaxial strain of about 7×10^{-4} along the growth direction was observed for GaAs/AlGaAs QDs (Sotomayor Torres *et al.*, 1994). A large spread of confinement energies in the range from 0 to 11 meV was reported for 40 nm wide InGaAs/InP wires. Annealing and wet etching steps carried out after dry etching were shown to remarkably alter the blue shift (MacLeod *et al.*, 1993). The observed variation of experimental results demonstrates the large influence of extrinsic factors.

The surface recombination velocity for the free GaAs surface is very high and close to the thermal velocity of nonequilibrium carriers ($\sim 10^7$ cm/s at 300 K) (Bimberg and Queisser, 1972). For small mesas this value is not dramatically smaller, even at low temperatures, as the nonresonantly created carriers can be trapped by the surface well before they are thermalized with the lattice. For the InGaAsP material system the surface recombination velocity is much smaller, being around 1×10^4 cm/s (Forchel *et al.*, 1988) at 300 K, and the emission from etched quantum dots can be observed in mesas having relatively small lateral sizes. By studying the luminescence efficiency versus mesa size at different temperatures, it was found (Maile *et al.*, 1989) that the surface recombination velocity of InGaAs increases from 1.6×10^3 cm/s at 4 K to 5.9×10^3 cm/s at 77 K and to 1.2×10^4 cm/s at 300 K.

Comparison of the experimental results on luminescence intensity versus lateral size (Forchel *et al.*, 1988; Andrews *et al.*, 1990; Beaumount, 1991; Sotomayor Torres *et al.*, 1994) demonstrates significant scatter of the experimental data of different groups which is related to different degrees of damage introduced by etching, different material purity and different interface or alloy structure of the precursor QW. High concentrations of localization centers, such as impurities or interface fluctuations, can suppress carrier lateral mobility and result in a relative increase in luminescence intensity, at least at low temperatures. This is particularly true for alloy quantum wells, where alloy scattering further reduces exciton mobility. As a general trend, a strong decrease in integrated photoluminescence (PL) intensity (by orders of magnitude) is manifested when the lateral size of the etched GaAs/AlGaAs QD is reduced to about or slightly below 100 nm at low temperatures. At high temperatures degradation of PL starts from the micrometer-range for the InGaAs/InP system and from dozens of micrometers for the AlGaAs/GaAs one (high purity samples). Lower surface recombination velocity still permits physical studies of etched InGaAs/GaAs QDs at low temperatures down to lateral sizes of 50 nm (Bayer *et al.*, 1995).

In view of the above discussion one can conclude that it is possible to fabricate free-standing etched quantum dots having a size down to ~ 10 nm. It seems to be difficult, however, to obtain bright luminescence from such small structures. Thus the present usefulness of such structures for devices is limited.

2.1.3 SELECTIVE INTERMIXING BASED ON ION IMPLANTATION

Selective intermixing of quantum wells has been used for fabrication of laterally buried heterostructures for many years (Werner *et al.*, 1989). Diffusion-induced disordering (Zahari and Tuck, 1985; Harrison *et al.*, 1989), intermixing under pulsed laser irradiation (Ralston *et al.*, 1987) (see Section 2.1.4), and implantation-induced disordering (Cibert et al., 1986; Venkateson et al., 1986; Kuttler et al., 1997) are some of the techniques used. Submicrometer lateral variations in the local composition were realized by intermixing superlattices using a scanned focused beam of Ga ions (Hirayama et al., 1985).

Selective intermixing by ion implantation and subsequent annealing is caused by a remarkable enhancement of diffusion coefficients by defects like vacancies or interstitials (Cibert et al., 1986; Kuttler et al., 1997) after a high concentration of impurities and point defects was created. Both impurities and point defects are ionized at elevated temperatures, leading to a large increase in the concentration of charge carriers. This increase changes the point defect equilibrium in the crystal (Kröger, 1964; Hurle, 1979). It results in a drastic increase in the concentration of more mobile species, such as, for example, charged interstitials (Tuck, 1985; Zahari and Tuck, 1985), causing an effective intermixing of superlattices. The unimplanted areas remain more or less unaffected.

2.1.4 SELECTIVE INTERMIXING BASED ON LASER ANNEALING

Laser-induced local interdiffusion of a GaAs/AlGaAs quantum well structure has been proposed to define QDs (Ralston et al., 1987; Brunner et al., 1992). Pulsed laser irradiation, e.g. of GaAs/AlGaAs superlattices, results in melting of the annealed area followed by subsequent ultrafast recrystallization. After recrystallization the composition of the superlattice is averaged. This method has been shown to provide very sharp interfaces between intermixed and nonintermixed regions of less than 5 nm (Ralston et al., 1987). Using an Ar laser beam with 500 nm focus, selective intermixing of a single 3 nm thick GaAs/AlGaAs quantum well (Brunner et al., 1992) has been demonstrated. A high-energy shift of the luminescence line by several tens of millielectronvolts has been observed. Due to the high activation energy of the interdiffusion process the resulting lateral potential barriers were five times steeper than the laser intensity profile (Schlesinger and Kuech, 1986).

A disadvantage of this method for applications is related to the quality of the material after laser annealing. Melting and subsequent fast recrystallization of the material results in a high concentration of nonequilibrium point defects and, correspondingly, the luminescence efficiency of the treated areas strongly degrades. It was shown (Brunner et al., 1992) that a reduction in the size of the unintermixed region from 1000 to 450 nm resulted in a drop in PL intensity by a factor of 20, and a further decrease in the size down to 300 nm resulted in an additional drop of one order of magnitude. Luminescence efficiency of uniformly intermixed areas was reported to be more than 300 times smaller than the luminescence of the untreated quantum well at low temperature. Another problem relates to the microscale uniformity of the recrystallized material.

2.1.5 STRAIN-INDUCED LATERAL CONFINEMENT

Formation of lateral confinement by strain gradients has been proposed for quantum wires and quantum dot fabrication to overcome problems of interface damage in free-standing quantum wires and dots (Kash, 1988, 1991). Here a highly strained film is deposited on top of a quantum well structure buried closely underneath the surface. If the film is not patterned, the strain is concentrated in the film and the quantum well is not affected, except for a slight bending of the substrate as a whole. If the film is patterned, the film strain partly relaxes at the expense of some strain in the substrate and quantum well. The lateral resolution of the method is only limited by the possibilities of electron lithography and very small mesas of stressors can be potentially formed. Recently, self-organized stressed islands have also been used as stressors (Lipsanen *et al.*, 1995) (see also Section 6.3). An initially biaxially compressed film induces in the substrate regions of compression in the vicinity of the stressor edges and regions of expansion under the stressor center (see Section 5.2.6). No problems with surface states on mesa side walls and related surface recombination exist in this case. The confinement potentials for carriers, however, are comparatively small.

2.1.6 QUANTUM DOTS GROWN ON PATTERNED SUBSTRATES

Growth of nanostructures on patterned substrates was originally initiated to avoid some of the problems characteristic of *free-standing* quantum wires and dots. The growth occurs on patterned surfaces, V-grooves or inverted pyramids with dimensions significantly larger than the nanostructure itself (Kapon *et al.*, 1987; Lebens *et al.*, 1990; Fukui *et al.*, 1991; Rajkumar *et al.*, 1992; Madhukar, 1993; Madhukar *et al.*, 1993; Sugawara, 1995; Hartmann *et al.*, 1997; Ishida *et al.*, 1998). The growth on patterned, sometimes higher-index substrates, is complex and several instabilities, such as spontaneous tilting of facets, spontaneous corrugation of facet surfaces, and non-uniformity of the growth rate in the vicinity of edges can prevent the realization of perfect structures. A high level of understanding of the growth process with complex surface and interfacet kinetics and energetics is required.

Three steps are required to fabricate QDs using this approach (see, for example, Rajkumar *et al.*, 1992):

(a) definition of lateral masks using lithographic techniques,
(b) wet chemical etching to produce the desired geometrical relief and an atomically clean surface suitable for growth,
(c) growth on the patterned or corrugated substrate.

Rajkumar *et al.* (1992) reported the fabrication of an array of mesas on an As terminated GaAs (111)B substrate. By photolithography, an array of 5 µm size resist patterns aligned along a ⟨1–10⟩ direction was defined, followed by wet chemical etching. Growth was stopped when each mesa consisted of a truncated triangular pyramid. The mesa top was an (111)B face and the three side facets were of the {100} type. The size of the mesa top can be decreased from the initial pattern defined size, eventually down to practically zero depending on the duration of etching. Even if the size of the mesa top is not sufficient to provide efficient lateral confinement one can reduce it further by overgrowth, due to lateral shrinkage of the mesa top. Figure 2.4 illustrates this process. Precondition of the shrinkage process is a higher growth rate on the mesa top rather than on the side walls. This condition is fulfilled if, for example, upward diffusion takes place.

Optimization of the growth temperature (Rajkumar et al., 1992) allows monotone dependence (shrinkage) to be obtained and reproducibility to be improved. At the same time, as growth continues, new facets emerge. A new facet was identified to be of the {211} type, contiguous to the (111)B mesa top. Using this or similar approaches QD structures with lateral dimensions down to 50 nm can be created (Rajkumar et al., 1992). A similar approach has been used by Nagamune et al. (1995).

Growth on patterned substrates and selective epitaxy in local openings of a dielectric mask gave qualitatively similar results. In the latter case, no further patterning of the substrate was required, as the growth was governed by the size and the shape of the opening in the dielectric mask. Moreover, as the size of the opening was defined by the resolution of the lithographic process, potentially very small (~ 10 nm) openings are possible. Such a formation of three-dimensional structures using selective metal-organic chemical vapor phase epitaxy in openings in a Si_3N_4

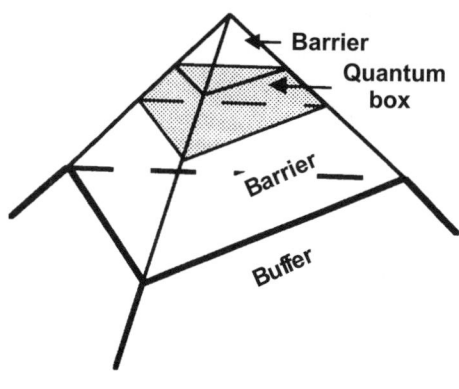

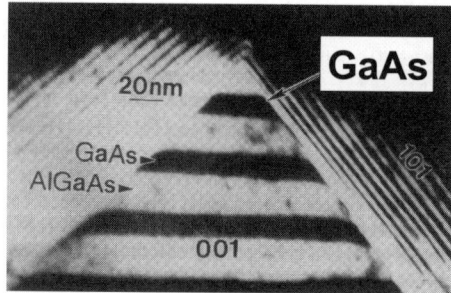

Fig. 2.4 Schematic diagram and cross-sectional TEM image of a growth profile resulting in quantum box mesas. (After Rajkumar et al., 1992)

mask was demonstrated, for example, by Lebens *et al.* (1990). The minimum GaAs dot diameter was 80 nm.

In a recent alternative approach selective electron beam exposure was used to define a pattern directly on to the GaAs surface, which can subsequently be used for selective nucleation of clusters (Sleight *et al.*, 1995). Also, contact printing (Krauss and Chou, 1997) presents a novel way of pattern formation.

When conventional lithographic techniques are used, the lateral density of QDs is defined by the pattern. The dot-to-dot distance is usually large (~ 0.3–$3\,\mu m$), preventing the realization of dense arrays of quantum dots. A general problem of growth on patterned substrates is its limited compatibility with the planar technology of double-heterostructure lasers. The mesoscopic size of the mesa opposes the possibility of formation of perfect waveguides. On the other hand, it might ease fabrication of distributed feedback lasers. In the case of growth in openings the existence of a dielectric mask limits certain applications.

2.1.7 CLEAVED EDGE OVERGROWTH

Twofold growth on the cleaved edge of a quantum well or superlattice (2CEO) had been predicted to lead to the formation of electronic quantum dots at the juncture of three orthogonal quantum wells (see Fig. 5.19 (Grundmann and Bimberg, 1997a). Such dots have been realized in the AlGaAs/GaAs system using MBE (Wegscheider *et al.*, 1997; Schedelbeck *et al.*, 1998).

2.2 QUANTUM DOTS FORMED BY INTERFACE FLUCTUATIONS

Interface fluctuations cause potential fluctuations in a quantum well which form efficient localization centers for excitons (Christen and Bimberg, 1990; Christen *et al.*, 1990). Optical properties of excitons localized at interface fluctuations are reviewed in Section 6.5. Such 'natural' quantum dots are characterized by small localization energy of the carriers and by inhomogeneity of lateral sizes and relatively small density of states. This causes quick saturation of the luminescence of localized excitons with increasing excitation density and depopulation of localized states due to thermal evaporation of excitons already at moderate excitation densities and observation temperatures, respectively (Kop'ev *et al.*, 1986).

2.3 SELF-ORGANIZED QUANTUM DOTS

The evolution of an initially two-dimensional growth into a three-dimensional corrugated growth front is a well-known phenomenon and has been observed for an abundance of systems. A comprehensive discussion of such growth modes, in particular in the presence of strain, is given in Section 3.6.1.

In a paper by Stranski† and Krastanow (1937) (SK) the possibility of island formation on an initially flat heteroepitaxial surface was proposed; actually, for the growth of lattice *matched* ionic crystals that had different charges. In the following years the term 'SK growth' was used in heteroepitaxy for the formation of islands on an initially two-dimensional layer (see Fig. 3.9), including growth of islands *relaxed* by misfit dislocations in strained heteroepitaxy (Bauer, 1958).

The formation of *coherent,* i.e. defect-free, islands as a result of SK growth of strained heterostructures is a fairly new concept and is now systematically exploited for the fabrication of quantum dots. Goldstein *et al.* (1985) observed for the first time that a regular pattern of islands forms (Fig. 2.5) in an InAs/GaAs superlattice. As a consequence of the formation of coherent islands from a strained epilayer, a significant amount (of the order of 50%) of its strain energy is relaxed (see Section 3.6.2). A number of early observations (Eaglesham and Cerullo, 1990; Guha *et al.*, 1990; Snyder *et al.*, 1991; Tabuchi *et al.*, 1992) stimulated further research in this direction. Important breakthroughs were reported almost simultaneously by four

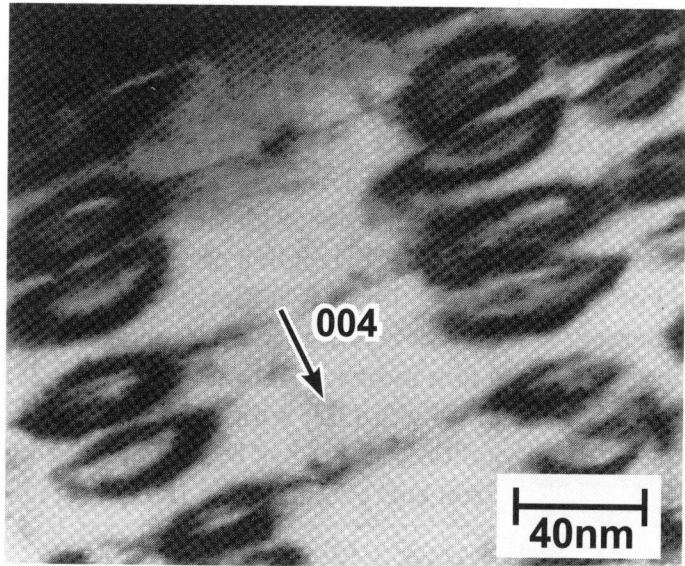

Fig. 2.5 Cross-sectional TEM image of fifteenfold 2.5 ML InAs/30 nm GaAs 'superlattice' consisting of stacked dot layers. Contrast is mainly due to strain fields and the diffraction vector [004] is indicated in the figure. (Reproduced by permission of American Institute of Physics from Goldstein *et al.*, 1985)

---

† Iwan N. Stranski (born 1897 in Sofia) obtained his PhD from Friedrich-Wilhelm-Universität (now Humboldt-Universität zu Berlin) in 1925. He held the position of a full professor at Technische Universität Berlin from 1945 to 1953, succeeding Max Volmer. In 1949 he became Prorector and served from 1951 to 1953 as Rector of TU Berlin.

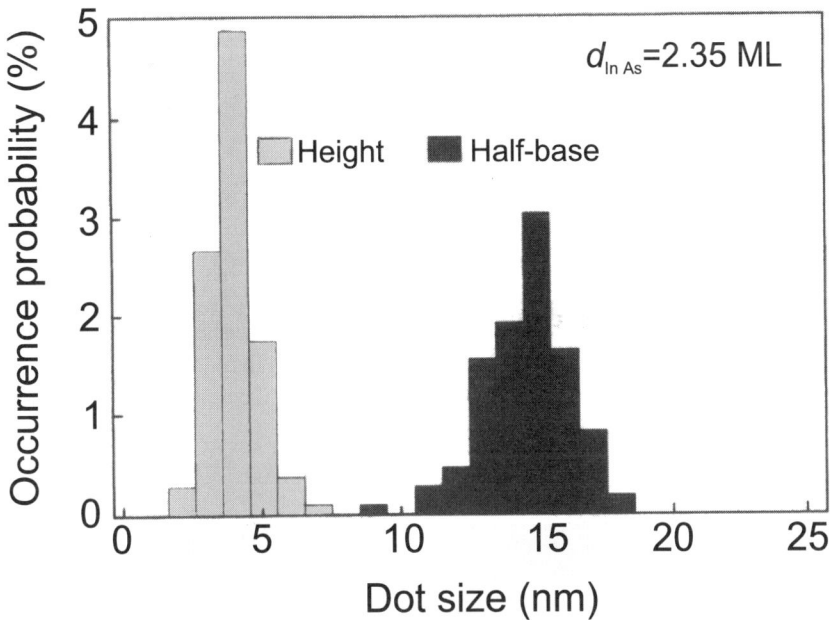

Fig. 2.6 Histogram of height and lateral size for MBE-grown InAs/GaAs quantum dots. (Reproduced by permission of American Institute of Physics from Moison *et al.*, 1994)

groups (Leonard *et al.*, 1993; Bimberg, 1994; Madhukar *et al.*, 1994; Moison *et al.*, 1994) where a narrow distribution of the island size was manifested (Fig. 2.6).

Structural and optical properties of self-organized quantum dots are the focus of the remainder of this book. As discussed in detail in Chapter 4, using suitable growth conditions in MBE or MOCVD, it is now possible to fabricate arrays and stacks of small QDs (~ 10 nm), ordered in size and shape, with high area density ($> 10^{11}$ cm^{-2}) and high optical quality. Such dot ensembles are suitable for creating devices such as low threshold lasers (see Section 8.2).

3 Self-Organization Concepts on Crystal Surfaces

3.1 INTRODUCTION

Self-organization of ordered nanostructures on crystal surfaces represents a unique phenomenon which opens exciting new possibilities both in fundamental research of low-dimensional objects and in their device applications. There is a variety of effects that can be used for fabrication of QDs:

(a) periodic surface faceting with subsequent heteroepitaxial growth on the faceted surface,
(b) formation of arrays of step bunches of equal height on vicinal surfaces,
(c) formation of ordered nanoscale two-dimensional islands, e.g. having mono- or bilayer height upon submonolayer deposition,
(d) formation of ordered arrays of three-dimensional coherently strained QDs.

In this section we consider theoretical concepts and their relation to experimental results on self-ordering phenomena on semiconductor surfaces and on formation mechanisms of self-ordered nanostructures (Shchukin and Bimberg, 1998). Thermodynamic arguments will mainly be considered for the various classes of self-organized semiconductor nanostructures. All of these structures are described as equilibrium structures of elastic domains. Despite the fact that the driving forces of the instability of a homogeneous phase are different in each case, the common driving force for the long-range ordering of the inhomogeneous phase is the elastic interaction. Also, kinetic theories of single-layer dot formation are reviewed and compared with the thermodynamic approach. A kinetic theory of the formation of multiple-sheet structures of QDs is considered. It includes equilibrium ordering of coherently strained islands in the first sheet and the strain-driven kinetic stacking of islands in the next sheets or strain-driven shape transformation effects induced by complex growth modes.

Spontaneous formation of periodically ordered domain structures in solids with the periodicity much larger than the lattice parameter is a general phenomenon which is at the origin of a large variety of different domain structures. This process, also called self-ordering, can occur if the homogeneous state of the system is thermodynamically unstable and the system undergoes a phase transition into an inhomogeneous state. In an inhomogeneous state, generally a coordinate-dependent order parameter is the source of a long-range field (electric, magnetic, strain). Therefore, a multidomain state of the solid is energetically more favorable than the single-domain state, since the former provides compensation of the long-range field at large distances outside the domain structure. The long-range field is responsible for the periodic ordering of the equilibrium domain structure that meets the conditions of the total Helmholtz free energy minimum. The free energy can be written as a sum of three distinct contributions:

$$F_{\text{total}} = F_{\text{domains}} + F_{\text{boundaries}} + E_{\text{long-range}} \qquad (3.1)$$

Here F_{domains} is the free energy of domains, $F_{\text{boundaries}}$ is the free energy of domain boundaries, and $E_{\text{long-range}}$ is the energy of a long-range field, which includes interaction between domains.

The conventional classification of domain structures is based upon the physical nature of the order parameter. An elastic strain field is of major importance for consideration since it exists in nearly all domain structures. The reason is that any phase transformation in solids is typically accompanied by a crystal lattice rearrangement. In most cases, neighboring domains have different lattice parameters of the bulk or surface unit cells. For the case of ferroelectric or ferromagnetic domain structures this phenomenon is well known as electrostriction or magnetostriction, respectively (Landau and Lifshits, 1960).

The focus of our consideration here will be given to so-called elastic domain structures where the elastic strain field is the major or only long-range field. Elastic domain structures only recently became the subject of monographs. The first systematic theoretical approach to elastic domain structures in bulk materials is given in the monographs by Khachaturyan (1974, 1983). Both composition-modulated structures in phase-separating metal alloys and domain structures induced by martensite transformation are treated and a review of experimental data is presented. The theory of elastic domain structures in epitaxial films of macroscopic thickness has been developed by Roitburd (1976) and Bruinsma and Zangwill (1986) for martensite-type domains and by Ipatova *et al.* (1993) for composition-modulated structures in alloys.

In recent years, due to its high importance for fabrication of semiconductor heterostructures with reduced dimensionality (quantum wires and quantum dots), self-ordering phenomena on crystal surfaces has become a subject of intense experimental and theoretical studies. The three classes of self-ordered surface structures displayed in Fig. 3.1 will be discussed here. These are periodically faceted surfaces of pure crystals, surface domain structures (e.g. ordered arrays of monolayer-height islands in heterophase systems at submonolayer coverage), and ordered arrays of three-dimensional coherently strained islands on lattice-mismatched substrates. Since typical linear dimensions of all these structures are of a nanometer length scale, they are commonly termed 'nanostructures'.

In Section 3.2 we consider the general problem of the equilibrium crystal shape in relation to surface faceting. In Section 3.3 periodic arrays of macroscopic step bunches are discussed. Section 3.4 focuses on the heteroepitaxial growth on corrugated surfaces. In Section 3.5 the formation of ordered arrays of planar surface domains is treated. The main section (3.6) is on ordered arrays of three-dimensional

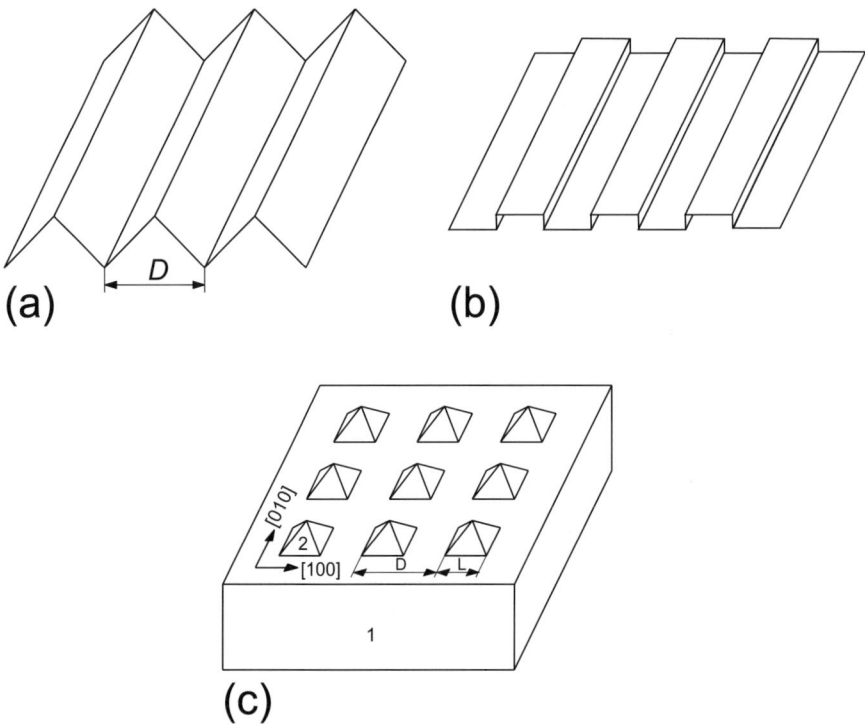

Fig. 3.1 Various classes of spontaneously ordered nanostructures: (a) periodically faceted surfaces, (b) surface domain structures, (c) ordered arrays of coherently strained islands (material 2) on a lattice-mismatched substrate (material 1)

islands. Finally, vertically correlated growth of nanostructures is discussed in Section 3.7.

3.2 SPONTANEOUS FACETING OF CRYSTAL SURFACES

3.2.1 THE PROBLEM OF EQUILIBRIUM CRYSTAL SHAPE

The phenomenon of equilibrium faceting plays the key role in the definition of the equilibrium crystal shape (ECS). Two problems are related to the ECS that correspond to two different physical situations. The first problem is the ECS of a single crystal which dates back at least to Wulff (1901); further developments of ECS theory can be found in Herring (1951a), Chernov (1961) and Mullins (1963). The exact thermodynamic formulation may be found, for example, in the review by Rottman and Wortis (1984), who discussed the shape of a single solid inclusion of a fixed volume ω which is in equilibrium with the liquid or with the gas phase. The surface free energy of the inclusion is the integral over the surface of ω:

$$F_{\text{surf}}(T, \omega) = \oint_{\partial \omega} \gamma(\hat{\mathbf{m}}; T) \, dA \qquad (3.2)$$

Here $\gamma(\hat{\mathbf{m}}; T)$ is the surface free energy per unit surface area dependent on the orientation $\hat{\mathbf{m}}$ of the surface element dA relative to the crystal axes. Thermodynamic theory of the ECS states that a macroscopic inclusion of a fixed volume

$$V(\omega) = \int_{\omega} dV \qquad (3.3)$$

takes at equilibrium the shape that minimizes the surface free energy (Eq. 3.2) subject to the constraint (Eq. 3.3).

The dependence of the surface free energy $\gamma(\hat{\mathbf{m}}; T)$ on orientation is expected to have cusps in symmetry directions, leading to facets in the crystal shape at sufficiently low temperatures. These cusps represent discontinuities in the angular derivatives of the surface free energies, the discontinuities being associated with the free energy of the steps on a given facet. With a temperature increase, the step free energies decrease, cusps blunt, and the corresponding facets shrink. Facets finally disappear at the roughening transition temperature T_R (which can be different for different directions) of the corresponding infinite planar surface. Above T_R the corresponding region of the ECS becomes smoothly rounded.

References to experimental data on equilibrium crystal shapes of micrometer-scale metal clusters are presented in Rottman and Wortis (1984). Transmission electron microscopy (TEM) measurements of the equilibrium shape of voids in Si enabled the orientation dependence of the surface free energy of Si to be revealed (Eaglesham et al., 1993). Statistical mechanics is the tool used for microscopic evaluation of the dependence of the surface free energy $\gamma(\hat{\mathbf{m}}; T)$ on orientation and for the determination of the ECS; an overview of the theoretical results may be found in Rottman and Wortis (1984).

The important issue of ECS theory is that there exist surface orientations that are not present in the crystal shape at a given temperature. At $T=0$ only a few high-symmetry surface orientations are present. All others are passive in the sense that they do not contribute to the ECS. With increasing temperature the domain of passive orientations shrinks as the crystal becomes more rounded.

The importance of this issue becomes even more visible as one considers the second problem of the ECS. This concerns a crystal in equilibrium with the liquid or with the gas phase and with the volume and all but the top surface fixed. This formulation of the problem is relevant to an experimental situation where only the top crystal surface is studied, e.g. thermal annealing of a crystal or growth interruptions introduced in crystal growth experiments.

The top crystal surface is not fixed and is allowed to rearrange into a hill-and-valley structure. Can rearranging the atoms into hills and valleys lower the free energy of a plane surface? When hills and valleys have a size large compared to the lattice parameter, new tilt facets can be defined, and the free energy of the hill-and-valley structure can be written as a surface integral over tilted facets:

$$F_{\text{surf}} = \int \frac{\gamma(\hat{\mathbf{m}}; T)}{(\hat{\mathbf{m}} \cdot \hat{\mathbf{n}})} \, dA \tag{3.4}$$

Here $\hat{\mathbf{m}}$ is the coordinate-dependent unit vector locally normal to the surface at each point and $\hat{\mathbf{n}}$ is the constant unit vector normal to the initially planar surface. Fixed side surfaces of the crystal imply that the average normal to the top surface coincides with the normal to the nominally planar surface, i.e.

$$\frac{1}{A} \int \hat{\mathbf{m}} \, dA = \hat{\mathbf{n}} \tag{3.5}$$

A being the total area of the nominally planar surface.

The theorem proved exactly by Herring (1951b) reads:

> If a given macroscopic surface of a crystal does not coincide in orientation with some portion of the boundary of the equilibrium shape, there will always exist a hill-and-valley structure which has a lower free energy than a flat surface, while if the given surface does occur in the equilibrium shape, no hill-and-valley structure can be more stable.

If the planar surface is unstable, the resulting hill-and-valley structure is determined by the minimum of the surface free energy (Eq. 3.4) subject to the constraint (Eq. 3.5). This minimization will yield the orientation of tilt facets as well as the fraction of the nominal planar surface on to which each facet is projected. Microscopic theory based on statistical mechanics can yield the orientational dependence of the surface free energy $\gamma(\hat{\mathbf{m}}; T)$ and thus allow the ECS to be determined. Recent developments in this area have been made by Williams *et al.* (1993) for surfaces vicinal to Si(111) and by Mukherjee *et al.* (1994) for surfaces vicinal to Si(001).

The theory formulated by Eqs. (3.4) and (3.5) does not yield, however, the facet *size*. This problem requires additional concepts of the *intrinsic surface stress* and of *capillary effects on solid surfaces*, which are addressed in the next two sections.

3.2.2 THE CONCEPT OF INTRINSIC SURFACE STRESS OF A CRYSTAL SURFACE

Since atoms in the surface layer are in a different environment than in the bulk, the surface layer energetically favors a lattice parameter different from the bulk value in the directions parallel to the surface. The surface layer is intrinsically stretched or compressed and therefore characterized by *intrinsic surface stress*.

The intrinsic surface stress of a solid is an analogue to the surface tension of a liquid (Fig. 3.2). However, there exists a fundamental difference in thermodynamic properties of a liquid surface and of a solid surface, pointed out long ago by Gibbs (1928); an explanation may also be found in Marschenko and Parshin (1980) and Needs (1987). The basic reason is that any liquid is *incompressible*. When a liquid

SELF-ORGANIZATION CONCEPTS ON CRYSTAL SURFACES 27

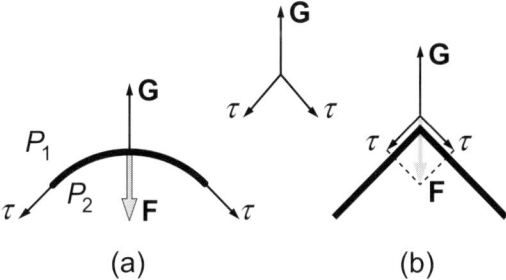

Fig. 3.2 Capillary effects on liquid and on solid surfaces. (a) Balance of forces acting on an element of the curved surface layer of a liquid. Forces caused by the surface tension τ are balanced by the force **G** acting from the bulk of the liquid. According to the third Newton law, the opposite force $\mathbf{F} = -\mathbf{G}$ is acting from the surface layer on the bulk of the liquid. This force results in the excess Laplace pressure $\Delta P = P_2 - P_1$ below the curved surface of the liquid, P_1 and P_2 being the values of the pressure below and above the curved liquid surface. (b) The balance of forces acting on a crystal edge. Forces caused by the intrinsic surface stress τ are balanced by the force **G** acting from the bulk of the crystal. The reaction $\mathbf{F} = -\mathbf{G}$ is acting from the surface layer on the bulk of the crystal, resulting in the strain field. The inset in the center depicts the force balance, which is similar for a liquid and for a crystal (all forces are per unit length in the direction perpendicular to the figure)

film is stretched, atoms or molecules move out from the bulk to form a new surface that is structurally identical to the existing surface. Thus, an attempt to stretch a liquid film by 1% results in a 1% increase in the number of surface atoms or molecules, whereas the spacing between surface atoms remains unchanged.

Thus the processes of the creation and deformation of a liquid surface are identical and are described by the *single* parameter γ, the energy required to create a unit area of the surface. However, when a crystal surface is stretched, the distance between atoms increases and the nature of the surface itself changes. The energy to create a unit area of the surface of a given orientation is characterized by *two* parameters: the scalar quantity γ, termed the surface energy, and the energy change due to deformation of the crystal surface, described by the intrinsic surface stress tensor $\tau_{\alpha\beta}$. The concept of the intrinsic surface stress tensor has been proposed by Gibbs (1928) and discussed later by Shuttleworth (1950), Herring (1951a) and Marchenko and Parshin (1980). The linearized change of the surface energy is written as the following integral over the surface:

$$\int \tau_{\alpha\beta}(\hat{\mathbf{m}}) \varepsilon_{\alpha\beta} \, \mathrm{d}A \tag{3.6}$$

where $\tau_{\alpha\beta}$ has nonvanishing components only in the surface plane, $\alpha, \beta = 1, 2$ (Marchenko and Parshin, 1980). The values of the intrinsic surface stress tensor can be either positive (tensile) or negative (compressive); a tensile surface stress is associated with the surface that favors contraction while a compressive surface stress favors expansion.

Values of the intrinsic surface stress for solids are known either from first-principle calculations or from comparison with indirect experimental data on parameters of surface domain structures. No direct experimental method for determining the intrinsic surface stress of a solid has been proposed so far. For most solid surfaces, the intrinsic surface stress is tensile (see, for example, Needs, 1987; Fiorentini *et al.*, 1993) whereas a compressive surface stress is known for Si(001) surfaces in the direction perpendicular to dimers (Garcia and Northrup, 1993). Results on the (2 × 4) GaAs(001) surface indicate tensile surface stress (Moll, 1998a). The order of magnitude of τ is 10–100 meV/Å^2.

Since the liquid surface is characterized by a single quantity γ which is both the energy for creation of a unit surface area (i.e. the surface energy) and the quantity responsible for capillary effects of the Laplace-pressure type, the term 'surface tension' is widely used for the surface energy γ. Mainly for historical reasons, usage of this term has been extended to surfaces and interfaces of solids. However, the use of 'surface tension' for solids turns out to be ambiguous since it may produce a confusion between the surface energy γ and the intrinsic surface stress $\tau_{\alpha\beta}$. Therefore, we intend to distinguish 'surface energy' and 'intrinsic surface stress' and will not use the term 'surface tension' for solids.

The change in surface energy due to strain (Eq. 3.6) indicates the interconnection between surface effects and strain-related phenomena. Within the framework of the linear theory of elasticity, where the bulk strain energy is a quadratic function of the strain, it is possible to expand the surface energy up to second-order terms in strain. The thermodynamics of solid surfaces and interfaces up to second-order terms in strain has been studied by Andreev (1981) and further developments have been made by Nozières and Wolf (1988). The dependence of the surface or interface free energy on strain is written as

$$\gamma(\hat{\mathbf{m}}; \varepsilon_{\alpha\beta}) = \gamma_0(\hat{\mathbf{m}}) + \tau_{\alpha\beta}(\hat{\mathbf{m}})\varepsilon_{\alpha\beta} + \tfrac{1}{2}S_{\alpha\beta\varphi\phi}(\hat{\mathbf{m}})\varepsilon_{\alpha\beta}\varepsilon_{\varphi\phi} + \tfrac{1}{2}h_{\alpha\beta i}(\hat{\mathbf{m}})\varepsilon_{\alpha\beta}\sigma_{ij}m_j \quad (3.7)$$

Here, $\hat{\mathbf{m}}$ is the normal to the surface or interface, Greek characters label two-dimensional indices in the surface plane whereas Latin indices are three-dimensional ones. Quadratic coefficients $S_{\alpha\beta\varphi\phi}$ and $h_{\alpha\beta i}$ have the meaning of surface excess elastic moduli and can be either positive or negative; σ_{ij} is the bulk elastic stress tensor, and the fourth term in Eq. (3.7) exists on solid–solid interfaces and vanishes on stress-free surfaces where the bulk stress tensor obeys the boundary conditions $\sigma_{ij}m_j = 0$.

Following the conventional approach of elasticity theory (Landau and Lifshits, 1959), we define all surface quantities defined per unit area of the undeformed surface. The dependence of the surface free energy on the strain (Eq. 3.7) can be interpreted as the *strain-induced renormalization* of the surface free energy.

3.2.3 FORCE MONOPOLES AT CRYSTAL EDGES

Both liquid and solid surfaces are intrinsically stressed. Solid surfaces exhibit capillary (or surface stress-induced) phenomena similar to the familiar Laplace capillary pressure near a curved liquid surface, and a strain field is generated at a curved surface of a solid. The existence of a surface stress-induced strain field was pointed out by Marchenko and Parshin (1980) who considered the elastic energy of a solid including both bulk and surface contributions:

$$E_{\text{elastic}} = \frac{1}{2} \int \lambda_{ijlm} \varepsilon_{ij} \varepsilon_{lm} \, dV + \int \tau_{\alpha\beta}(\hat{\mathbf{m}}) \varepsilon_{\alpha\beta} \, dA \qquad (3.8)$$

The elastic strain field in the crystal is created by the force

$$F_i = \frac{\partial \tau_{i\beta}}{\partial r_\beta}. \qquad (3.9)$$

The ultimate case of a curved solid surface is a sharp edge between neighboring facets. Figure 3.2 depicts the force balance on the curved surface of a liquid and at the crystal edge and thus demonstrates the similarity between capillary effects in liquids and solids. The surface tension on a curved liquid surface results in an excess pressure below the curved surface, and, similarly, the intrinsic surface stress of crystal surfaces causes an effective force applied to the edge of the crystal. This force monopole generates the strain field that affects significantly the energetics of faceted surfaces and promotes the formation of a periodic structure of facets.

2.3.4 SPONTANEOUS FORMATION OF PERIODICALLY FACETED SURFACES

If a planar crystal surface is unstable and breaks up into a system of tilt facets, the conservation of the average orientation of the normal to the surface (Eq. 3.5) implies the coexistence of alternating facets (Fig. 3.1a). At the intersection of neighboring facets there appear either sharp crystal edges or narrow rounded parts of the surface. Both types of intersection may be described as linear defects at the surface. These linear defects give a short-range contribution into the surface free energy and a long-range contribution due to elastic strain energy.

If the order parameter related to the phase transition at the surface is linearly coupled to the strain field, i.e. if a linear striction effect exists, formation of a periodic structure, whose period can be macroscopically large, is favored (Andreev, 1980). For the faceting phase transition the possibility of the formation of a faceted surface with a macroscopic period has been pointed out by Andreev and Kosevich (1981).

The theory of a periodically faceted surface has been developed in Marchenko (1981a, 1981b), predicting the period of equilibrium structures. Following Marchenko (1981a, 1981b), we consider the faceted surface with a one-dimensional periodic saw-tooth profile depicted in Fig. 3.3. The free energy per unit projected area equals

$$F = F_{\text{surf}} + E_{\text{edges}} + \Delta E_{\text{elastic}} \qquad (3.10)$$

Here F_{surf} is the free energy of tilted facets, E_{edges} is the short-range energy of the edges, and $\Delta E_{\text{elastic}}$ is the elastic energy due to the discontinuity of the surface stress tensor τ_{ij} at the crystal edges.

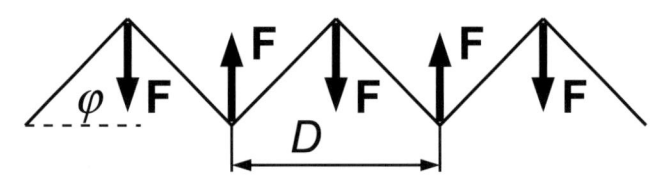

Fig. 3.3 Saw-tooth profile of the faceted surface. Effective forces of alternating sign are applied to neighboring edges

The free energy of tilted facets per unit projected area depends only on the orientation of the facets, $E_{\text{surf}} = \gamma(\varphi)\sec(\varphi)$, and does not depend on the period of the structure D. The short-range energy of the edges per unit projected area equals

$$E_{\text{edges}} = \frac{\eta}{D} \equiv \frac{\eta^+(\varphi) + \eta^-(\varphi)}{D} \qquad (3.11)$$

where $\eta^+(\varphi)$ is the short-range energy of the convex edge per unit length of the edge and $\eta^-(\varphi)$ denotes the same energy for the concave edge. The short-range energy can be roughly considered as a contribution of additional dangling bonds due to the edge formation. This term becomes dominant for a very small facet period.

E_{elastic} is given by Eq. (3.8), which contains both linear and quadratic terms as a function of strain and yields zero energy in the absence of strain. Therefore, the elastic strain energy in the equilibrium is *negative*, which corresponds to the relaxation of the surface stress at crystal edges. This negative elastic energy will be termed below the elastic relaxation energy due to crystal edges. The effective elastic forces acting at the edges are displayed in Fig. 3.3. *Force monopoles* are acting at each edge and forces applied to neighboring edges are balanced, so that the total force applied to the system vanishes. Elastic strains generated by linear crystal edges propagate into the crystal across a distance of the order of D and decay at larger distances from the surface. Since the strain field is generated at linear defects at the surface, namely at the linear crystal edges, the elastic relaxation energy depends logarithmically on the period D of the structure:

$$\Delta E_{\text{elastic}} = -\frac{\tilde{C}(\varphi)F^2}{YD}\ln\left(\frac{D}{a}\right) = -\frac{C(\varphi)\tau^2}{YD}\ln\left(\frac{D}{a}\right) \qquad (3.12)$$

Here τ is the characteristic value of the intrinsic surface stress tensor, Y is Young's modulus, a is the lattice parameter, $C(\varphi)$ is the geometric factor which accounts for the particular symmetry of the tensor τ_{ij}, the elastic anisotropy of the crystal, etc.

By substituting Eqs. (3.11) and (3.12) into Eq. (3.10), one obtains the following expression for the free energy of the faceted surface per unit projected area:

$$F = \frac{\gamma(\varphi)}{\cos\varphi} + \frac{\eta(\varphi)}{D} - \frac{C(\varphi)\tau^2}{YD}\ln\left(\frac{D}{a}\right) \qquad (3.13)$$

The dependence of the free energy versus the period of the faceted surface D is displayed in Fig. 3.4. Due to the logarithmic dependence of the elastic relaxation

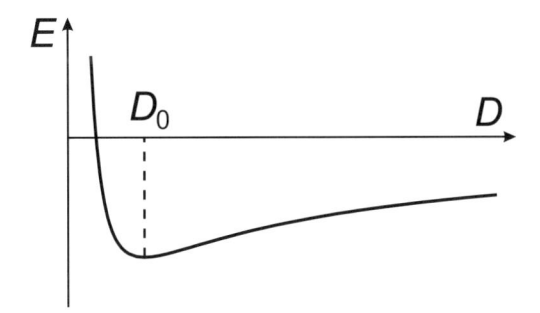

Fig. 3.4 Energy of a periodically faceted surface versus the period D. There always exists an optimum period of faceting D_{opt} due to the logarithmic dependence of the elastic relaxation energy on the period D

energy on the period D, there always exists an optimum period of faceting D_{opt} equal to

$$D_{opt} = a \exp\left[\frac{\eta(\varphi)Y}{C(\varphi)\tau^2} + 1\right] \quad (3.14)$$

All material parameters that enter the exponential in Eq. (3.14) have typical atomic values. Therefore the argument of the exponential is of the order of 1. Since the exponential function is steep, the argument that can eventually be equal to about 3 or more results in a macroscopic period D which exceeds the lattice parameter a at least by an order of magnitude. In this case a macroscopic approach is indeed justified.

It should be noted that the free energy of facets, i.e. the first term in Eqs. (3.10) and (3.13), contains the entropy contribution which describes the dependence of the faceting itself and of the facet orientation on temperature. However, there exists another entropy contribution to F associated with possible deviations of the structure from a perfect periodic structure (the configuration entropy). For periodically faceted surfaces, the configuration entropy has not been considered in the literature so far. Therefore the present discussion refers, strictly speaking, to the case $T=0$. In the following we will omit everywhere the entropy contribution to F and will discuss only the total energy of the system.

Periodic faceting with a macroscopic period has been observed on various surfaces of different materials, e.g. on surfaces vicinal to Si(111) (Hibino *et al.*, 1993; Williams *et al.*, 1993), on surfaces vicinal to GaAs(001) (Kasu and Kobayashi, 1993; Golubok *et al.*, 1994; Ledentsov *et al.*, 1994a, 1994b), on surfaces vicinal to Pt(001) (Watson *et al.*, 1993), on high-index surfaces of GaAs (Lötzel *et al.*, 1991) and Si (Baski and Whitman, 1995), on the low-index surface TaC(110) (Zuo *et al.*, 1993), and on Ir(110) (Koch *et al.*, 1991). All the above-cited experimental observations of faceting refer to conditions of thermal annealing and point to equilibrium structures.

3.3 PERIODIC ARRAYS OF MACROSCOPIC STEP BUNCHES

An important example of self-organized nanostructures on crystal surfaces is the spontaneous formation of a quasi-periodic array of step bunches on a vicinal surface. A homogeneous vicinal surface of a crystal consists of planar terraces with low Miller indices, neighboring terraces being separated by equidistant monoatomic or monomolecular steps. The phenomenon of step bunching has been known for a long time as one of various possible kinetic instabilities during crystal growth on vicinal surfaces (see, for example, Chernov, 1984). Recent experiments on surfaces vicinal to Si(111) (Williams et al., 1993) or GaAs(001) (Kasu and Kobayashi, 1993; Golubok et al., 1994; Ledentsov et al., 1994a, 1994b) have revealed step bunching, which appears after the annealing of a cleaved surface or during growth interruption. This indicates an *equilibrium* mechanism driving step bunching. The height of step bunches was found to be homogeneous throughout the sample. The observation of equal-width terraces (or equidistant step bunches) was reported for GaAs(001) (Kasu and Kobayashi, 1993; Golubok et al., 1994; Ledentsov et al., 1994a, 1994b). The height of step bunches (7–15 monolayers) allows them to be considered as macroscopic step bunches forming new facets (Fig. 3.5). Another example of faceting of a vicinal surface is its breaking into two vicinal surfaces with different concentration of steps (Williams et al., 1993).

Figure 3.5 depicts a periodic array of step bunches of a macroscopic height. The discontinuity of the intrinsic surface stress tensor τ_{ij} at the edges results in effective force monopoles acting at the edges. Thus, the elastic interaction between the two edges of the same step bunch is a monopole–monopole interaction. Force monopoles acting at the two edges of the same bunch compensate each other, and the strain field due to a single step bunch is equal at large distances to a strain field of an elastic dipole. Therefore, the elastic interaction between step bunches is the dipole–dipole interaction which decreases with the separation L as L^{-2}, which is similar to the elastic interaction between steps of a microscopic height (Marchenko and Parshin, 1980). This means a very fast decrease of the strength of the interaction between step

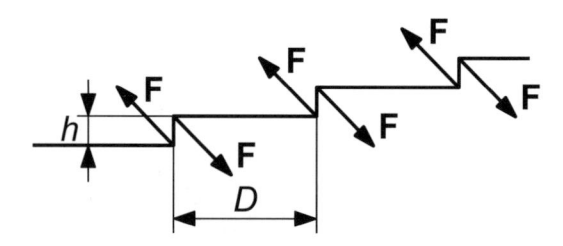

Fig. 3.5 A periodic array of macroscopic step bunches resulting from faceting of a vicinal surface. Force monopoles are acting at the edges of the structure. The elastic interaction between the two edges of one step bunch is a monopole–monopole interaction, whereas the interaction between different step bunches has a dipole–dipole character

bunches with period. The total energy per unit surface area equals (dipole–dipole interaction and higher terms are truncated)

$$E = \gamma_0 + \gamma_1 \frac{h}{D} + \frac{C_1}{D} - \frac{C_2}{D} \ln\left(\frac{h}{a}\right) \quad (3.15)$$

Since the average orientation of the faceted surface in Fig. 3.5 coincides with the orientation of the initially homogeneous vicinal surface, there exists a relation between the height of macroscopic step bunches h and the period D of the structure, $h = D\varphi$, where φ is the miscut angle of the initially homogeneous vicinal surface. By using this relation, it is possible to rewrite Eq. (3.15) in a form similar to Eq. (3.13), $E = \gamma_0 + \gamma_1 \varphi + \varphi[C_1 h^{-1} - C_2 h^{-1} \ln(h/a)]$. Due to the logarithmic dependence of the elastic relaxation energy on the height of the step bunch, there always exists an optimum equilibrium height of step bunches. At the same time, for weakly misoriented surfaces having a large period, contrary to the step height control, the periodicity control is weak in view of the relatively weak dipole–dipole interaction between neighboring step bunches and one can speak rather of 'quasi-periodic' arrays.

3.4 HETEROEPITAXIAL GROWTH ON CORRUGATED SUBSTRATES

Periodic surface faceting gives a possibility for direct fabrication of ordered arrays of quantum wires or quantum dots depending on the type of the array of facets (1D or 2D). The growth of a deposited material 2 on a faceted surface of material 1 allows, in principle, the fabrication of quantum wires provided the growth proceeds in the grooves of the faceted substrate. Here we focus on the heteroepitaxial growth in the system GaAs/AlAs where the two materials are nearly lattice-matched. Examples of the heteroepitaxial growth on faceted surfaces in the GaAs/AlAs system are the growth on a faceted surface vicinal to (001) (Kasu and Kobayashi, 1993) on faceted (311) surfaces (Alferov et al., 1988; Nötzel et al., 1991), and faceted (110) surfaces (Inoue et al., 1993; Takeuchi et al., 1995). Also the Ge/Si system has been explored (Brunner et al., 1998).

The theory of quasi-equilibrium heterophase growth on periodically corrugated substrates has been developed by Shchukin et al. (1995b). The present consideration is focused on the practical case where both the surface of material 1 and the surface of material 2 are unstable against faceting. This is true for both GaAs and AlAs surfaces vicinal to (001) (Kasu and Kobayashi, 1993) and for (311) surfaces of GaAs and AlAs (Nötzel et al., 1991). Several possibilities for the morphology of the heterophase system are depicted in Fig. 3.6.

The total energy of the heterophase system equals (Shchukin et al., 1995b)

$$E = E_{\text{surf}} + E_{\text{interface}} + E_{\text{edges}} + \Delta E_{\text{elastic}} \quad (3.16)$$

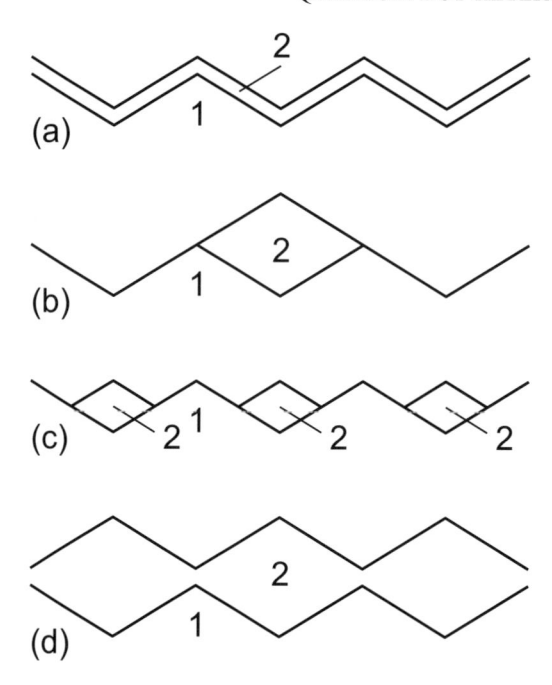

Fig. 3.6 Structures of a heterophase system: (a) homogeneous coverage, (b) system with separated 'thick' clusters, (c) system of 'thin' clusters, (d) heterophase system at large coverage, where the periodic surface corrugation is restored and the hills of the top surface of the heterophase system appear over the valleys of the substrate, and vice versa. The heterophase system contains a continuous layer of material 2 with periodically modulated thickness

Here, in addition to the three contributions to the energy of the faceted surface (see Eq. 3.10), the contribution of the interface energy has to be included. A comparison of total energies for several distinct types of the heterophase structures depicted in Fig. 3.6 has been carried out by Shchukin *et al.* (1995b), leading to the following conclusions. The selection between two possible growth modes is determined by the fact of whether the deposited material wets or does not wet the substrate. If the deposited material wets the substrate, then the homogeneous coverage of the periodically corrugated substrate occurs (Fig. 3.6a). An example is the growth of AlAs on a periodically corrugated vicinal surface of GaAs(001) 3° off towards [1-10] (Kasu and Kobayashi, 1993).

If the deposited material does not wet the substrate, then isolated clusters of the deposited material appear on periodically corrugated substrate (Fig. 3.6b and c). This situation is likely to be realized for the growth of GaAs on vicinal surfaces of AlAs(001) 3° off towards [1-10] (Kasu and Kobayashi, 1993) and for both GaAs/AlAs and AlAs/GaAs heterophase growth on (311)A surfaces (Nötzel *et al.*, 1991; Alferov *et al.*, 1992).

In the case of inhomogeneous cluster coverage the periodic surface corrugation is restored after the deposition of the first several monolayers. Then the hills of the top surface of the heterophase system appear over the valleys of the substrate, and vice versa, and a continuous layer of the deposited material with periodically modulated thickness is formed (Fig. 3.6d). Thus, the formation of clusters allows direct fabrication of quantum wires and quantum wire superlattices in heterophase semiconductor systems. Since any periodically faceted surface represents a structure of elastic domains, its geometrical parameters (e.g. the period) can be tuned in a controlled fashion by applying external stress (Shchukin *et al.*, 1995a).

To conclude the present section, we emphasize that the spontaneous periodic faceting of semiconductor surfaces and the cluster growth in grooves gives the possibility of direct fabrication of isolated quantum wires and dots, quantum wire and quantum dot superlattices, and quantum well superlattices with modulated thickness of quantum wells.

3.5 ORDERED ARRAYS OF PLANAR SURFACE DOMAINS

Another important class of self-ordered nanostructures is associated with periodically ordered structures of planar surface domains. Surface domain structures occur if different phases can coexist on the surface, e.g. phases of (2×1) and (1×2) surface reconstruction of Si(001), monolayer-height islands in heterophase systems, etc. Neighboring domains possess different values of the intrinsic surface stress tensor τ_{ij}, which gives rise to elastic relaxation. The existence of surface domain structures governed by the discontinuity of τ_{ij} has been predicted by Marchenko (1981a, 1981b). Although the geometry of these structures is very different from that of periodically faceted surfaces, the energetics is basically the same. Figure 3.7 depicts force monopoles that are acting at domain boundaries. These force monopoles lead to elastic relaxation. The total energy of the domain structure per unit surface area equals

$$E = E_{\text{surf}} + E_{\text{boundaries}} + \Delta E_{\text{elastic}} \quad (3.17)$$

The surface energy E_{surf} does not depend on the period of the structure D. The energy of domain boundaries equals $E_{\text{boundaries}} = C_1 \eta D^{-1}$. The elastic relaxation

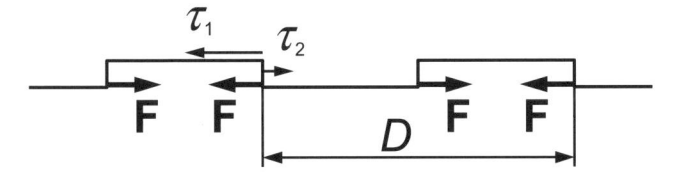

Fig. 3.7 Effective forces applied to domain boundaries of a system of planar surface domains

energy is $\Delta E_{\text{elastic}} = -C_2(\Delta\tau)^2 Y^{-1} D^{-1} \ln(D/a)$. Due to the logarithmic dependence of the elastic relaxation energy on the period of the structure D, the total energy (Eq. 3.16) always has a *minimum* at a certain optimum period:

$$D_{\text{opt}} = a \, \exp\left[\frac{C_1 \eta Y}{C_2(\Delta\tau)^2} + 1\right] \qquad (3.18)$$

The best-known system exhibiting surface stress domain structures is the Si(001) surface and corresponding vicinal surfaces. Coexisting domains are domains of (2 × 1) and (1 × 2) surface reconstruction separated by single-height atomic steps. Such a domain structure was first observed by Men *et al.* (1988), and the explanation was given by Alerhand *et al.* (1988). The comparison of the measured period of the domain structure with Eq. (3.17) allows the value of $\Delta\tau$ to be extracted from experiment (provided the short-range energy of single-height atomic steps is known) and $\Delta\tau$ to be compared with results from first-principle calculations. A detailed discussion may be found, for example, in Dabrowski *et al.* (1994). Force monopoles applied to single-height atomic steps lead to a variety of structures of vicinal surfaces as a function of miscut angle. These structures include transition from single-layer atomic steps to double-layer steps, wavy steps, and hill-and-valley structures. The review of both experimental and theoretical works on these structures may be found, for example, in Mukherjee *et al.* (1994). The transition from single-layer to double-layer steps (Grundmann *et al.*, 1991a) was used to grow antiphase-domain free layers of group III–V materials on Si(001) (Grundmann *et al.*, 1991b).

Another class of surface-stress domain structures includes monolayer-height islands formed in heterophase systems at submonolayer coverage. Periodic domains of stripes have been observed in the O/Cu(110) system (Kern *et al.*, 1991) or the InAs/GaAs(001) system (Bressler-Hill *et al.*, 1994).

Force monopoles acting at opposite boundaries of a given domain are balanced, and the interaction between domains at large distances again has a dipole–dipole character. Similar to elastic domain structures, domain structures are known for systems with dipole–dipole electrostatic interactions (Langmuir monolayers at the water/air interface) or for systems with dipole–dipole magnetic interactions in ferromagnetic films or in planar-confined ferrofluid/water mixtures in magnetic fields. Corresponding references may be found in Kwok On and Vanderbilt (1995).

Although the scaling behavior of the energy is similar for all systems mentioned, the particular pattern of the domain structure depends strongly on the orientational dependence of the long-range interaction and on the energies of domain boundaries. For an isotropic interaction, there exists a transition from a 1D array of stripes to a 2D hexagonal array of discs (Vanderbilt, 1992; Kwok On and Vanderbilt, 1995). Discs are more favorable if the coverage of the surface is close to 0 or to 1. Anisotropic interactions, e.g. elastic interactions (Fig. 3.8), and anisotropic domain boundaries, e.g. crystal surfaces, favor striped domains over disc-shaped domains.

A relation between the size of a single surface domain (for low domain concentration) and the minimum separation between them at an intermediate

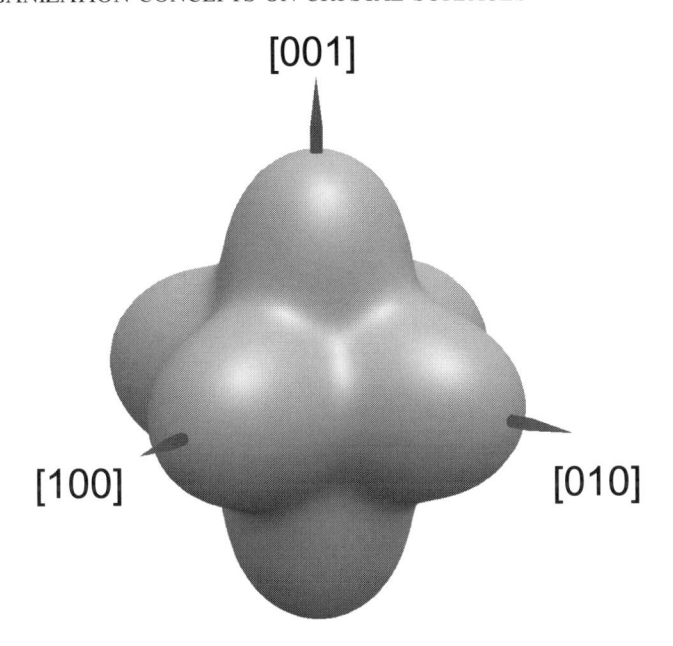

SELF-ORGANIZATION CONCEPTS ON CRYSTAL SURFACES

Fig. 3.8 $1/Y$, Y being Young's modulus for different directions in GaAs, visualizing the anisotropy of elastic properties in zincblende crystals

concentration has been established by Zeppenfeld *et al.* (1994). In the particular case of monolayer-height islands, this result means that there exists an optimum island size for a dilute array of islands. The existence of an optimum size of islands implies that two-dimensional strained islands in the heterophase system do not undergo Ostwald (1900) ripening.

Similar results are valid for a dilute array of strained islands of macroscopic height for the case where islands have a planar top surface, and the height of the islands is *kinetically limited* to a value considerably smaller than the lateral size (Tersoff *et al.*, 1993). The global geometry of such islands is similar to the geometry of two-dimensional islands. The important result is that there exists an optimum size in a system of planar strained islands and no Ostwald ripening occurs.

3.6 ORDERED ARRAYS OF THREE-DIMENSIONAL COHERENTLY STRAINED ISLANDS

3.6.1 HETEROEPITAXIAL GROWTH IN LATTICE-MISMATCHED SYSTEMS

There are three well-known modes of heteroepitaxial growth: Frank–van der Merwe (1949), Volmer–Weber (1926) and Stranski–Krastanow (1937). They represent layer-by-layer growth (FvdM, 2D), island growth (VW, 3D), and layer-by-layer

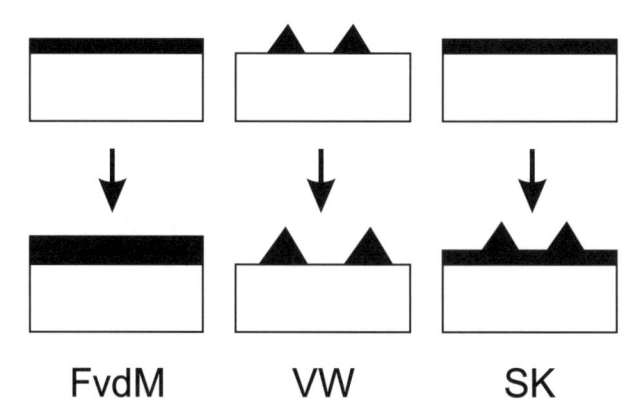

Fig. 3.9 Schematic diagrams of the three possible growth modes: Frank–van der Merwe (FvdM), Volmer–Weber (VW), and Stranski–Krastanow (SK)

plus islands (SK) (Fig. 3.9). The particular growth mode for a given system depends on the interface energies and on the lattice mismatch.

In lattice-matched systems, the growth mode is solely governed by the interface and surface energies. If the sum of the epilayer surface energy γ_2 and of the interface energy γ_{12} is lower than the energy of the substrate surface, $\gamma_2 + \gamma_{12} < \gamma_1$, i.e. if the deposited material *wets* the substrate, the FvdM mode occurs. A change in $\gamma_2 + \gamma_{12}$ alone may drive a transition from the FvdM to the VW growth mode. For a strained epilayer with small interface energy, initial growth may occur layer by layer, but a thicker layer has a large strain energy and can lower its energy by forming isolated islands in which strain is relaxed. Thus the SK growth mode occurs. It was traditionally believed that islands formed in the SK growth mode are dislocated. However, experiments on InAs/GaAs (001) (Goldstein et al., 1985) and on Ge/Si(001) (Eaglesham and Cerullo, 1990; Mo et al., 1990) have demonstrated the formation of three-dimensional *coherently* strained islands.

The relaxation of the elastic energy due to the formation of coherently strained islands is related to the Asaro–Tiller–Grinfield instability (Asaro and Tiller, 1972; Grinfield, 1986; Srolovitz, 1989; Spencer et al., 1991) of a strained layer against a long-wavelength corrugation of the surface. To illustrate the physical mechanism of the elastic relaxation, it is convenient to consider a strongly pronounced corrugation. Among examples of such strongly pronounced corrugations are islands (Vanderbilt and Wickham, 1991), troughs (Vanderbilt and Wickham, 1991), surface cusps (Jesson et al., 1993) and cracks (Yang and Srolovitz, 1993). The formation of troughs, cusps, and cracks can occur in a strained epitaxial film of a certain macroscopic thickness under annealing. For the first stages of heteroepitaxial growth on a substrate, the formation of islands seems to be the only coherent mechanism of elastic relaxation.

Figure 3.10 demonstrates two islands of a different shape. A flat island with a small height/width ratio is practically nonrelaxed, whereas a hypothetical island

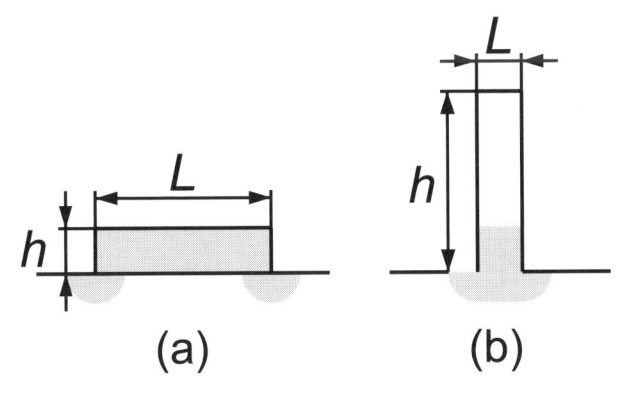

Fig. 3.10 Effect of the island shape on the volume elastic relaxation of a coherently strained island. The grey area represents the part with a large strain energy density. (a) The island with a height/width ratio $h/L \ll 1$ is not relaxed, while (b) the island with a height/width ratio $h/L \gg 1$ is almost completely relaxed

having the shape of a whisker with a large height/width ratio is almost completely relaxed. Thus the elastic relaxation depends strongly on the island shape. For a given shape, the elastic relaxation energy is proportional to the volume of the island.

Thus the volume elastic relaxation of coherently strained islands is an alternative mechanism of relaxation which competes with the formation of dislocations. The theory developed by Vanderbilt and Wickham (1991) compares the two mechanisms of elastic relaxation and yields a phase diagram of a lattice-mismatched system where all possible morphologies are present, i.e. uniform films, dislocated islands, and coherent islands (Fig. 3.11). The formation of an island from a uniform film is accompanied by a relaxation of the elastic energy, $\Delta E^V_{elastic} < 0$, and by a change of the surface area, $\Delta A < 0$. The size of the corresponding change of the surface energy depends on the formation of side facets of the islands and on the disappearance of certain areas of the planar surface. It is usually believed that the change of the surface energy caused by the formation of islands is *positive*, $\Delta E_{surf} > 0$. It was shown by Vanderbilt and Wickham (1991) that the morphology of the mismatched system is determined by the relation between ΔE_{surf} and the energy of the dislocated interface E^{disl}_{interf}. The ratio of these two energies, denoted $\Gamma = E^{disl}_{interf}/\Delta E_{surf}$, is the control parameter that governs the morphological phase diagram of Fig. 3.11.

If ΔE_{surf} is positive and large, or the energy of the dislocated interface is relatively small, the corresponding value Γ on the phase diagram of Fig. 3.11 is smaller than Γ_0. Then formation of coherently strained islands is not favorable. With the increase of the amount Q of the deposited material, a transition occurs from a uniform film to dislocated islands (or a dislocated film), and coherently strained islands are not formed.

If ΔE_{surf} is positive and small, or the energy of the dislocated interface is relatively large, the corresponding value Γ on the phase diagram of Fig. 3.11 is larger than Γ_0.

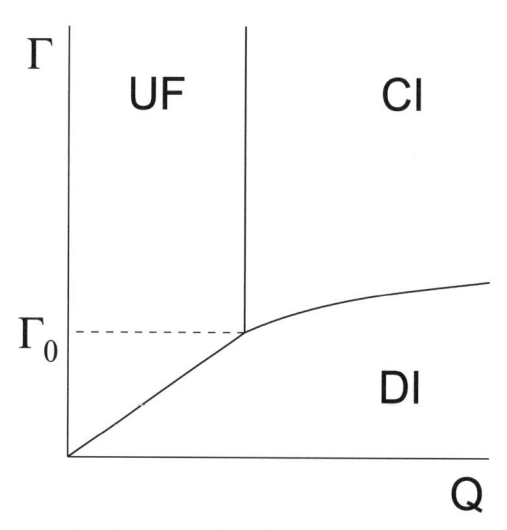

Fig. 3.11 Phase diagram showing the preferred morphology as a function of the amount of deposited material Q (horizontal axis) and of the quantity $\Gamma = E_{\text{interf}}^{\text{disl}}/\Delta E_{\text{surf}}$, where ΔE_{surf} is the change of the surface energy due to island formation and $E_{\text{interf}}^{\text{disl}}$ is the energy of a dislocated interface. Labels UF, CI, and DI refer to 'uniform film', 'coherent island', and 'dislocated island', respectively. (After Vanderbilt and Wickham, 1991)

With the increase of the amount of the deposited material, a transition from a uniform film to *coherent islands* occurs. Further deposition or growth interruption may cause the onset of dislocations in the islands. The theory of Vanderbilt and Wickham (1991) deals with islands having the shape of elongated prisms ('ridges'). The theory of Ratsch and Zangwill (1993) yields the existence of these morphologies, i.e. uniform films, coherent islands, and dislocated islands in the different case of pyramid-shaped islands.

If the change in the surface energy due to island formation is positive and small, coherent islands can be formed and the total energy of the island can be written as $E_{\text{island}} = E_{\text{edges}} + \Delta E_{\text{surf}} - E_{\text{relaxation}}^{\text{V}}$. E_{edges} denotes the short-range energy of the island edges, $\Delta E_{\text{surface}}$ the excess surface energy of the island, and $-E_{\text{relaxation}}^{\text{V}}$ the energy of elastic relaxation of the volume strain caused by lattice mismatch with the substrate. The scaling behavior is $E_{\text{edges}} \sim L (E_{\text{edges}} = AL)$, $E \sim L^2$ ($E_{\text{surf}} = BL^2$) and $E_{\text{relaxation}}^{\text{V}} \sim L^3$ ($E_{\text{relaxation}}^{\text{V}} = CL^3$), where L is the size of the island. If the contribution of the surface energy dominates over the short-range edge energy, island formation is energetically favorable only for cases where the island size exceeds a critical value $L_{\text{crit}} = 2B/3C$. Smaller islands even occasionally formed due to statistical fluctuations will dissolve.

When coherent islands with a size larger than L_{crit} are formed, the common belief is that they undergo Ostwald (1900) ripening in order to reduce the overall surface of the islands, thus reducing the total surface energy of the system. The process of ripening implies the growth of large islands at the expense of the evaporation of small islands. Thus islands undergoing ripening should become dislocated at some point.

We note that in the case $\Delta E_{\text{surf}} < 0$ a critical island size exists, but is defined mostly by the interplay between the short-range edge energy term and the negative surface energy term ($L_{\text{crit}} \approx A/B$)). The critical island size is generally much smaller in this case. After islands exceed a critical size they undergo ripening to reduce the short-range edge energy. However, if the total amount of the deposited material is fixed, at some point ripening becomes energetically unfavorable as the growth of the size of an island will reduce the total area of the wetting layer not covered by islands and having a higher surface energy. Therefore at some characteristic island size, defined by the balance of short-range edge energy of all islands and the total negative surface energy, ripening stops.

Experimental studies of coherent islands of InGaAs/GaAs(001) and InAs/GaAs(001) systems have revealed a surprisingly narrow size distribution of the islands (Leonard *et al.*, 1993; Ledentsov *et al.*, 1994a, 1994b; Moison *et al.*, 1994; Grundmann *et al.*, 1995b), which does not follow from the SK growth mode itself. Besides that, in Ledentsov *et al.* (1994a), Bimberg *et al.* (1995), Cirlin *et al.* (1995) and Grundmann *et al.* (1995b) have reported that coherent islands of InAs form, under certain conditions, a quasi-periodic square lattice on the GaAs(001) surface. The periodicity and the island size do not change with time upon growth interruptions. A new class of self-ordered nanostructures, namely ordered arrays of coherently strained three-dimensional islands, was discovered.

3.6.2 ENERGETICS OF A DILUTE ARRAY OF 3D ISLANDS

The energetics is considered of an array of three-dimensional coherent strained islands under the constraint of a fixed amount of deposited material Q assembled in all islands, where Q is defined in numbers of monolayers. Both the substrate and the deposited material are treated as elastically anisotropic cubic media with equal elastic moduli λ_{ijlm}, and the lattice mismatch between the two materials being equal, $\varepsilon_0 = \Delta a/a$, where a is the lattice spacing. The energy of the uniformly strained film of the thickness (Qa) on the (001) substrate is given by $E_{\text{EL}}^{(0)} = \lambda \varepsilon_0^2 A(Qa)$, where the elastic modulus λ is $\lambda = (C_{11} + 2C_{12})(C_{11} - C_{12})/C_{11}$, C_{11} and C_{12} are elastic compliances (see Section 5.1.1), and A is the surface area.

The energy of the heterophase lattice-mismatched system with islands is equal:

$$\tilde{E}_{\text{total}} = \tilde{E}_{\text{elastic}} + \tilde{E}_{\text{surf}} + \tilde{E}_{\text{edges}} \tag{3.19}$$

where

$$\tilde{E}_{\text{elastic}} = \frac{1}{2} \int \lambda_{ijlm} [\varepsilon_{ij}(\mathbf{r}) - \varepsilon_0 \delta_{ij} \vartheta(\mathbf{r})][\varepsilon_{lm}(\mathbf{r}) - \varepsilon_0 \delta_{lm} \vartheta(\mathbf{r})] \mathrm{d}V$$

$\varepsilon_{ij}(\mathbf{r})$ being the strain tensor, $\vartheta(\mathbf{r}) = 1$ in the deposited material, and $\vartheta(\mathbf{r}) = 0$ in the substrate. The surface energy per unit area γ is *renormalized* in the strain field:

$$\gamma(\varepsilon_{\alpha\beta}) = \gamma_0 + \tau_{\alpha\beta}(\varepsilon_{\alpha\beta} - \varepsilon_0\delta_{\alpha\beta}\theta) + \frac{1}{2}S_{\alpha\beta\mu\nu}(\varepsilon_{\alpha\beta} - \varepsilon_0\delta_{\alpha\beta}\theta)(\varepsilon_{\mu\nu} - \varepsilon_0\delta_{\mu\nu}\theta) + \cdots$$

where $\theta = 1$ for the surface of the deposited material and $\theta = 0$ for the substrate surface, $\tau_{\alpha\beta}$ is the intrinsic surface stress tensor, $S_{\alpha\beta\mu\nu}$ is the tensor of the surface excess elastic moduli (Andreev, 1981), and α,β,μ,ν are 2D indices in the local facet plane. The total *renormalized* surface energy of the heterophase system equals

$$\tilde{E}_{\text{surf}} = \int \gamma_0(\hat{\mathbf{m}}) + \tau_{\alpha\beta}(\hat{\mathbf{m}})[\varepsilon_{\alpha\beta}(\mathbf{r}) - \varepsilon_0\delta_{\alpha\beta}\theta(\mathbf{r})]$$
$$+ \frac{1}{2}S_{\alpha\beta\mu\nu}(\hat{\mathbf{m}})[\varepsilon_{\alpha\beta}(\mathbf{r}) - \varepsilon_0\delta_{\alpha\beta}\theta(\mathbf{r})][\varepsilon_{\mu\nu}(\mathbf{r}) - \varepsilon_0\delta_{\mu\nu}\theta(\mathbf{r})](\hat{\mathbf{m}}\cdot\hat{\mathbf{n}})^{-1}\,dA \quad (3.20)$$

Here $\hat{\mathbf{m}} = \hat{\mathbf{m}}(\mathbf{r})$ is the local normal to the facet, $\hat{\mathbf{n}} = (0, 0, 1)$ is the normal to the flat surface, and the integration in Eq. (3.20) is carried out over the reference flat surface. The surface energy $\gamma_0(\hat{\mathbf{m}})$ usually has cusped local minima for low-index facets (Herring, 1951a). We consider here the situation where the quantity $\gamma_0(\hat{\mathbf{m}})(\hat{\mathbf{m}}\cdot\hat{\mathbf{n}})^{-1}$ has, besides the cusped absolute minimum for the (001) surface, also cusped local minima for four equivalent facets: $(k0l)$, $(0kl)$, $(\bar{k}0l)$, and $(0\bar{k}l)$. Then, under a range of conditions, the island will be bounded by $(k0l)$, $(0kl)$, $(\bar{k}0l)$, and $(0\bar{k}l)$ facets, e.g. it will have the shape of a pyramid or of a prism, which are shown in the inset of Fig. 3.12. The third term in Eq. (3.19) is the short-range energy of edges.

For lattice-mismatched systems with edges, the total strain field is the sum of two contributions, one due to the lattice mismatch and the other due to the discontinuity of the intrinsic surface stress tensor $\tau_{\alpha\beta}$ at the edges. By evaluating the change of the energy of the heterophase system due to the formation of a single island of a characteristic size L, one obtains (Shchukin *et al.*, 1995c)

$$\tilde{E}(L) = \left[-f_1(\varphi_0)\lambda\varepsilon_0^2 L^3 + (\Delta\Gamma)L^2 - \frac{f_2(\varphi_0)\tau^2}{\lambda}L\ln\left(\frac{L}{2\pi a}\right) + f_3(\varphi_0)\,\eta L\right] \quad (3.21)$$

Here $\varphi_0 = \tan^{-1}(k/l)$ is the tilt angle of island facets. The first term in Eq. (3.21) is the energy of the volume elastic relaxation $\Delta\tilde{E}^{\text{V}}_{\text{elastic}}$; it is always negative. The second term is the change of the renormalized surface energy of the system due to island formation, $\Delta\tilde{E}^{\text{renorm}}_{\text{surf}} = (\Delta\Gamma)L^2$, where $(\Delta\Gamma) = \gamma_0(\varphi_0)\sec\varphi_0 - \gamma_0(0) - g_1(\varphi_0)\tau\varepsilon_0 - g_2(\varphi_0)S\varepsilon_0^2$. The bare difference of the surface energies of the tilted facet and the planar surface, $\gamma_0(\varphi_0)\sec\varphi_0 - \gamma_0(0) > 0$, indicates that an unstrained planar surface is stable against faceting. Due to the renormalization terms, $(\Delta\Gamma)$ can be *of either sign*. The third term in Eq. (3.21) is the contribution of the edges of the island to the elastic relaxation energy, $\Delta\tilde{E}^{\text{edges}}_{\text{elastic}} \sim -L\ln L$; it is always negative. The fourth term in Eq. (3.21) is the short-range energy of edges, where η is the characteristic energy per unit length of the edge. The coefficients f_1, f_2, f_3 depend only on the tilt angle of facets φ_0.

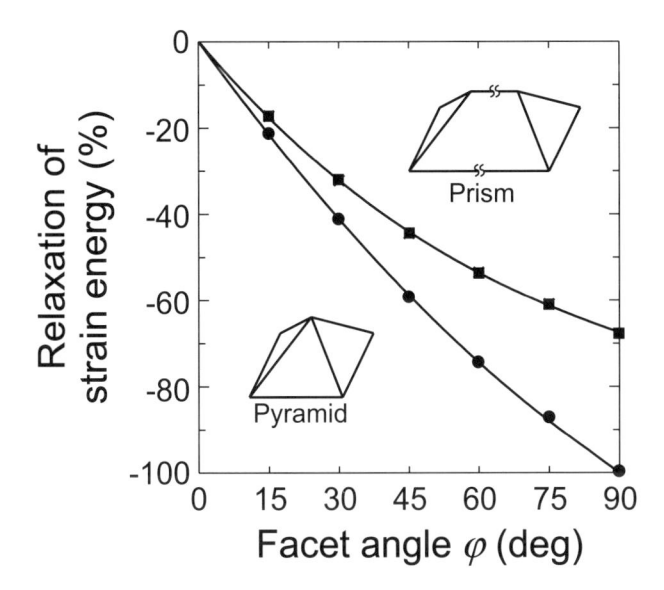

SELF-ORGANIZATION CONCEPTS ON CRYSTAL SURFACES

Fig. 3.12 Relaxation of elastic strain energy for 3D coherently strained islands versus the tilt angle of island facets for pyramids (lower curve) and for elongated prisms (upper curve). Circles and squares represent results of calculations by the finite element method, which are fitted by the solid lines

3.6.3 ORDERING OF 3D ISLANDS IN SHAPE

For a dilute system of islands, where the average distance between the islands is large compared to the island size L, the equilibration of the island shape by atomic migration on the island is faster than material exchange between islands. Then for any given volume of an island, there exists an equilibrium shape. For sufficiently large islands, the first two terms in the island energy (Eq. 3.21), $\Delta \tilde{E}^V_{elastic}$ and $\Delta \tilde{E}^{renorm}_{surf}$, are the two dominant ones.

We focus on the situation here where the surface free energy of tilted facets of the deposited material per unit projected area, $\gamma_0(\hat{m})(\hat{m} \cdot \hat{n})^{-1}$, has cusped local minima for four equivalent facets: $(k0l)$, $(0kl)$, $(\bar{k}0l)$, and $(0\bar{k}l)$. It yields cusped minima in $\Delta \tilde{E}^{renorm}_{surf}$ which fix the tilt angle of island facets for a certain interval of island volumes (or sizes). Then, we consider islands bounded by four facets with the same tilt angle relative to the substrate, i.e., by facets $(k0l)$, $(0kl)$, $(\bar{k}0l)$, and $(0\bar{k}l)$.

Comparison of pyramids and prisms is presented in Fig. 3.12 where the volume elastic relaxation energy versus the tilt angle of facets is displayed for a square-based pyramid and for an infinitely elongated prism. For the prism, the component of the strain field in the direction along the prism remains equal to ε_0 throughout the entire volume of the island, whereas for the pyramid, all components of the strain undergo relaxation and decrease with the height. Thus, Fig. 3.12 demonstrates that the volume elastic relaxation is more efficient for pyramids ('quantum dots') than for infinitely elongated prisms ('quantum wires'), which explains the preferred pyramid-like shape of the islands. In a more detailed treatment, if the surface energy $\gamma(\hat{\mathbf{m}})$ does not have cusped local minima, then the tilt angle of facets φ will vary with the size of the island.

The dependence of the island shape on the island volume has been studied by Pehlke et al. (1996, 1997). Using calculated *ab initio* surface energies of different InAs facets and elastic relaxation energies for different shapes of islands, the equilibrium shape for InAs/GaAs(001) is revealed to be strongly dependent on the volume. In Fig. 3.13 the elastic energy of the islands is shown for different shapes. Within the configuration space the optimum shape has been found to be a mesa-type hill bounded by {101} and {111} facets and a (001) surface on the top. However, the strain-induced renormalization of surface energies was not taken into account. If the shape of the islands is a function of their volume, then the scaling analysis of the total energy of island arrays is no longer possible. However, ordering of islands *in size* generally occurs if formation of the islands leads to the *decrease* of the renormalized surface energy.

3.6.4 ORDERING OF 3D ISLANDS IN SIZE VERSUS OSTWALD RIPENING

The equilibrium in the array of islands can be reached by material exchange between islands which occurs via surface migration. For a *dilute* system of islands, the elastic interaction between islands via the strained substrate may be neglected. Then the energy of the array is the sum of contributions of single islands (Eq. 3.21). Here we consider a dilute array of identical pyramid-shaped islands with the base length L.

The equilibrium is defined by the condition of the total energy minimum under the constraint of a fixed amount of the deposited material. Then, instead of minimization of the total energy, it is possible to use an equivalent procedure and to minimize the energy per atom in the island $E(L)$. Dividing $\tilde{E}(L)$ from Eq. (3.21) by the volume of a single island $\frac{1}{6} \tan \varphi_0 L^3$ and multiplying by the atomic volume Ω, one obtains (Shchukin et al., 1995c)

$$E(L) = 6 \cot \varphi_0 \Omega \left[-f_1(\varphi_0)\lambda\varepsilon_0^2 + \frac{(\Delta\Gamma)}{L} - \frac{f_2(\varphi_0)\tau^2}{\lambda L^2} \ln\left(\frac{L}{2\pi a}\right) + \frac{f_3(\varphi_0)\eta}{L^2} \right] \quad (3.22)$$

The volume elastic relaxation energy $\Delta \tilde{E}_{\text{elastic}}^V$ (the first term in Eq. 3.22) does *not* depend on the island size L. To search the minimum of $E(L)$ from Eq. (3.22), we introduce the characteristic length

$$L_0 = 2\pi a \, \exp\left[\frac{f_3(\varphi_0)\eta\lambda}{f_2(\varphi_0)\tau^2} + \frac{1}{2} \right] \quad (3.23)$$

and the characteristic energy per one atom

$$E_0 = \frac{1}{2} \frac{\Omega f_2(\varphi_0)\tau^2}{\lambda L_0^2} \quad (3.24)$$

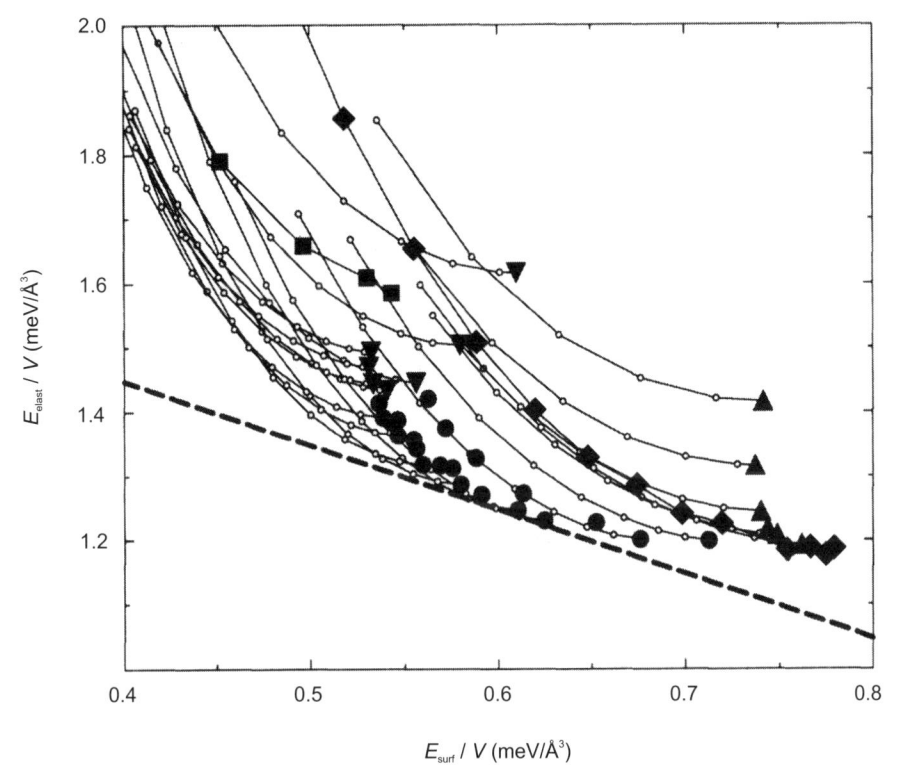

SELF-ORGANIZATION CONCEPTS ON CRYSTAL SURFACES

Fig. 3.13 Elastic energy per volume versus surface energy per volume for InAs islands with volume 2.88×10^5 Å^3 for different dot shapes. Squares: square-based pyramid with $\{101\}$ facets and (001) truncated $\{101\}$ pyramids. Diamonds: square-based pyramids with $\{111\}$ and $\{-1-1-1\}$ facets and (001) truncated $\{111\}$ pyramids. Triangles up: 'huts' with $\{111\}$ and $\{-1-1-1\}$ facets. Triangles down: square-based $\{101\}$ pyramids with $\{-1-1-1\}$ truncated edges. Dots: islands with $\{101\}$, $\{111\}$ and $\{-1-1-1\}$ facets. Filled symbols denote numerical results, while open circles correspond to simple analytical approximations for (001) truncated 'mesa-shaped' islands. It is assumed that the elastic energy does not change when the (almost fully relaxed) top part is cut off. Full lines connect islands that are created in this way, varying the height of the (001) top surface plane. The dashed line is the curve of constant total energy $E_{\text{elast}} + E_{\text{surf}}$ that selects the equilibrium shape. Generalization to arbitrary volume can be done using $E_{\text{elast}} \sim V$ and $E_{\text{surf}} \sim V^{2/3}$; for larger volumes the slope of the total energy line becomes smaller. (After Pehlke et al., 1996)

Then we may write the sum of all L-dependent terms in $E(L)$ as follows:

$$\hat{E}(L) = E_0 \left[-2\left(\frac{L_0}{L}\right)^2 \ln\left(\frac{e^{1/2}L}{L_0}\right) + \frac{2\alpha}{e^{1/2}}\left(\frac{L_0}{L}\right) \right] \quad (3.25)$$

The function $\hat{E}(L)$ is governed by the control parameter

$$\alpha = \frac{e^{1/2}\lambda L_0}{f_2(\varphi_0)\tau^2}(\Delta\Gamma) \quad (3.26)$$

which is the ratio of the renormalized surface energy and of the contribution of the edges to the elastic relaxation energy, $|\Delta E_{\text{elastic}}^{\text{edges}}|$. The energy of a dilute array of islands per atom versus the size of the island L is displayed in Fig. 3.14 for different values of α. There exists an optimum size of islands L_{opt}, corresponding to the absolute minimum of the energy, $\min \hat{E}(L) \equiv E(L_{\text{opt}}) \leq 0$. Ripening of islands would correspond to $L \Rightarrow \infty$, then the energy $\hat{E}(L) \to 0$. Thus an array of identical islands of optimum size L_{opt} is a stable array, and islands do not undergo ripening. If $1 < \alpha < 2\mathrm{e}^{-1/2} \approx 1.2$, there exists only a local minimum of the energy, corresponding to a meta-stable array where $\hat{E}(L) > 0$. If $\alpha > 1.2$, the local minimum in the energy $\hat{E}(L)$ disappears. For both cases where $\alpha > 1$, there exists a thermodynamic tendency to ripening. The energy minimum corresponds to one large cluster where all deposited material is collected. If $\Delta \Gamma < 0$ (and $\alpha < 0$) the formation of 3D islands does not only lead to a decrease of strain energy due to the relaxation but also to the *decrease* of the renormalized surface energy.

For an InAs pyramid with {101} side facets on top of an InAs wetting layer deposited on a GaAs(001) surface, the calculation of $\Delta\Gamma$ yields

$$\begin{aligned}\Delta\Gamma = &1.41\gamma_0^{(101)} + \gamma_{\text{interface}} - \gamma_{\text{WL}}^{(001)} \\ &- [0.72\tau_{\mu\mu}^{(101)} + 0.40\tau_{\nu\nu}^{(101)} - 0.15(\tau_{\zeta\zeta}^{(001)} + \tau_{\eta\eta}^{(001)})]\varepsilon_0 \\ &+ [0.22S_{\mu\mu\mu\mu}^{(101)} + 0.08S_{\nu\nu\nu\nu}^{(101)} + 0.25S_{\mu\mu\nu\nu}^{(101)} + 0.10S_{\mu\nu\mu\nu}^{(101)} \\ &- 0.01(S_{\zeta\zeta\zeta\zeta}^{(001)} + S_{\eta\eta\eta\eta}^{(001)}) + 0.03S_{\zeta\zeta\eta\eta}^{(001)} + 0.01S_{\zeta\eta\zeta\eta}^{(001)}]\varepsilon_0^2\end{aligned} \quad (3.27)$$

where the axes μ, ν, ζ, η are defined in Fig. 3.15. The change of the surface energy due to the formation of a pyramid contains contributions due to the appearance of tilted $\langle 101 \rangle$ facets of InAs (the first term in Eq. 3.27), the appearance of the InAs/GaAs

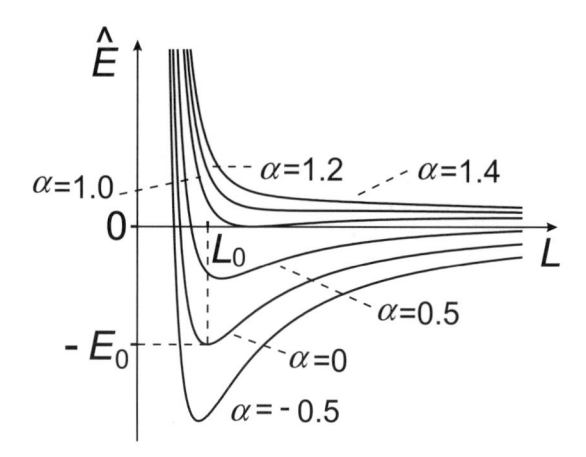

Fig. 3.14 Energy of a dilute array of 3D coherently strained islands per atom versus the size of the island. The parameter α is the ratio of the renormalized surface energy of the island, $\Delta E_{\text{surf}}^{\text{renorm}}$, and of the contribution of the edges to the elastic relaxation energy, $|\Delta E_{\text{elastic}}^{\text{edges}}|$

SELF-ORGANIZATION CONCEPTS ON CRYSTAL SURFACES

interface under the pyramid (the second term in Eq. 3.27), the disappearance of the L^2 area of the wetting layer (the third term in Eq. 3.27), linear renormalization terms (the second line of Eq. 3.27), and quadratic renormalization terms (the third and fourth line in Eq. 3.27).

The change of the surface energy due to the formation of an island is governed mainly by the interplay of two major contributions in $\Delta\Gamma$ from Eq. 3.27, the one due to the creation of the side facets of InAs and the other due to disappearance of a certain area of the wetting layer, all other terms in Eq. 3.27 being small corrections. It should be noted that a decrease of the surface energy in the InAs/GaAs(001) system ($\Delta\Gamma < 0$) may occur in spite of the fact, that the (001) surface of the bulk InAs is stable against faceting. The reason is that the appearance of the tilted {101} facets of InAs competes with the disappearance of a certain area of the *wetting layer*. The wetting layer of InAs having a microscopic thickness of one or two monolayers, should be regarded as a surface with *its own surface energy*, the latter has not to be equal to the surface energy of InAs(001)

The quantity $\Delta\Gamma$ could be evaluated if *all* quantities entering Eq. 3.27 were known from *ab initio* calculations. By using *so far known* values for $\gamma_{InAs}^{(101)} = 41$ meV/Å^2 (Pehlke *et al.*, 1996), for the interface energy, for the intrinsic surface stress $\tau_{\alpha\beta}$, and for the excess surface elastic moduli $S_{\alpha\beta}\Phi\Psi$ from Moll *et al.*, (1998b), it was shown that the criterion $\Delta\Gamma < 0$ is equivalent to $\gamma_{WL}^{(001)} > 55$ meV/Å^2. Whether this condition holds could be established by *ab initio* calculations.

However, one should be very careful while applying *ab initio* surface energies to evaluate crucial quantities entering the macroscopic theory. Firstly, preferred surface reconstructions and corresponding values of surface energies have been obtained in above cited papers only for temperatures $T = 0$ while typical growth temperatures are 450 °C and higher. It is known, however, that the surface reconstruction can change and does change with temperature, and so does the surface energy. Secondly, calculated surface energies refer to infinitely large surfaces while the reconstruction

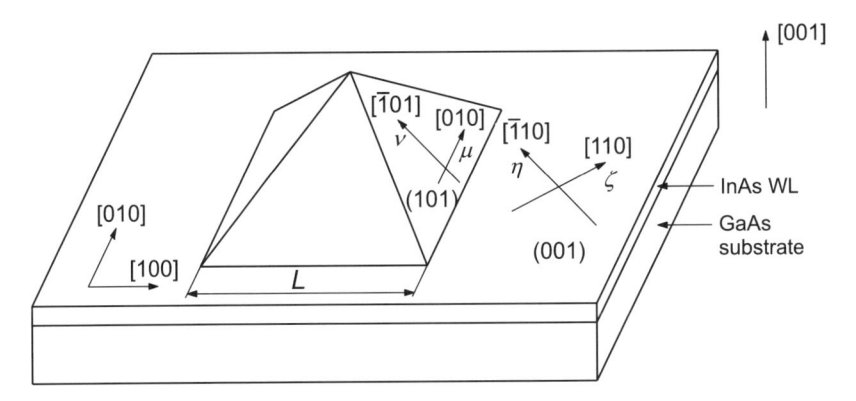

Fig. 3.15 Geometry of an InAs pyramid on an InAs wetting layer deposited on a GaAs(001) surface

on nanometer-scale facets could be different from that on a plane. Thirdly, to apply the macroscopic theory to finite temperatures, one has to take into account entropy contribution to the free energy which was so far neglected. Thus, substantial efforts from both *ab initio* and macroscopic theories are required in order to get an ultimate theoretical answer on the nature of ordering in a given particular system.

3.6.5 LATERAL ORDERING OF 3D ISLANDS

For a *dense* system of islands, elastic interaction between islands via the strained substrate becomes essential. The system of interacting islands can then be considered as a system of elastic domains where the minimum of the strain energy corresponds to a periodic domain structure Khachaturyan, 1974; Marchenko, 1981a, 1981b; Vanderbilt, 1992; Ipatova *et al.*, 1993; Zeppenfeld *et al.*, 1994). In the approximation of a small tilt angle φ_0 of the facets, the sum of the energy per unit surface area $\Delta E_{\text{elastic}}^{(1)}$ for isolated islands and the interaction energy per unit surface area $E_{\text{interaction}}$ equals $\Delta E_{\text{elastic}}^{(1)} + E_{\text{interaction}} = (Qa)\tilde{\lambda}\varepsilon_0^2\varphi_0[-g_i(\varsigma)]$.

Here $\tilde{\lambda} = (C_{11} + 2C_{12})^2(C_{11} - C_{12})^3 C_{11}^{-3}(C_{11} + C_{12})^{-1}$ and ς is the fraction of the surface covered by islands. The functions $-g_i(\varsigma)$ for different arrays of islands are displayed in Fig. 3.16.

Among different 2D arrays of pyramid-shaped islands on the (001) surface of an elastically anisotropic cubic medium, the minimum energy corresponds to a 2D square lattice of islands with primitive lattice vectors along the 'soft' directions [100] and [010] (see Fig. 3.8). There exist two factors that favor the square lattice. First is

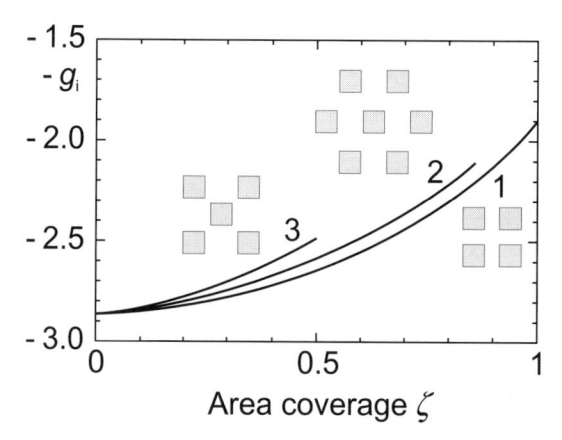

Fig. 3.16 The functions $-g_i(\zeta)$ (see Section 3.6.5) for different arrays of coherently strained islands versus the fraction ζ of the surface covered by islands. Different curves are for different Bravais lattices: (1) 2D square lattice of pyramids with primitive lattice vectors (1,0,0) and (0,1,0); (2) 2D hexagonal lattice of pyramids with primitive lattice vectors $(-1/2, -\sqrt{3}/2, 0)$ and (1,0,0); (3) 2D 'checkerboard' square lattice of pyramids with primitive lattice vectors $(1, -1,0)$ and (1,1,0). Curves 2 and 3 terminate at the maximum possible coverage for given arrays

the cubic anisotropy of elastic moduli of the medium and second is the square shape of the base of a single island.

In the approximation of small tilt angles used in deriving $\Delta E_{\text{elastic}}^{\text{V}} + E_{\text{interaction}}$ in Fig. 3.16, significant elastic relaxation energy has already been obtained in agreement with Vanderbilt and Wickham (1991). For large values of φ_0, this approximation overestimates the elastic relaxation energy. However, the difference between the exact and the approximate values of $\Delta E_{\text{elastic}}^{\text{V}}$ is small, even for the angle $\varphi_0 = 45°$. In the approximation of small tilt angles the energy of the volume elastic relaxation is found to be -64% of the total strain energy of the flat film. Calculations using finite element methods yield an elastic relaxation energy of -60%.

The close similarity between the approximation for small tilt angles and the exact numerical solution remains for arrays of interacting islands. An evaluation of the elastic energy of interacting arrays of islands by the finite element method for a tilt angle $\varphi_0 = 45°$ shows that the cubic anisotropy of the elastic moduli in this case also favors a 2D *square* lattice of islands with primitive lattice vectors along the 'soft' directions [100] and [010].

3.6.6 PHASE DIAGRAM OF TWO-DIMENSIONAL ARRAY OF 3D ISLANDS

The main part of the interaction energy is the energy of dipole–dipole elastic repulsion between islands, $E_{\text{interaction}} = (Qa) \times 6 \cot\varphi_0 f_5(\varphi_0) \lambda \varepsilon_0^2 \zeta^{3/2}$, where ζ denotes again the surface area coverage by islands. Thus, the curve $-g_1(\zeta)$ in Fig. 3.16 in the region $0 \leq \zeta \leq 0.5$ can be approximated by $-g_1(\zeta) = -2.87 + 0.61\zeta^{3/2}$ with an accuracy of 1%. To express ζ in terms of Q and L, the volume of the pyramid is set equal to the volume of the initially uniform film per unit period $D \times D$ of the square lattice: $\frac{1}{6}L^3 \tan\varphi_0 = D^2(Qa)$. Hence, $\zeta \equiv L^2/D^2 = 6Qa \cot\varphi_0/L$. Now, adding $E_{\text{interaction}} \sim \zeta^{3/2} \sim L^{-3/2}$ to the energy of Eq. (3.22), we obtain the energy of the array of interacting islands. To consider L-dependent terms in the energy per one atom $E\prime(L) = E_{\text{dilute}}(L) + E_{\text{interaction}}(L)$ in detail, we introduce the characteristic length L_0 from Eq. (3.23) and the characteristic energy E_0 from Eq. (3.24). Then the sum of all L-dependent terms in $E(L)$ may be written as the function of the dimensionless length L/L_0:

$$E\prime(L) = E_0 \left[-2 \left(\frac{L_0}{L} \right)^2 \ln\left(\frac{e^{1/2}L}{L_0} \right) + \frac{4\beta}{e^{3/4}} \left(\frac{L_0}{L} \right)^{3/2} + \frac{2\alpha}{e^{1/2}} \left(\frac{L_0}{L} \right) \right] \quad (3.28)$$

This function is governed by two control parameters, where α is defined in Eq. (3.26), and

$$\beta = (Qa)^{3/2} \frac{e^{3/4} f_5(\varphi_0)(6 \cot\varphi_0)^{3/2} (\lambda \varepsilon_0)^2 L_0^{1/2}}{2 f_3(\varphi_0) \tau^2} \quad (3.29)$$

The parameter β is the ratio $E_{\text{interaction}}/|\Delta E_{\text{elastic}}^{\text{edges}}|$, which increases with the amount of the deposited material as $Q^{3/2}$.

Searching the minima of the energy $\hat{E}(L)$ from Eq. (3.28) for different α and β, we obtain the phase diagram of Fig. 3.17. For region 1 of the phase diagram, there exists an optimum size of islands L_{opt}, corresponding to the absolute minimum of the energy, min $\hat{E}(L) \equiv \hat{E}(L_{opt}) < 0$. On the other hand, the ripening of islands would correspond to $L \to \infty$ where the energy $\hat{E}(L) \to 0$. Thus a 2D periodic square lattice of islands of the optimum size L_{opt} is a stable array, and islands do not undergo ripening. For region 2 of the phase diagram, there exists only a local minimum of the energy, corresponding to a metastable array where $\hat{E}(L) > 0$. For region 3, the local minimum in the energy $E'(L)$ disappears. In both regions 2 and 3, there exists a thermodynamic tendency to ripening. If the system initially corresponds to the region 1 and the amount of the deposited material Q increases, then the point in the phase diagram moves to regions 2 and 3, and islands undergo ripening.

If $\alpha \leqslant 0$, there exists an absolute minimum of the energy $\hat{E}(L)$ for an arbitrary value of β, and min $\hat{E}(L) \equiv \hat{E}(L_{opt}) < 0$. Besides the absolute minimum of $\hat{E}(L)$,

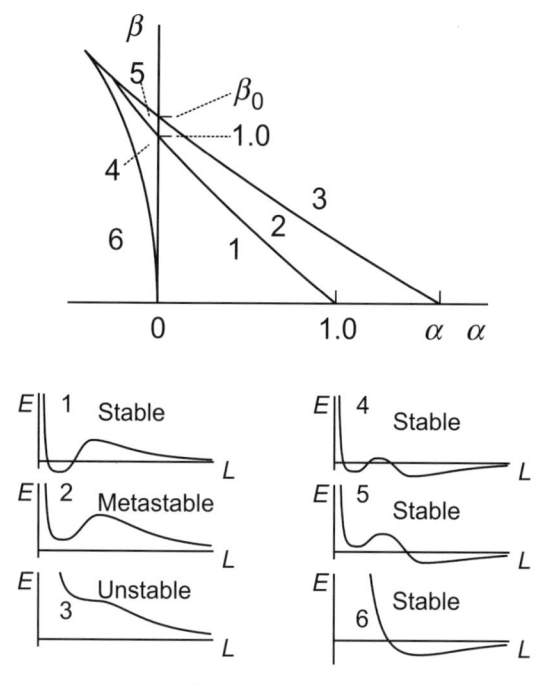

Fig. 3.17 Schematic phase diagram of the stability of a square lattice of coherently strained islands in the plane of control parameters α, β. Here α and β are defined in Eqs. (3.26) and (3.29). In the lower part of the figure, the energy of the interacting array of islands versus the size of the islands is plotted. Plots labeled 1 to 6 correspond to different regions of the phase diagram. In regions 1, 4, 5, and 6, stable arrays of islands exist that do not undergo ripening. In regions 2 and 3, all arrays of islands undergo ripening. $\alpha_0 = 2e^{-1/2} \approx 1.213$, $\beta_0 = \frac{4}{3}e^{-1/4} \approx 1.038$.

there may also exist a local minimum with an energy <0 in region 4 and >0 in region 5. No metastable state exists in region 6.

In order to estimate characteristic values α and β, $\tau \approx 100\,\text{meV}/\text{Å}^2$, $\lambda \approx 500\,\text{meV}/\text{Å}^3$, $L_0 \approx 100\,\text{Å}$, and $a = 2.8\,\text{Å}$ were substituted in Eq. (3.29). This gives $\beta \approx 1$ for $|\varepsilon_0| \approx 1\%$ and $Q = 0.5$ monolayers. Therefore, if $\alpha > 0$, the array of islands corresponds to region 1 of the phase diagram only for low coverage Q. On the other hand, the typical difference in the surface energy between facets of different orientation is of the order of ≈ 5–$10\,\text{meV}/\text{Å}^2$ (Eaglesham et al., 1993). It follows then from Eq. (3.26) that the parameter α for a strongly mismatched system may become negative due to mismatch-induced renormalization of the surface energy of island facets. If $\alpha < 0$, then the increase of $|\varepsilon_0|$, e.g. by the increase of x for the heterophase system $\text{In}_x\text{Ga}_{1-x}\text{As}/\text{GaAs}(001)$, results in a decrease of L_{opt}.

3.6.7 ORDERING-TO-RIPENING PHASE TRANSITION FOR 3D ISLANDS

The key difference between arrays of three-dimensional coherently strained islands and periodically faceted surfaces or periodic structures of surface domains is the existence of both an ordering regime and a ripening regime in the island case. The possible transition between these two regimes is governed by the surface energy of the island facets. A remarkable fact is that such a phase transition driven by the As pressure has been observed experimentally (Ledentsov et al., 1996b).

The MBE growth of four monolayers of InAs at a substrate temperature $T = 480\,°\text{C}$ and an As pressure $P_{\text{As}}^{(0)} = (1.5\text{–}3) \times 10^{-6}\,\text{Torr}$ results in an array of 140 Å pyramid-shaped QDs arranged in a 2D square lattice with principal axes along the [100] and [010] directions. This array is stable and does not undergo ripening upon growth interruption.

Significant changes of MBE growth conditions ($T = 480\,°\text{C}$ and $P_{\text{As}}^{(0)} = 10^{-5}\,\text{Torr}$) lead to a ripening regime which results in the formation of large macroscopic islands (Ledentsov et al., 1996b). This shows that kinetics is sufficient to drive the system to a state with lower energy, i.e. to a stable array for the ordering regime or to macroscopic clusters for the ripening regime. The shift from the ordering regime to the ripening regime might be caused by the As pressure induced change of the surface reconstruction and the consequent change of the surface energy (Qian et al., 1988).

3.6.8 KINETIC THEORIES OF ORDERING

Several *kinetic* models of island formation and ordering have been proposed (e.g. Chen et al., 1995; Chen and Washburn, 1996; Jesson et al., 1996b; Barabási, 1997; Dobbs et al., 1997). Such models take into account microscopic processes of adatoms on the crystal surface, such as deposition, diffusion, and attachment and detachment to islands.

In Barabási (1997) a one-dimensional model has been established using a Monte Carlo method including strain relaxation of the lattice at each step. The strain distribution around an island has an important impact on the kinetics (Chen and Washburn 1996; Barabási, 1997). The decrease of strain energy with increasing distance from a 3D island biases the random motion of ad-atoms away from the island. For large islands the bonding energy of an ad-atom at its edge becomes comparable to the strain energy at the edge, enhancing its detachment. These effects slow the growth rate of larger islands and increase the ad-atom density away from them, thus enhancing the nucleation of new islands. This eventually leads to a narrow size distribution for sufficiently large strain ($\geqslant 5\%$) (Barabási, 1997). Increasing deposition leads mainly to a higher density of islands of the same size.

In Dobbs et al. (1997) a mean field theory for the densities of ad-atoms, 2D and 3D islands has been formulated, attempting to explain data from Seifert et al. (1996a). As soon as they nucleate, 3D islands act as traps for ad-atoms and atoms detaching from (flat) 2D islands. The increase of the density of 3D islands with increasing coverage is steep and saturates quickly. Additional material leads to no further increase of the island density, but an increase of the 3D islands. This model, however, does not predict the development of a particular island size.

3.6.9 EQUILIBRIUM ORDERING VERSUS KINETIC-CONTROLLED ORDERING OF 3D ISLANDS

Discussing the ordering phenomena in a system of coherently strained islands on a lattice-mismatched substrate, one faces inevitably the question of whether the ordering is an equilibrium phenomenon or a kinetic-controlled one. It should be remembered that thermodynamic equilibrium of course relies microscopically on kinetic processes in a detailed balance. One way to investigate kinetic effects on dot formation is rapid freezing of a certain configuration, e.g. by cooling or burying the structure underneath a cap layer before it can reach thermodynamic equilibrium. 'True' kinetic effects in dot formation and ordering are only those that do *not* lead to the thermodynamically preferred state with lowest free energy. This can occur by inhibiting the thermodynamic equilibrium structure or by creating shapes, sizes, and ordering dynamically (Ruggerone et al., 1997).

Some experimental results discussed in Section 4.2.3 are in favor of equilibrium ordering, at least in the system of InAs/GaAs or InGaAs/GaAs strained islands in certain precisely defined growth windows. In particular, the evolution of a particular dot size up to an equilibrium value upon growth interruption probably indicates that the dominating ordering effects for the system in question are of a thermodynamic nature.

From a theoretical point of view, the possibility of equilibrium ordering of islands in size depends dramatically on the effect of strain-induced renormalization of the surface energies of the wetting layer and the island facets. We note here that $\Delta\Gamma$ from Eq. (3.22) is very sensitive to the orientation of island facets. First, the bare surface energy is expected to be lower for low-index facets (e.g. (101)) and, second, the renormalization terms, which are governed by the volume elastic relaxation, strongly depend on the tilt angle of island facets. For example, calculations of Priester and Lannoo (1995) based on the VFF model (see Section 5.1.3) were performed for islands in the InAs/GaAs(001) system, which have (104)-type side facets. These calculations did not reveal any minimum in the total energy versus the size of the island L.

So far, the thermodynamic approach has focused on the calculation of the total energy and did not consider the entropy term in the free energy. This entropy term, first, will lead to a finite concentration of the residual gas of ad-atoms on the surface and to a reduction of the total amount of material assembled in islands. Second, it is known for stepped vicinal surfaces (see, for example, Joós et al., 1991) that the entropy term prevents the 'collision' of steps and results in an effective repulsion between steps. Similarly, one might expect that for an array of islands, the entropy term will also lead to an effective repulsion between islands.

3.7 VERTICALLY CORRELATED GROWTH OF NANOSTRUCTURES

3.7.1 VERTICALLY CORRELATED GROWTH OF NANOSTRUCTURES DUE TO A MODULATED STRAIN FIELD

An array of coherently strained islands (e.g. InAs), being again covered, creates a strain field in the surrounding matrix. If the thickness of the deposited material exceeds the height of the islands, the islands are typically completely buried. When the next layer of InAs is grown, the growth proceeds in a completely new growth mode, i.e. growth on a surface with a modulated surface unit cell size, similar to the case of artificial stressors discussed in Section 2.1.5.

The modulated strain field on the surface leads to modulation of the chemical potentials of surface ad-atoms (Srolovitz, 1989). The modulation of the chemical potential makes a large impact on the growth mode. The migration of ad-atoms on the surface consists of diffusion and drift in the strain field (sometimes the term 'directional migration' is used for the drift of ad-atoms). This drift is a driving force of the kinetic self-ordering in this complex growth mode. Such a kinetic mechanism is responsible for the instability of the epitaxial growth of the homogeneous alloy and may result in the formation of composition-modulated structures (Malyshkin and Shchukin, 1993; Ipatova et al., 1996).

The strain field on the surface generated by buried islands can be calculated using an analytical approximation of continuum elasticity theory (Maradudin and Wallis, 1980) or numerically (Grundmann et al., 1996d) (see Section 5.1.2.5). For In ad-atoms migrating on the GaAs surface, the minimum of the chemical potential corresponds to the maximum of the tensile strain that is achieved, directly above buried islands (Eq. 5.19). Therefore, In ad-atoms are attracted to buried islands of InAs, resulting in vertical stacking of InAs islands in multisheet structures, a theoretical consideration first used by Xie et al. (1995). The theory by Tersoff et al. (1996) considers preferential nucleation of islands at the surface position of maximum strain, thus explaining vertical stacking in the Ge/Si(001) system. Experimental data and theoretical modeling (Tersoff et al., 1996) show that lateral ordering in subsequent sheets of islands is better pronounced than in the first sheet. Self-ordering effects, which are likely to be purely equilibrium ones for the first sheet of islands, become more pronounced for the growth of multisheet structures. In the latter case, self-ordering effects are controlled both by energetics and by kinetics.

Since such vertically stacked islands (quantum dots) are essentially separated, both electron and hole wave functions are effectively localized inside each individual island. Therefore the regularity of the above arrangement does not result in a modification of the basic electronic properties of the structures, such as radiative lifetime, energy spectrum, carrier capture and relaxation mechanisms or material gain. This fact initiated attempts to fabricate electronically coupled QDs (Ledentsov et al., 1996d; Solomon et al., 1996; Heinrichsdorff et al., 1997d) with smaller vertical separation.

3.7.2 ENERGETICS OF STRANSKI-KRASTANOW GROWTH FOR SUBSEQUENT INAS AND GAAS DEPOSITION CYCLES

Single-cycle deposition of InAs on the GaAs (001) surface above the critical thickness (≈ 0.45 nm) leads to the formation of pyramid-shaped InAs islands on the InAs wetting layer. Further growth of GaAs on a surface with a wetting layer and locally formed islands is affected by the inhomogeneous strain field due to the pyramids (Srolovitz, 1989; Xie et al., 1994). The strain field modulates the surface chemical potential for Ga ad-atoms as follows (see, for example, Srolovitz, 1989):

$$\mu^{Ga}(\mathbf{r}) = \left[\mu_0^{Ga} + \frac{Y\varepsilon_t^2(\mathbf{r})\Omega}{2}\right] + \gamma\Omega\kappa(\mathbf{r}) \qquad (3.30)$$

Here μ^{Ga} is the chemical potential of Ga ad-atoms on the reference flat surface with the lattice parameter equal to that of bulk GaAs. The second term in the brackets is the elastic energy correction to μ_0^{Ga}, where $\varepsilon_t(\mathbf{r})$ is the tangential component of the local strain defined with respect to unstrained GaAs, Y is Young's modulus, and Ω is the atomic volume. The third term is the surface energy contribution to the chemical potential, where γ is the surface energy and $\kappa(\mathbf{r})$ is the local curvature of the surface. Due to the elastic energy correction to μ_0^{Ga}, the incorporation of GaAs on the facets of elastically relaxed InAs pyramids is energetically unfavorable. The gradient of the surface chemical potential leads to a locally directional migration of Ga ad-atoms away from the InAs islands (Fig. 3.18a to d). The latter results in a reduction of the growth rate and in a curved growth front in the vicinity of the islands (Xie et al., 1994). When the InAs islands are partly covered by GaAs, the effect of the strain inhomogeneity on the profile of the growing surface decreases, and a planar growth

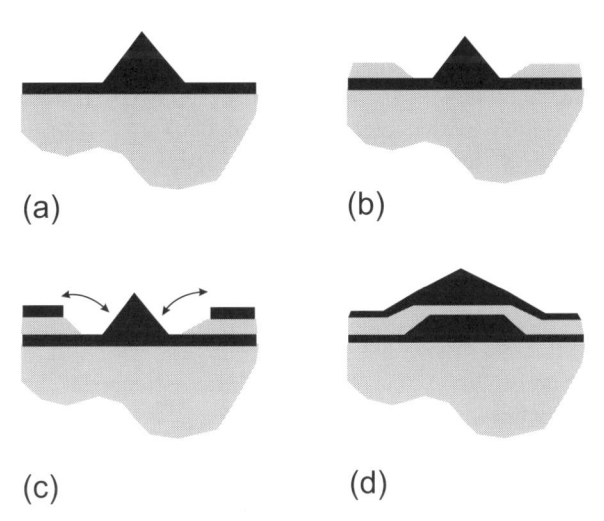

SELF-ORGANIZATION CONCEPTS ON CRYSTAL SURFACES 55

Fig. 3.18 Progressive stages of the formation of vertically coupled quantum dot structures, depicted schematically. Efficient exchange reactions on the surface due to the energetics of Stranski–Krastanow growth occur during growth interruption on stage (c).

front is completely re-established after the deposition of ≈ 6 nm (Ruvimov et al., 1995).

The situation changes drastically if the GaAs growth is interrupted well before the dots are completely covered by GaAs and another InAs deposition cycle is introduced (or just the growth interruption time is chosen to be long enough to let the system come to equilibrium). According to the SK InAs/GaAs growth mode, it is energetically favorable for InAs to evaporate from the InAs islands and to cover the free surface of GaAs, forming a second wetting layer. The surface chemical potential of In atoms is equal in this case to

$$\mu_0^{\text{In}}(\mathbf{r}) = \left\{ \mu_0^{\text{In}} + \frac{Y[\varepsilon_t(\mathbf{r}) - \varepsilon_0]^2 \Omega}{2} \right\} + \gamma \Omega \kappa(\mathbf{r}) - \frac{\zeta \Omega \vartheta(\mathbf{r})}{a} \quad (3.31)$$

Here μ_0^{In} is the chemical potential of In ad-atoms on the unstressed (completely relaxed) surface, where the lattice parameter is equal to that of InAs. The elastic energy correction is minimum for a completely relaxed surface, where $\varepsilon_t(\mathbf{r}) = \varepsilon_0$. The third term in Eq. (3.31) is the same as in Eq. (3.30). The last term represents the effect of wetting, where ζ is the energy benefit per unit area due to the formation of the wetting layer, a is the lattice parameter, and $\vartheta(\mathbf{r}) = 1$ on the GaAs surface and $\vartheta(\mathbf{r}) = 0$ on the InAs surface. The last term in Eq. (3.31) plays the dominant role in the directed migration of indium atoms before the second wetting layer is formed. According to this term there exists a thermodynamically favorable tendency for indium atoms to be detached from the InAs island and to cover the free GaAs surface. As the InAs pyramids are only partly covered with GaAs, InAs of the upper part of the InAs island is available and the top of the pyramid can be completely dissolved at this stage. Further evaporation of indium from the InAs pyramid will occur from the laterally confined part, unless the enhanced curvature of the nearby GaAs region and, consequently, enhanced surface energy (the last term in Eq. 3.30) make the planarization (partial or complete) of the GaAs surface by directional migration of Ga ad-atoms energetically more favorable. Then the rest of the InAs island will be completely confined by GaAs.

Formation of split islands is possible only if the surface exchange reactions are fast enough. After GaAs is deposited and a new InAs deposition cycle has just started, most of the surface is InAs-free, and the directional migration of In and Ga ad-atoms continues. At our typical substrate temperatures (480 °C) and for a growth mode with 0.3 ML (monolayer) InAs deposition cycles with 10 s growth interruptions after each cycle, the exchange reactions are fast enough to produce severe morphological modifications, even on a much larger geometrical scale (Ledentsov et al., 1996d). Simultaneously, with the shape transformation of the InAs islands deposition of extra InAs occurs, finally resulting in the formation of a complete InAs wetting layer (≈ 1.5 ML). Then, after the second wetting layer is formed, the second term in Eq. (3.31) provides a tendency for excess In atoms to attach to the region of locally modulated lattice parameter due to existing islands. According to this energy term, the formation of InAs islands at new positions is energetically unfavorable. As a result, a vertical arrangement of two InAs islands separated by a several monolayer thick GaAs layer (split pyramid or VCQD) can be formed. Then the process can be continued. Schematically, the formation of the VCQD structure is illustrated in Fig. 3.18a to d.

3.7.3 STRAIN ENERGY OF VERTICALLY COUPLED STRUCTURES

The possibility of fabricating such VCQD structures depends mainly on the energetics of the split pyramid with respect to other possible arrangements. Several possible final states are compared in Fig. 3.19. Figure 3.19a depicts the case where no splitting occurs. Figure 3.19b to d refers to the situation where the 'buried' part of the pyramid floats up partially (Fig. 3.19b and c or Fig. 3.19d) and is replaced by GaAs. We estimate here the gain in the elastic energy for structures of Fig. 3.19b to d with respect to that of Fig. 3.19a. This gain is due to the fact that a certain volume of InAs is transferred from the buried region, where it is not relaxed, to the uncovered pyramid, where it is partially relaxed. The elastic energy for each of the structures of Fig. 3.19 is determined by

$$E_{\text{elastic}} = \frac{1}{2}\lambda^{(2)}_{ijlm}\int_{(2)}[\varepsilon_{ij}(\mathbf{r}) - \varepsilon_0\delta_{ij}][\varepsilon_{lm}(\mathbf{r}) - \varepsilon_0\delta_{lm}]\,dV + \frac{1}{2}\lambda^{(1)}_{ijlm}\int_{(1)}\varepsilon_{ij}(\mathbf{r})\,dV \quad (3.32)$$

Here, indices (1) and (2) denote GaAs and InAs, respectively; λ_{ijlm} is the tensor of elastic moduli. The displacement vector $u_i(\mathbf{r})$ in the heterophase system obeys equilibrium equations of the elasticity theory in each material,

$$\lambda^{(1,2)}_{ijlm}\frac{\partial^2 u_m(\mathbf{r})}{\partial r_j \partial r_i} = 0 \quad (3.33)$$

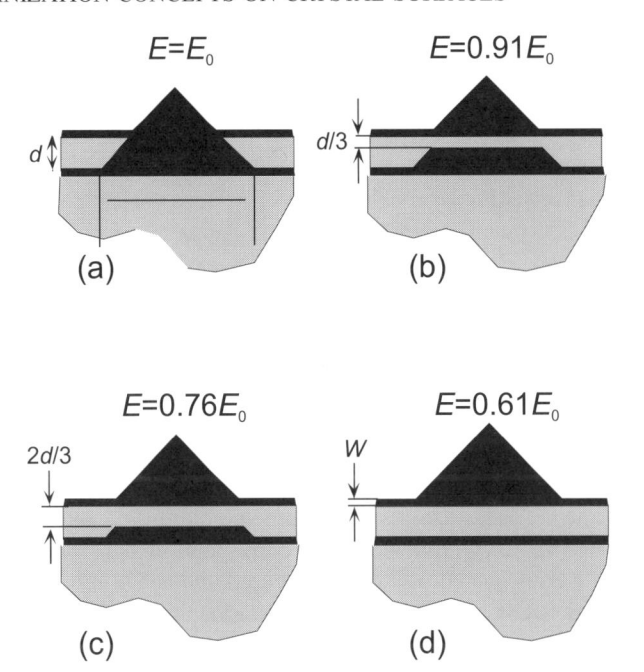

SELF-ORGANIZATION CONCEPTS ON CRYSTAL SURFACES

Fig. 3.19 Different possible final states for multilayer InAs (black) and GaAs (grey) deposition. Relative dimensions used in calculations of the elastic energy are as follows: $d = 3$ W, $L = 24$ W. The tilt angle of the pyramid facets is $45°$. We show that a reduction occurs in the elastic energy for structures (b to d) with respect to that of (a) caused by the InAs transfer from the buried part to the uncovered part of the pyramid. No total energy reduction is induced by simple splitting without InAs transfer

obeys boundary conditions at the interface and at the free surface and vanishes deep in the substrate. The boundary conditions at the InAs/GaAs interface are as follows (Khachaturyan, 1974):

$$\lambda_{ijlm}^{(2)} n_i(\mathbf{r}) \left[\frac{\partial u_m(\mathbf{r})}{\partial r_l} - \varepsilon_0 \delta_{lm} \right] \Bigg|^{(2)} = \lambda_{ijlm}^{(1)} n_j(\mathbf{r}) \left[\frac{\partial u_m(\mathbf{r})}{\partial r_l} \right] \Bigg|^{(1)} \quad (3.34)$$

where $n_j(\mathbf{r})$ is the outer normal to the InAs region at the interface. The boundary conditions at the free surface of InAs are as follows (Ipatova et al., 1993):

$$\lambda_{ijlm}^{(2)} m_i(\mathbf{r}) \left[\frac{\partial u_m(\mathbf{r})}{\partial r_l} - \varepsilon_0 \delta_{lm} \right] \Bigg|^{(2)} = 0 \quad (3.35)$$

where $m_j(\mathbf{r})$ is the outer normal to the free surface. We solve Eq. (3.32) subject to boundary conditions Eqs. (3.33) and (3.34) by the finite element method, calculate the strain tensor $\varepsilon_{ij}(\mathbf{r})$, evaluate the elastic energy of the two wetting layers, and thus obtain the net elastic energy of each of the islands displayed in Fig. 3.19.

If the net energy of the non-split pyramid of Fig. 3.19a is denoted as E_0, the net energies of split pyramids are as follows: $0.89E_0$ for (b), $0.76E_0$ for (c), and $0.61E_0$ for (d). To analyze other possible shape transformations of the pyramid, we have calculated E_{elastic} for several split structures, where InAs from the buried part of the pyramid is transferred, not to the uncovered part of the pyramid, but is redistributed between two planar wetting layers. Such a splitting does not lead to a reduction in the elastic energy, but, to the contrary, results in an increase in the elastic energy up to $1.20E_0$.

If the process described is repeated in a multicycle deposition mode, the volume of each island progressively increases with successive deposition cycles of InAs due to the transfer of InAs from the buried part of the structure to the uncovered part. The elastic energy for structures like those of Fig. 3.19b to d is relaxed with respect to that of Fig. 3.19a when at the final stage a certain volume of InAs is transferred from the buried region (where it is not relaxed) to the uncovered pyramid (where it is partially elastically relaxed). This general conclusion is not affected by possible deviations of the real shape of islands from the simplified shapes of Fig. 3.19. Although the maximum energy gain would correspond to a complete transfer of the buried InAs to the uncovered part of the pyramid, the actual resulting shape of the island strongly depends on the kinetics of the growth process involved. Thus the strain energy calculations also support the total energy reduction due to kinetically induced splitting of the pyramids, in addition to the energetics of the Stranski–Krastanow growth itself. On the other hand, the splitting can be suppressed because of the slower kinetics of surface exchange reactions for high deposition rates and low substrate temperatures.

4 Growth and Structural Characterization of Self-Organized Quantum Dots

4.1 INTRODUCTION

Self-organized growth of quantum dots has been successfully demonstrated using both molecular beam epitaxy (MBE) (Mo *et al.*, 1990; Guha *et al.*, 1990; Yao *et al.*, 1991; Grandjean *et al.*, 1993; Leonard *et al.*, 1993; Moison *et al.*, 1994; Nabetani *et al.*, 1994; Cirlin *et al.*, 1995; Grundmann *et al.*, 1995b; Guryanov *et al.*, 1995a, 1995b; Hausler *et al.*, 1996; Kobayashi *et al.*, 1996; Kurtenbach *et al.*, 1995; Ledentsov *et al.*, 1996b; Ngo *et al.*, 1996) and metal-organic chemical vapor deposition (MOCVD) (Carlsson *et al.*, 1994; Nötzel *et al.*, 1994a, 1994b; Oshinowo *et al.*, 1994; Heinrichsdorff *et al.*, 1996a). For in-depth information on MBE techniques we refer the reader to Gossard (1981), Ploog (1988), Kop'ev and Ledentsov (1988), and Herman and Sitter (1989) and for the MOCVD method to Razeghi (1989), Stringfellow (1989) and Razeghi (1995).

Primary methods for structural characterization can be divided into:

(a) *direct imaging* methods such as scanning tunneling microscopy (STM), atomic force microscopy (AFM) (Wiesendanger, 1994), and transmission electron microscopy (TEM) (Reimer, 1984; Cerva and Oppolzer, 1990; Ourmazd *et al.*, 1990; Bimberg *et al.*, 1992; Neumann *et al.*, 1996).

(b) *diffraction* methods such as reflective high-energy electron diffraction (RHEED) (Joyce *et al.*, 1984; Larsen and Dobson, 1988) and its ellipsometric equivalent reflectance anisotropic spectroscopy (RAS) (Aspnes, 1985; Rinaldi *et al.*, 1996) and X-ray diffraction (XRD) (Bartels and Nijman, 1978; Tapfer and Ploog, 1986; Segmüller *et al.*, 1989; Krost *et al.*, 1996a).

Scanning tunneling microscopy has the advantage of being able to reveal directly the morphology of a surface on an atomic level and to enable manipulation of surface atoms, e.g. to form lines and figures on the surface. Atomic force microscopy has in principal atomic resolution. Typically a lateral resolution of a few nanometers and a much finer z resolution of 0.1 nm is achieved. The actual resolution depends on the specific size and shape of the tip. Tip effects can modify slightly the apparent height and to a larger extent the lateral size of the QD. Scanning techniques can mostly not distinguish between coherent and dislocated clusters.

One should keep in mind, however, that STM and AFM plan-view measurements of uncovered dots are usually not performed at the growth temperature. The surface morphology actually investigated can thus be completely different from that directly after growth. For example, QD size, facet angle, and density have all been shown to be a function of the details of the growth conditions like the deposition temperature (Ledentsov et al., 1996b). At the same time, the disappearance of InAs QDs due to switching off the arsenic flux and depositing only 0.15 ML of pure indium has been reported (Ledentsov et al., 1996b), to mention another example. These problems are not encountered if STM cross-sectional experiments are performed on covered samples (Wu et al., 1997; Legrand et al., 1998).

The problems mentioned can be avoided by using transmission electron microscopic techniques like high-resolution electron microscopy (HREM) or electron energy loss spectroscopy (EELS), which provide information on quantum dot morphology which is frozen-in by direct coverage (Ruvimov et al., 1995a; Kosogov et al., 1996). These techniques are, however, very time consuming and the apparent shape of the dots is affected by strain fields. Further mathematical modeling of the HREM and TEM images is necessary to understand contrast generation (Androussi et al., 1994; Ruvimov and Scheerschmidt, 1995).

RHEED is a highly surface-sensitive ultra-high vacuum technique used to monitor growth in MBE systems. Upon transformation of an initially two-dimensional ordered surface with monolayer high islands into a corrugated structure the RHEED pattern changes from streaky to spotty (e.g. see Nabetani et al., 1994; Guryanov et al., 1996b). RHEED is thus a very valuable tool for *in situ* monitoring of the dot formation.

RAS allows the asymmetry of the dielectric properties of the surface to be monitored. Surface reconstructions (Kamiya et al., 1992) and oscillations due to monolayer growth (Reinhardt et al., 1993) can be identified. The formation of larger objects can be found in scattered (stray) light intensity (Olson and Kibbler, 1986). RAS has been applied to the *in-situ* monitoring of MOCVD growth of self-organized QDs by Steimetz et al. (1996).

X-ray diffraction techniques are useful for structural investigation after growth. Results for single dot layers (Krost, 1996b) and stacks (Darhuber et al., 1997a, 1997b) have been reported for the InAs/GaAs and Ge/Si systems. The diffraction signal due to dots is rather weak since the dots are much larger (~ 10 nm) than the probing wavelength (~ 0.1 nm). Thus high-brightness X-ray sources and experimental set-ups yielding a large dynamic range like double-, triple- or four-crystal spectrometers are of importance.

In the following we will discuss first MBE (Section 4.2) and MOCVD (Section 4.3) growth of quantum dots in the InGaAs/GaAs material system. Most work has been devoted to this model system. In Section 4.4 other material systems such as Ge/Si, InP/GaInP, Sb compounds, group III nitrides, and II–VI heterostructures are presented for which similar and exciting results on dot formation have been found. In Section 4.5 the vertical ordering of dots in stacks is discussed. Attempts to achieve artificial alignment of self-organized dots are compiled in Section 4.6.

4.2 MBE OF InGaAs/GaAs QUANTUM DOTS

4.2.1 DEPOSITION BELOW THE 2D–3D TRANSITION

Submonolayer InAs deposition on GaAs results in the formation of wire-like one-monolayer high islands, as proposed from optical anisotropy data by Wang *et al.* (1994). Islands oriented along the [0-11] direction having a uniform width of ~ 4 nm were visualized using STM (Fig. 4.1) (Bressler-Hill *et al.*, 1994, 1995). The island length/width ratio increases with the increase of the total amount of InAs deposited. Such arrays of islands may be considered as broken quantum wires or quantum dot arrays depending on the coverage with InAs (Ledentsov *et al.*, 1996b). These submonolayer structures show no ripening. There exists an optimal width of the islands. Very uniform arrays yielding narrow luminescence linewidths (down to FWHM (full width at half-maximum) $= 0.15$ meV) were reported (Belousov *et al.*, 1995). The orientation of islands is a function of the surface symmetry defined by the surface reconstruction and by the anisotropy of ad-atom diffusion. Typically the diffusion length in [1-10] is larger than along [110]. Ordering of two-dimensional islands in size leads to a reduction of total energy due to elastic relaxation of intrinsic surface stress as predicted by Marchenko (1981a, 1981b) (see Section 3.2.2).

For the intermediate stage of InAs growth (1–1.5 ML deposition), before the 2D–3D transition starts, the RHEED pattern is still streaky, indicating a flat character of the surface features, similar to the submonolayer deposition stage. This range of depositions was studied using STM by Cirlin *et al.* (1995) and Guryanov *et al.*

Fig. 4.1 STM image of a 1° A-type GaAs(001) (2 × 4) surface after an InAs deposition of 0.75 ML. (After Bressler-Hill *et al.*, 1994)

(1996a, 1996b). It was shown that an InAs monolayer on the (001) GaAs surface is stable for short growth interruptions Δt of ~ 10 s, but decomposes under ultra-long (~ 1000 s) growth interruptions to multilayer InAs islands having an anisotropic shape and uniform 'submonolayer' InAs coverage (Ledentsov et al., 1994b).

In Fig. 4.2a to c, STM images of the surface morphology of InAs on singular GaAs (001) are shown (Guryanov et al., 1996a). For 1 ML deposition and subsequent growth interruption (GRI) of 30 s (Fig. 4.2a) a dense array of InAs quantum wires directed along the [100] direction with a characteristic period of 30 nm is found. Increase in the mean thickness up to 1.5 ML InAs (Fig. 4.2b and c) results in a complex parquet structure with quasi-periodic modulation of the thickness in both [001] and [010], already for a 2 s growth interruption time. In these cases the characteristic period is increased to 40–50 nm. Such parquet structures were predicted theoretically (Marchenko, 1981a, 1981b) where complex structures of surface domains are considered. For clarification, a simple model of a parquet structure is depicted in Fig. 4.2d. With increasing GRI the parquet

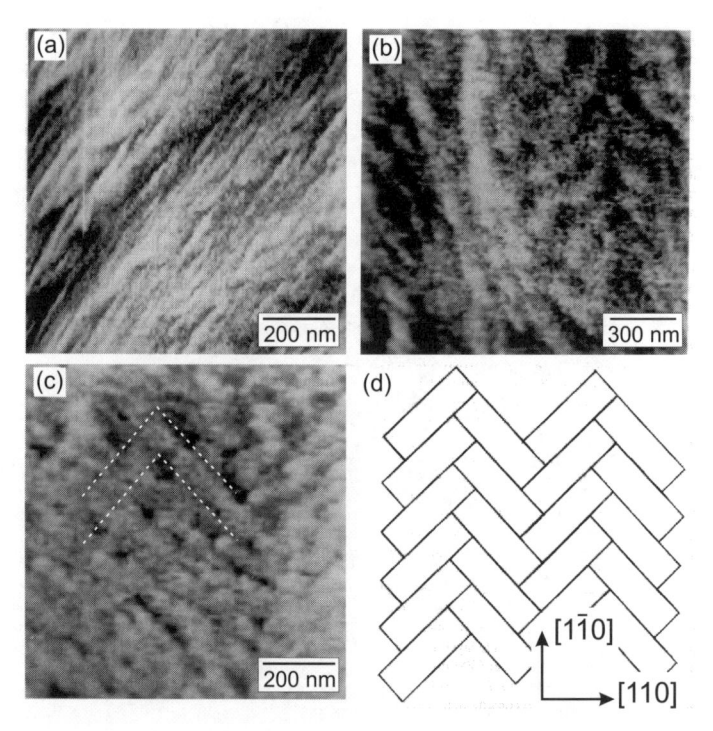

Fig. 4.2 STM images of GaAs(100) singular surface after InAs deposition: (a) 1 ML InAs with a growth interruption time of 30 s, scan area of $900 \times 900\,\text{nm}^2$; (b) 1.5 ML InAs with a growth interruption time of 30 s, scan area of $550 \times 550\,\text{nm}^2$; (c) 1.5 ML InAs with a growth interruption time of 2 s, scan area $900 \times 900\,\text{nm}^2$; (d) two-dimensional model of parquet structure. Sides of the images are parallel to the [110] and [1-10] directions. (After Guryanov et al., 1996a)

GROWTH OF SELF-ORGANIZED QDS   63

structure becomes more pronounced (Fig. 4.2b and c). These observations indicate the dependence of the pre-transformational morphology of InAs on minor growth details and demonstrate that ordering and formation of nanostructures occurs at very early stages of growth.

As shown by Marchenko (1981a, 1981b), the 2D domain structure on surfaces is caused by the discontinuity of the intrinsic *surface stress* tensor τ_{ij} on the boundary between neighboring domains (Section 3.2.2). The orientation of these boundaries is determined by the direction of the main axes of the tensor τ_{ij}, which are [110] and [1–10] directions on the (001) surface of zincblende semiconductors. On the other hand, the bulk anisotropy of elastic moduli of these semiconductors (see Fig. 3.8) favors domain structures in the directions of the lowest stiffness, i.e. in [100] and [010] directions. The interplay of these two factors may explain the different orientations of ordering observed by Bressler-Hill *et al.* (1994, 1995) and Cirlin *et al.* (1995) and may cause, under certain conditions, the complex 'parquet' structures. Additional stress due to steps on vicinal surfaces may also affect the domain structures and drastically change the surface morphology.

4.2.2 2D–3D TRANSITION

In a simplified scenario, three-dimensional islands start to develop on top of a two-dimensional wetting layer above a critical thickness d_c. As is clear from the previous section, the structure of the 2D layer can be rather complex; however, the corrugation is only in the monolayer (ML) range. Leonard *et al.* (1994) determined the area density of 3D islands ρ_D from AFM images versus the amount d of InAs deposited (Fig. 4.3). The samples were prepared using MBE at a deposition

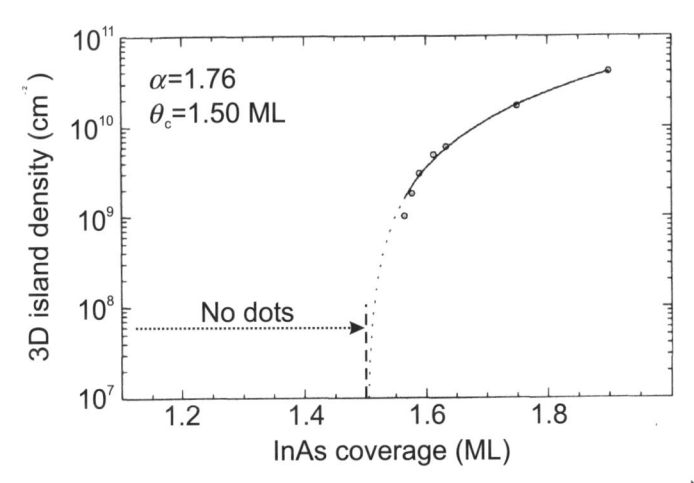

Fig. 4.3 Area density of 3D InAs quantum dots as a function of average thickness of deposited InAs at 530 °C by MBE. Line is fit curve described in the text. (After Leonard *et al.*, 1994)

temperature of 530 °C. The dependence can be described by a relation similar to that of a first-order phase transition:

$$\rho_D = \rho_0(d - d_c)^\alpha \tag{4.1}$$

where d_c is the critical InAs coverage, α the exponent and ρ_0 the normalization density. The fit parameters from Fig. 4.3 are $\alpha = 1.76$, $\rho_0 = 2 \times 10^{11}$ cm^{-2}, and $d_c = 1.5$ ML.

A much more complicated picture has been reported in Heitz et al. (1997a) for the 2D–3D morphology transition during MBE at a temperature of 500 °C. InAs features up to 5 ML high appear at ~ 1.25 ML deposition, disappear, and reappear prior to the onset of well-developed 3D islands at 1.57 ML. In Fig. 4.4 the morphology for InAs coverage increasing from 0.87 to 1.61 ML is shown. In Fig. 4.5 the area density of small and large 2D clusters, small and large 3D clusters, and 3D islands as determined from such STM images are plotted versus the deposited amount of InAs. The dependence of the density of 3D islands on InAs coverage is similar to Fig. 4.3.

A comparative STM study of InAs grown on exactly oriented GaAs (001) and vicinal (001) surfaces (3° toward [1-10]) during submonolayer molecular beam epitaxy in the vicinity of the 2D–3D transition has been reported in Cirlin et al. (1995). The evolution of a dense array of closely packed InAs 'wires', similar to that observed for 1–1.5 ML deposition, is observed (Fig. 4.6a). An increase of GRI to 10 s (Fig. 4.6b) results in the formation of 3D islands, closely packed in chains along the [100] and also [010] directions. For longer annealing time (30 s, Fig. 4.6c) the dots are well separated from each other in both lateral directions and aligned on a statistically distorted square lattice (Cirlin et al., 1995). For 2 ML InAs deposition on the vicinal surface, the parquet structure is more stable and a longer GRI (~ 60 s) is required for transformation into dots. A similar transformation of extended wire-like segments in dots has been reported for Ge islands on Si(001), indicating that the phenomenon is of quite general origin (Cullis, 1996; Jesson et al., 1996a).

The absence of a critical thickness for quantum dot formation has been reported by Polimeni et al. (1996) for InAs deposited at 420 °C using MBE.

4.2.3 QUANTUM DOT STRUCTURE

In Fig. 4.7 plan-view TEM images and PL spectra of samples with 2 ML (a, c) and 4 ML InAs (b) deposited at 480 °C with $\sim 2 \times 10^{-6}$ Torr arsenic pressure (P_{As}) are shown. No growth interruptions were introduced in cases (a) and (b). Sample (c) was deposited with submonolayer (0.3 ML InAs) growth cycles separated by 100 s long growth interruptions. The dots formed after the critical layer thickness is just exceeded (Fig. 4.7a) are small, mostly do not show well-resolved crystalline shape, and exhibit large size dispersion. TEM contrast modulation with stripes along the [100] direction resembles the pre-transformational InAs layer morphology (Section 4.2.1) and indicates that the wetting layer is modulated in thickness. Similar effects have been reported for InGaAs/GaAs growth, while the stripes were oriented

GROWTH OF SELF-ORGANIZED QDS 65

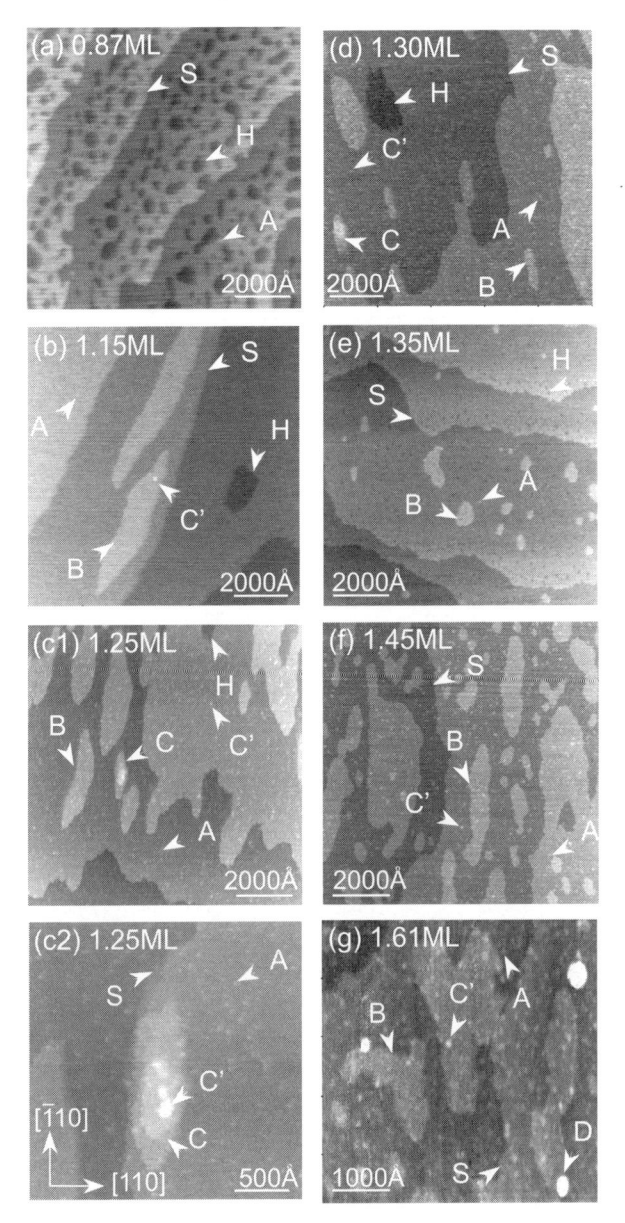

Fig. 4.4 STM images showing the evolution of InAs morphology on GaAs(001) for different thicknesses: (a) 0.87 ML, (b) 1.15 ML, (c1, c2) 1.25 ML, (d) 1.30 ML, (e) 1.35 ML, (f) 1.45 ML, and (g) 1.61 ML. The labels in the figure denote small clusters (A), large 2D clusters (B), small quasi-3D cluster (C'), large quasi-3D clusters (C), 3D islands (D), 1 ML high steps (S) and 1 ML deep holes (H). (After Heitz *et al.*, 1997a)

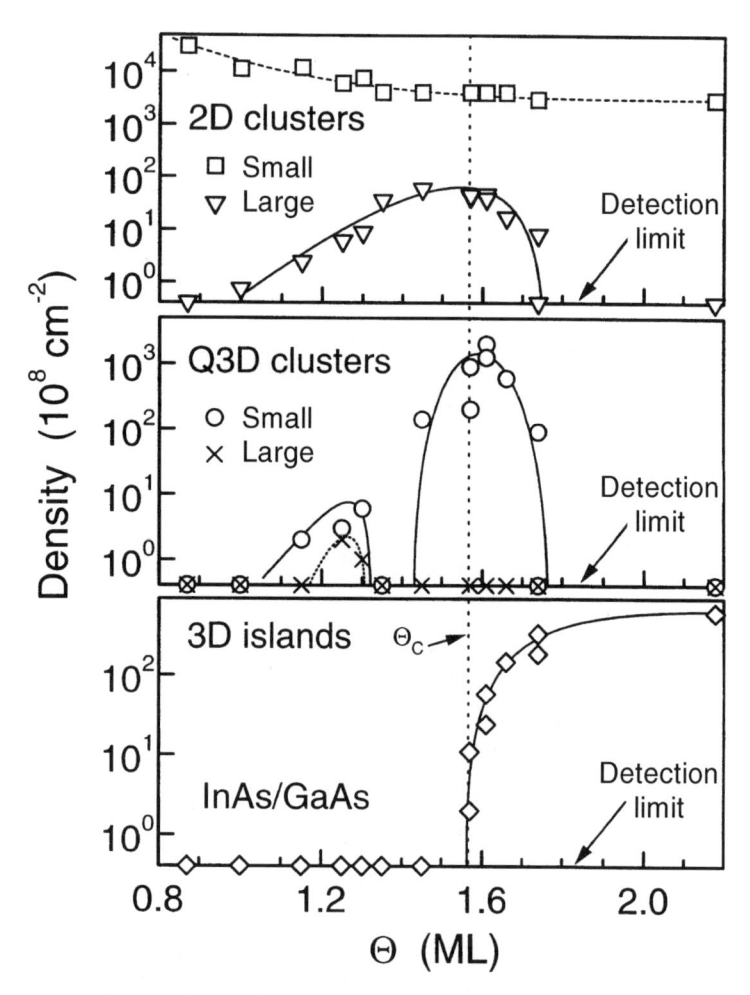

Fig. 4.5 Area density of 2D and quasi-3D clusters as well as 3D InAs islands on GaAs(001), determined from the STM images like those in Fig. 4.4, as a function of average deposition thickness. (After Heitz et al., 1997a)

along the [1-10] direction (Ming et al., 1995; Ruvimov et al., 1995). An increase to 4 ML InAs (Fig. 4.7c) results in a dense array of well-developed dots having a size of ~ 14 nm. The sides of the square-shaped base of the dot are aligned parallel to the [100] and [010] directions. The larger size of the dots in the case of the 4 ML sample agrees well with a strong shift of the corresponding PL line toward smaller photon energies.

Since the shape of the dots is quite different for 2 ML and for 4 ML InAs deposition one might question the existence of an *equilibrium shape* of the dots. The introduction of growth interruption of 40 s (10 s) after InAs deposition is enough to

GROWTH OF SELF-ORGANIZED QDS 67

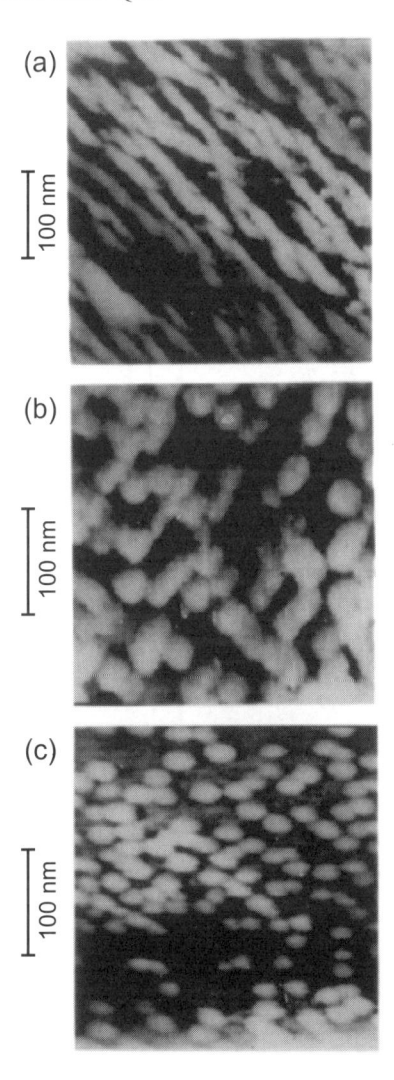

Fig. 4.6 STM images of GaAs (001) samples after deposition of 2 ML InAs with a growth interruption time of (a) 2 s, Å, (b) 10 s and (c) 30 s. (After Cirlin *et al.*, 1995)

let the dots reach the same equilibrium size for 2.5 ML (3 ML) InAs deposition as for 4 ML deposition. Large clusters and dislocations do not appear. With very long growth interruptions of 10 min total (Fig. 4.7b) it is possible to let even 2 ML dots reach equilibrium size (of course, with a lower area density due to the small amount of deposited InAs) and the PL peak coincides in energy with the 4 ML InAs PL peak at ~ 1.1 eV. For this interruption time the wetting layer starts to decompose (Ledentsov *et al.*, 1995a), resulting in the additional appearance of large clusters.

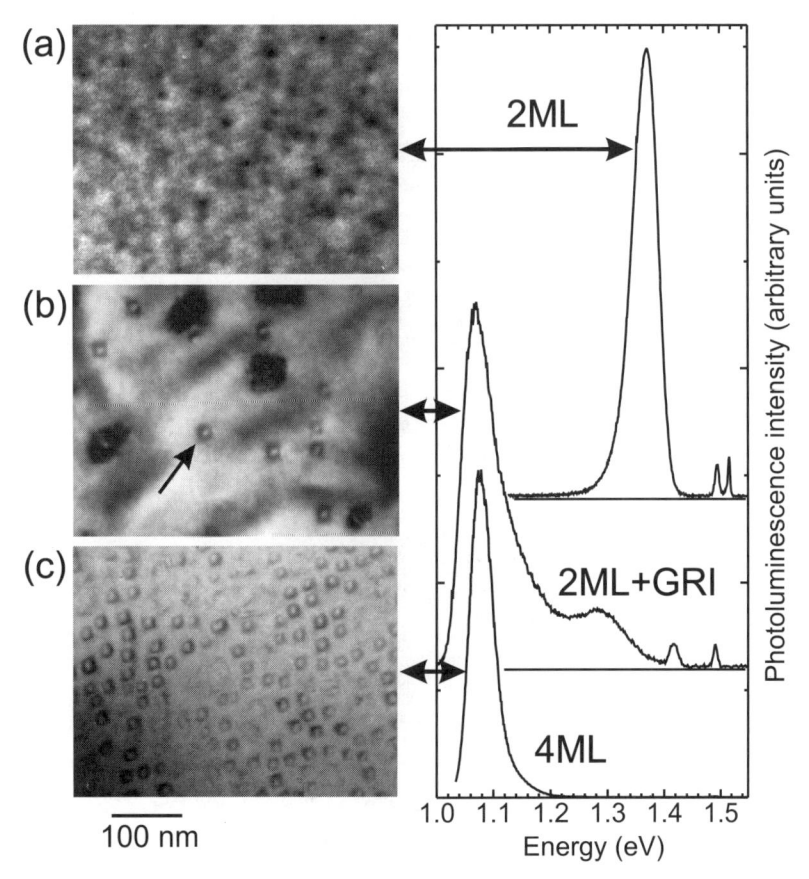

Fig. 4.7 PL spectra and plan-view transmission electron microscopy (TEM) images of structures with 2 ML (a,c) and 4 ML InAs (b) deposited at 480 °C with standard $\sim 2 \times 10^{-6}$ Torr arsenic pressure. No growth interruptions were introduced in cases (a) and (b). Sample (c) was deposited with submonolayer (0.3 ML InAs) growth cycles separated by 100 s long growth interruptions. (After Ledentsov et al., 1996b)

Many different shapes have been reported for InAs dots grown on GaAs using MBE. Moison et al. (1994) found in AFM investigations at room temperature facets ranging from {014} to {011} for dots deposited at 500 °C. Nabetani et al. (1994) concluded that large dots formed on {113} facets based on RHEED analysis for 13.3 nm upon 2 ML indium deposition at 480 °C. From a detailed RHEED analysis Lee et al., (1998) concluded that dots formed on {316} facets for 1.68 ML of InAs deposited at 500 °C.

Leonard et al. (1993) reported lens-shaped dots without particular facets found in cross section TEM analysis for InAs deposition at 530 °C. Plan-view TEM images of 2 and 4 ML InAs quantum dots (Fig. 4.7) grown at 480 °C show square-shaped objects with base sides along the ⟨100⟩ directions (Bimberg et al., 1995; Grundmann

GROWTH OF SELF-ORGANIZED QDS

et al., 1995b; Ruvimov *et al.*, 1995). Cross-sectional HREM images of such dots together with numerical simulations of the TEM and HREM contrast formation reveal a pyramidal shape (Ruvimov and Scheerschmidt, 1995; Ruvimov *et al.*, 1995) with {011}-like facets. In Fig. 4.8 a cross-sectional TEM image of a InAs quantum dot is shown. From the TEM contrast simulation in Fig. 4.9 it is obvious that only special imaging conditions yield an image that resembles the actual geometry. Conventional TEM is predicted to yield lens-shape images for pyramids for any parameters. Image analysis using a combination of methods performed on cross-sectional HREM images of InGaAs/GaAs quantum dots grown with MBE at 500 °C revealed rather flat islands (Woggon *et al.*, 1997). In Liao *et al.*, 1998 it is argued that near-square contrast in in-plane [001] on-zone bright-field TEM images can also arise from spherically symmetric QDs.

4.2.4 HIERARCHY OF SELF-ORGANIZATION MECHANISMS

In Fig. 4.10 the hierarchy of experimentally observed self-organization mechanisms of quantum dots is schematically shown.

Facets with distinct orientation close to {011} and a uniform shape (square base) as well as ordering in size (relative standard deviation ≤ 10%) have been observed for InGaAs/GaAs quantum dots grown by MBE (and also MOCVD; see Section 4.3). The lateral arrangement of dots depends on the growth mode and detailed conditions. MBE-grown dense arrays of QDs in the InAs/GaAs system, deposited at 480 °C, exhibit lateral ordering in a square lattice with primitive unit vectors in the [100] and [010] directions (Bimberg *et al.*, 1995; Grundmann *et al.*, 1995b) (e.g. see

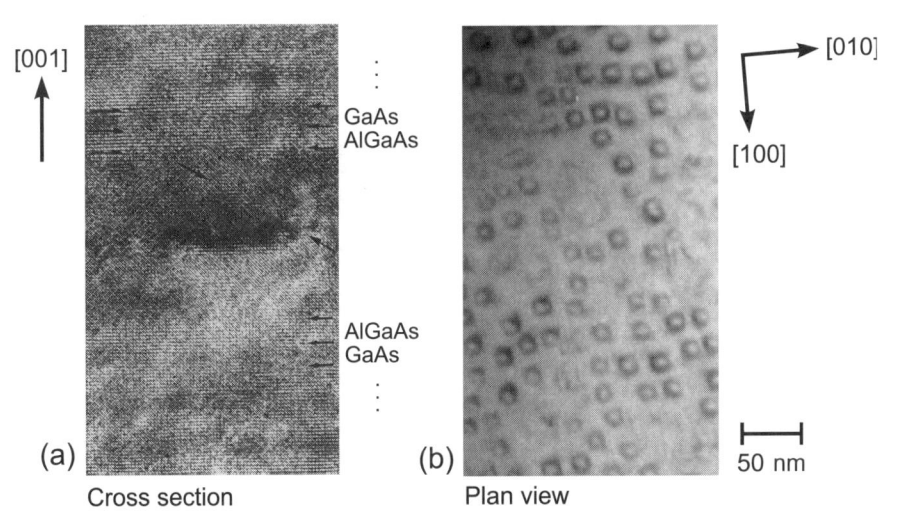

Fig. 4.8 (a) Cross-sectional and (b) plan-view TEM image of 4 ML InAs quantum dots on GaAs(001). The QDs are within a GaAs quantum well surrounded by a AlGaAs/GaAs superlattice as indicated in (a). (After Ruvimov *et al.*, 1995)

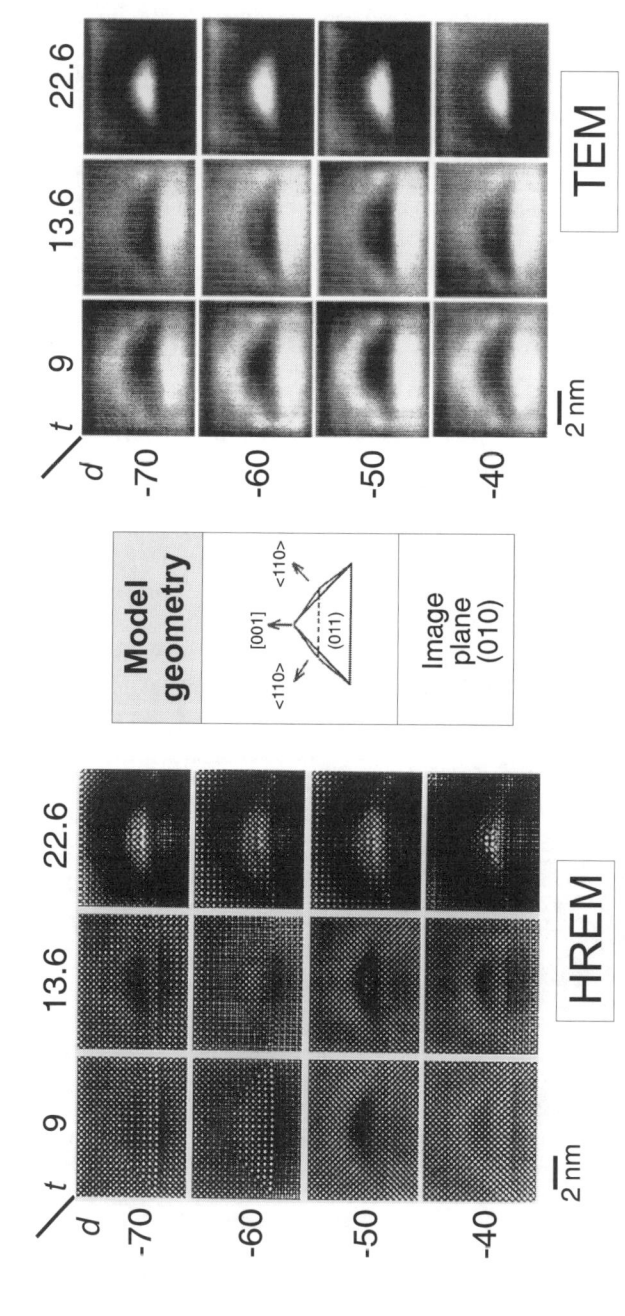

Fig. 4.9 Simulated high-resolution electron microscopy (HREM) and conventional (010) plane cross-sectional TEM images of strained InAs pyramidal quantum dots in a GaAs matrix for different sample foil thickness t and de-focus d (both in nm). (After Ruvimov *et al.*, 1995)

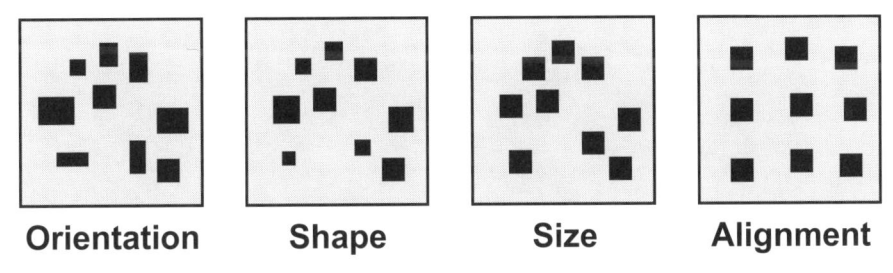

Orientation   **Shape**   **Size**   **Alignment**

Fig. 4.10 Schematic representation of the hierarchy of self-ordering mechanisms for quantum dots

Figs. 4.8b and 4.22). A hexagonal arrangement of InGaAs/GaAs QDs grown by MBE at 500 °C was reported by Moison et al. (1994). Hexagonal and square arrangements of islands are very close in their total energies, as can be seen from Fig. 3.16 (Shchukin et al., 1996). A hexagonal arrangement of islands certainly presents a possible metastable state of the system.

4.2.5 INFLUENCE OF DEPOSITION CONDITIONS

4.2.5.1 Deposition mode and misorientation

A comparative systematic study of InAs growth on singular and misoriented (3° toward the [011] direction) GaAs(001) surfaces for two different deposition modes during submonolayer molecular beam epitaxy has been reported by Guryanov et al. (1996b). In the simultaneous deposition mode (SDM) continuous As flux impinging on the surface and simultaneous In deposition using submonolayer growth cycles are used. In the alternate deposition mode (ADM) In and As are alternately deposited. All other growth parameters, such as substrate temperature, arsenic flux impinging on the surface, and the amount of the InAs deposited, are kept constant. Using RHEED and STM it was found that the deposition mode dramatically affects the final arrangement of dots. Surface misorientation significantly decreases the dot density for the SDM and stimulates ordering of dots in chains along the [001] direction. The ADM results in a much higher density of dots on vicinal substrates, while ordering of dots along the [001] *and* [010] directions is pronounced for singular substrates.

For the SDM a dense array of InAs quantum dots ($1.0 \times 10^{11}\,\text{cm}^{-2}$) is formed after 2 ML of InAs are deposited on the singular surface (Fig. 4.11a). The dot size is $\sim 30\,\text{nm}$ and the dots are locally aligned along the [100] and [010] directions. Dots formed using the SDM on the vicinal substrate are much smaller ($\sim 18\,\text{nm}$) and have a lower area density (around $0.3 \times 10^{11}\,\text{cm}^{-2}$). These dots are arranged in chains only along the [100] direction. The larger lateral size and higher density of dots formed on the singular substrate (Fig. 4.11b) is in agreement with a smaller InAs critical thickness in this case. Increase of the average InAs thickness to 3 ML increases the density of dots for both singular and vicinal substrates. The density of

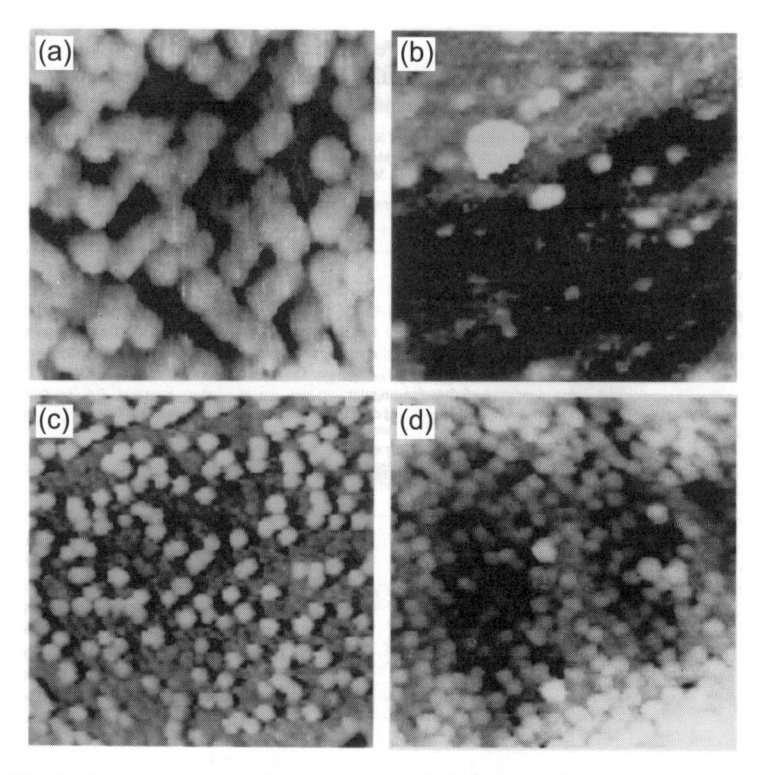

Fig. 4.11 Surface morphology of the samples with 2 ML InAs deposited on singular (a, c) and vicinal (b, d) surfaces using the simultaneous In and As deposition mode (SDM) (a, b) and using the alternate In and As deposition mode (ADM) (c, d). The growth interruption time (Δt) is equal to 10 s for (a, b) and to 2 s for (c, d). The scan area is 30 nm × 30 nm for (a,b), 100 nm × 100 nm for (c), and 60 nm × 60 nm for (d). Sides of the images are parallel to [011] or [0-11] direction. (After Guryanov et al., 1996b)

dots, however, is still lower for the misoriented surface ($1.3 \times 10^{11}\,\mathrm{cm}^{-2}$ and $0.9 \times 10^{11}\,\mathrm{cm}^{-2}$, respectively). There is a marked ordering of dots along the [010] direction in the latter case as well.

The results obtained for the ADM are very different. In Fig. 4.11c and d, STM images of the samples with 2 ML InAs deposited on singular (c) and vicinal (d) surfaces using ADM ($\Delta t = 2\,\mathrm{s}$) are shown. For singular substrates the resulting density of dots is much lower than in the SDM case (0.18×10^{11} and $1.0 \times 10^{11}\,\mathrm{cm}^{-2}$, respectively). Average lateral sizes of the dots in the ADM case (Fig. 4.11c and d) are ~ 35 and ~ 18 nm, respectively. The remarkable result of Fig. 4.11d is a high density of dots in the case of the vicinal surface ($0.8 \times 10^{11}\,\mathrm{cm}^{-2}$). Another interesting observation for the ADM is a completely isotropic distribution of dots on both singular and vicinal substrates.

One can speculate that in the case of the ADM, the thickness of the wetting layer is smaller, as no complete arsenic monoatomic layer is formed on top of the In atomic layer. This results in a smaller driving force for corrugations of the InAs wetting layer before the 2D–3D transformation starts. After the As$_4$ flux is switched on, a fast transformation to dots starts and the resulting dot distribution is laterally isotropic. On the other hand, if the InAs layer is formed by the SDM, layer corrugation and deformation of the substrate are pronounced as the 'effective' wetting layer thickness is higher in this case by at least an As monoatomic layer. Intentional substrate misorientation provides in this case an additional source for elastic strain relaxation via monolayer InAs/GaAs interface steps of high density. The dot formation starts at a larger critical thickness and the dot density is smaller. The high density of dots observed for the ADM growth on the vicinal substrate is in agreement with the relatively small elastic relaxation via interface steps due to a much 'softer' wetting layer. This is also in agreement with RHEED studies indicating no delay in the dot formation for vicinal substrates in this case. On the other hand, the increased concentration of interface steps along the [011] direction can suppress the long-wavelength corrugation along the [010] direction and increase the dot density.

4.2.5.2 As pressure

The stability of an array of InAs dots depends largely on the arsenic pressure (Ledentsov et al., 1996b). In Fig. 4.12 samples are compared that have been grown at a number of different arsenic pressures, varied around the standard MBE arsenic pressure $P_{As}^{(0)} \approx 2 \times 10^{-6}$ Torr, for otherwise identical conditions ($T_G = 480\,°C$). For growth at $P_{As}^{(0)}$, the typical equilibrium array of dots of high density (5×10^{10} cm^{-2}) develops. This type of dot array is stable upon variation of the As pressure by about 50%. An increase in As pressure by a factor of 3 ($3P_{As}^{(0)}$) results in a dramatic change in the morphology. The size of dots reduces, and a large concentration of dislocated InAs clusters ~ 50–100 nm in size appears. A reduction of the arsenic pressure affects the dots in a different way. At an arsenic pressure $\frac{1}{6}P_{As}^{(0)}$, the dots completely disappear in favor of macroscopic two-dimensional InAs islands (~ 100 nm). The transition from 3D islands to 2D islands with a decrease in the arsenic pressure is *reversible* whereas the ordering-to-ripening transition found for an increase in As pressure is accompanied by the onset of misfit dislocations and is *irreversible*. A similar reversible transition between an InGaAs dot structure and an InGaAsP flat surface in chemical beam epitaxy was reported by Ozawa et al. (1997). Supply of phosphorus induces the morphology change from dots to a flat InGaAsP surface. When the arsine beam is reapplied, a return to InGaAs dots is observed.

The stoichiometry of (001) surfaces of group III–V semiconductors that are in equilibrium with the gas is known to depend strongly on the partial pressure of the group V element. Thus a change in As pressure leads to a change in the surface energy of a GaAs(001) surface and to a change in the surface reconstruction (Qian et al., 1988). The wetting layer of InAs on a GaAs (001) surface is expected to exhibit a similar behavior with arsenic pressure. A decrease in the surface energy results in an increase in the control parameter α in Eq. (3.26) and favors a transition from the

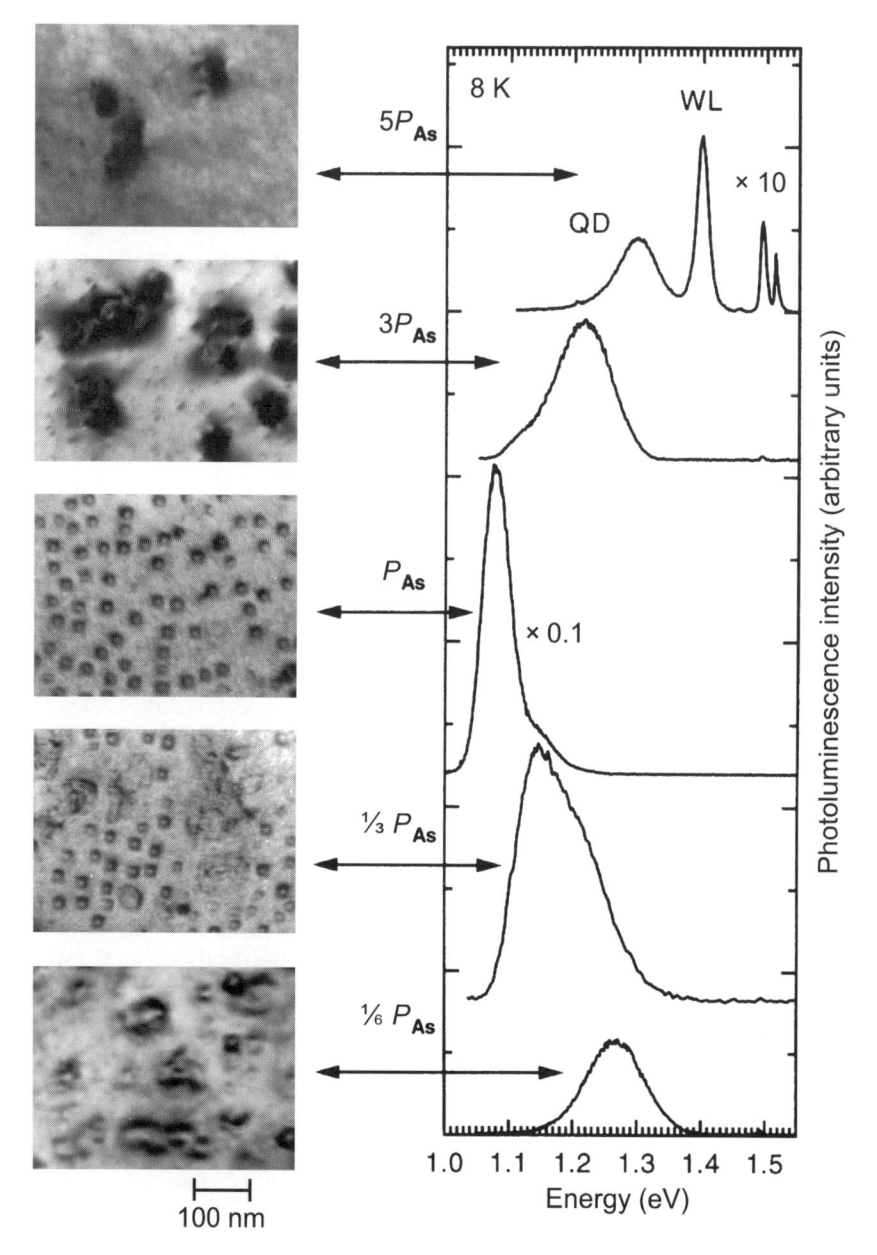

Fig. 4.12 Plan-view TEM images of InAs morphology on GaAs(001) and corresponding PL spectra ($T = 8$ K) for structures grown by MBE at different arsenic pressures. The pressure P_{As} corresponds to 2×10^{-6} Torr. (After Ledentsov *et al.*, 1996b)

ordering regime to the ripening regime. This is likely to be the case when the arsenic pressure increases from $P_{As}^{(0)}$ to $3P_{As}^{(0)}$. At a lower As pressure, on the other hand, In is known to segregate on the surface. Such an In layer can be regarded as a quasi-liquid phase (Ivanov *et al.*, 1990) that shows no strain-induced renormalization of the surface energy, since a liquid is incompressible. Therefore, an increase in the surface area due to As pressure reduction induced island formation may lead to a large increase in the surface energy making the formation of 3D islands unfavorable.

The growth regime of cation-rich conditions has been explored by Tournié and Ploog (1993), Trampert *et al.* (1994) and Tournié *et al.* (1995), who find that two-dimensional nucleation occurs and no 3D islands are formed, in agreement with the above general discussion.

4.2.5.3 High index substrates

All parameters governing quantum dot formation, like strain energy, surface energy, etc., depend on the substrate orientation in the systems discussed here. Quantum dot formation on high index substrates is thus different from formation on (001) substrates and has been investigated by various groups. Yamaguchi *et al.* (1996) observed that the MBE growth front remains two-dimensional on a (111)A-oriented substrate. Large strain relaxation is found from RHEED patterns after 2 ML deposition of InAs and 3D island formation is inhibited. A possible explanation is given by the high surface energy or the small lifetime of arsenic molecules on the (111)A surface corresponding to cation-rich conditions favoring 2D nucleation (see the previous section).

Formation of $In_{0.3}Ga_{0.7}As$ dots on GaAs (311)B substrates during hydrogen-assisted MBE was reported by Chun *et al.* (1996). On GaAs (311)B substrates ordered islands have also been fabricated using gas source MBE by Nishi *et al.* (1997). A typical base diameter of the dots was about 120 nm. The height varied from 3 to 13 nm for an increase in the nominal thickness of the $In_{0.25}Ga_{0.75}As$ layer from 4 to 8 nm. Ordering of islands along lines inclined about 60° from the [110] direction was observed.

Arrays of GaAs/AlGaAs quantum dots formed at atomic hydrogen induced step bunches on [011] oriented mesas on (311) substrates have been observed by Ploog (1988).

Fabrication of InGaAs/GaAs quantum dots on (111)B GaAs substrate was claimed by Tsai and Lee (1998). The optimum growth temperature was found to be 480 °C. The lateral extension of the QDs was estimated smaller than 50 nm, the height around 5 nm. The island formation was attributed to 3D growth mode rather than strain relaxation.

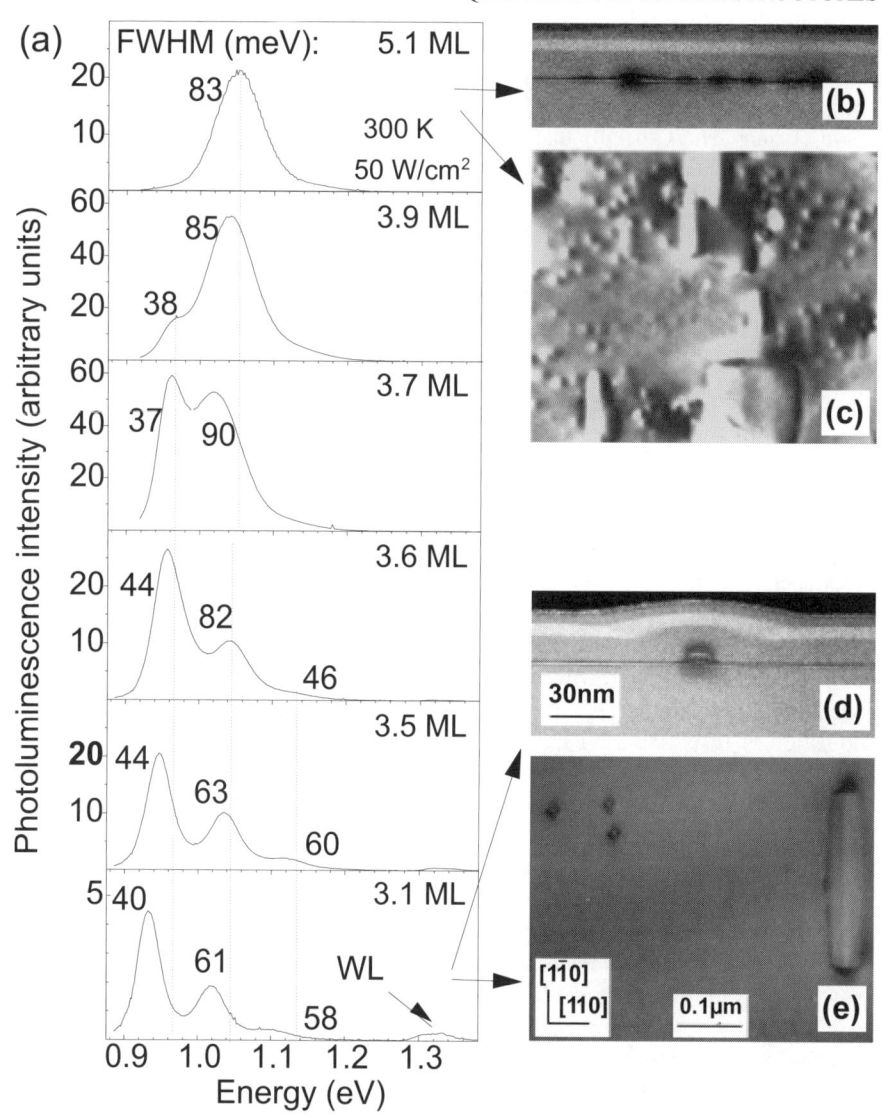

Fig. 4.13 Dependence of PL spectra (a) on deposited $In_{0.4}Ga_{0.6}As$ thickness during MOCVD growth ($T_G = 450\,°C$) and corresponding TEM cross-sectional (b, d) and plan-view (c, e) images. (After Heinrichsdorff *et al.*, 1996b)

4.3 MOCVD GROWTH OF InGaAs/GaAs QUANTUM DOTS

4.3.1 2D-3D TRANSITION

The 2D–3D transition of InGaAs on GaAs(001) during MOCVD has been investigated for different In concentrations by Heinrichsdorff *et al.* (1996b).

Typically observed morphologies and corresponding PL spectra are shown in Fig. 4.13 for In$_{0.4}$Ga$_{0.6}$As of different thicknesses deposited on GaAs. For deposition up to 3.6 ML the spectra consist of a triplet assigned to luminescence from the QD ground and two excited states. With increasing thickness this triplet gradually shifts to a higher energy and disappears. Instead, one broader peak at 1.05 eV comes up and dominates the spectrum for 3.9 ML deposition. The corresponding TEM images show that the dot density increases with layer thickness (Fig. 4.13e and c), whereas the dot height decreases (Fig. 4.13d and b), as deduced from the reduced corrugation of the AlGaAs cap layer. Figure 4.14 summarizes the types of PL spectra obtained for samples with varying In content and layer thickness. Below 33% In concentration no dot formation is observed. For higher concentrations the 2D–3D transition (QW to triplet spectra) occurs for increasing In content at decreasing thickness. The data are not explained by a curve of constant strain energy, as one might suspect, but by the total amount of deposited indium (Heinrichsdorff *et al.*, 1996b). These results allow the conclusion for In segregation toward the surface and subsequent formation of QDs (Fig. 4.13d and e), underlining the importance of kinetic effects in MOCVD.

Upon formation of QDs a decrease of indium content from 55 to 45% in the wetting layer was found (Krost *et al.*, 1996b). Since strain relaxation is most efficient in the 3D islands, the driving force behind this process can be understood in terms of minimization of strain energy.

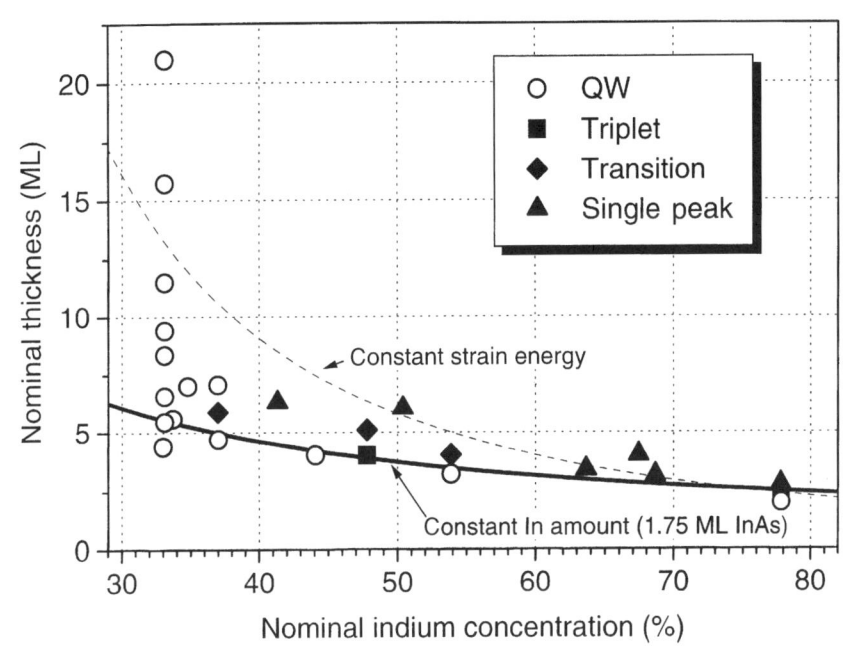

Fig. 4.14 Phase diagram of dot formation as a function of In content and deposited thickness, deducted from PL spectra according to Fig. 4.13a. (After Heinrichsdorff *et al.*, 1996b)

4.3.2 QUANTUM DOTS

The growth of self-organized InGaAs quantum dots on GaAs using MOCVD has been reported by Mukai *et al.* (1994, 1996a), Oshinowo *et al.* (1994), Heinrichsdorff *et al.* (1996a, 1996b, 1997c, 1997d), Ledentsov *et al.* (1996a) and Geiger *et al.* (1997). Using cross-sectional SEM, Oshinowo *et al.* (1994) observed a triangular shape. An in-plane square shape of the QD base was visualized in plan-view TEM images (Fig. 4.15a) (Heinrichsdorff *et al.*, 1996b). The growth of binary InAs dots had limited success initially due to occurrence of large three-dimensional dislocated In-rich clusters (see Fig. 4.15b), which coexisted with the quantum dots (Heinrichsdorff *et al.*, 1996a; Geiger *et al.*, 1997) at standard growth conditions.

Formation of clusters, which also act as nonradiative defects, can be completely avoided by switching off arsine flux during growth interruption (Heinrichsdorff *et al.*, 1997d). Defect-free and dense arrays of InAs QDs can now be fabricated (Fig. 4.16) (Heinrichsdorff *et al.*, 1997d). The local arrangement of these QDs in short chains along ⟨110⟩-type directions (Heinrichsdorff *et al.*, 1996b) might indicate the dominance of kinetic ordering effects such as anisotropic diffusion of ad-atoms. A further degree of freedom to tune the emission wavelength and a further increase of quantum efficiency was achieved by deposition of a thin InGaAs layer on top of the InAs QDs (Heinrichsdorff *et al.*, 1997d). Room temperature emission at wavelengths up to 1.4 µm has been reported (Heinrichsdorff, 1998).

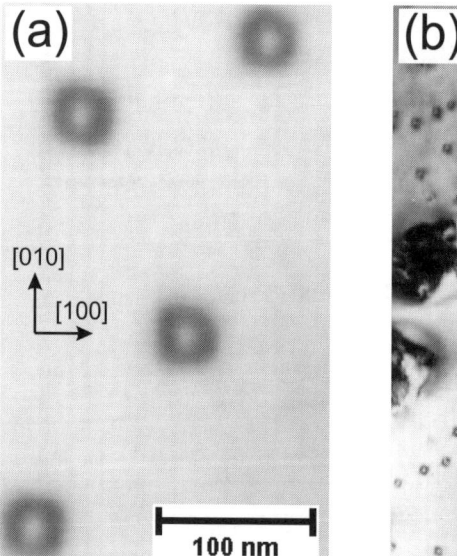

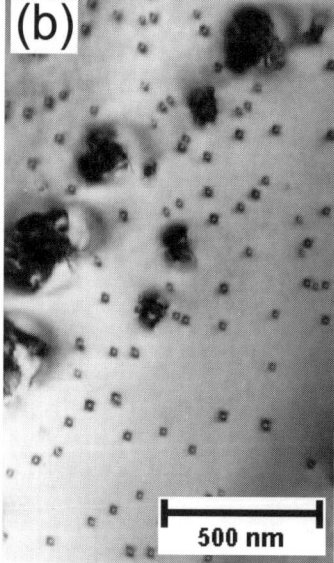

Fig. 4.15 Plan-view TEM image of InGaAs/GaAs quantum dots grown by MOCVD at 475 °C in two different magnifications: (a) reveals square-shaped dot base, (b) shows quantum dots and coexistent dislocated In-rich clusters. (After Heinrichsdorff *et al.*, 1996a)

GROWTH OF SELF-ORGANIZED QDS 79

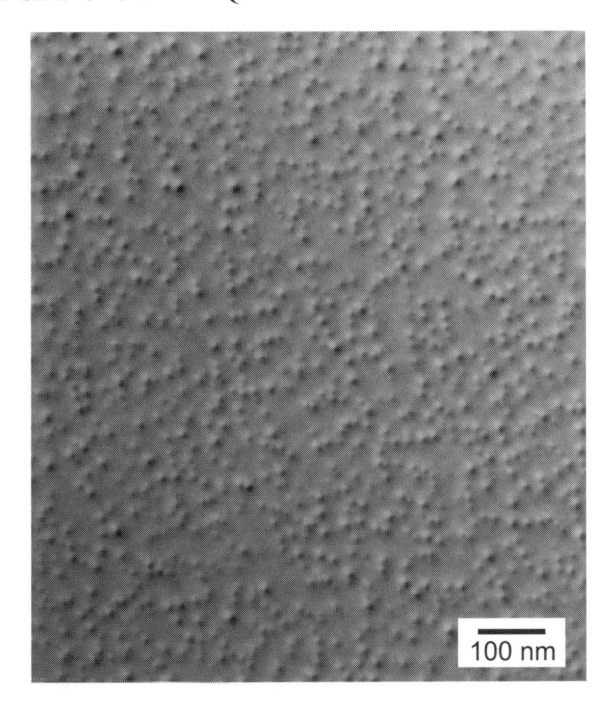

Fig. 4.16 Large area plan-view image of defect-free single layer of InAs quantum dots on GaAs grown by MOCVD

4.3.3 HIGH INDEX SUBSTRATES

The MOCVD growth of InGaAs islands on GaAs ($n11$) substrates has been investigated by Nötzel et al. (1994b) and Temmyo et al. (1996). On A-type surfaces step bunching was observed, leading to the formation of wire-like structures for the (311)A substrate orientation. On B-type surfaces the InGaAs film deposited on top of an AlGaAs layer transforms during growth interruption into disks that are directly buried within AlGaAs microcrystals (Nötzel et al., 1994b). In Fig. 4.17 the morphologies of AlGaAs microcrystals containing $In_{0.4}Ga_{0.6}As$ islands on GaAs (211)B, (311)B, and (511)B are compared (Nötzel et al., 1994c). The uniformity and ordering of the disks are apparently largest on the (311)B surface. The direction of the alignment is about 45° off the [1-10] azimuth. The size of the microcrystals depends strongly on growth temperature, where a 200 nm base width is found for growth at 800°C and 70 nm for deposition at 720°C. The base width of the 70 nm microcrystals is almost independent of the InGaAs coverage between 2 and 3.5 nm. With increasing coverage the island density increases.

Similar results have been reported for MOCVD growth of GaInAs/AlInAs and GaInAs/InP heterostructures on InP (311)B (Nötzel et al., 1995). InP islands on GaInP (311)B were fabricated by Reaves et al. (1996).

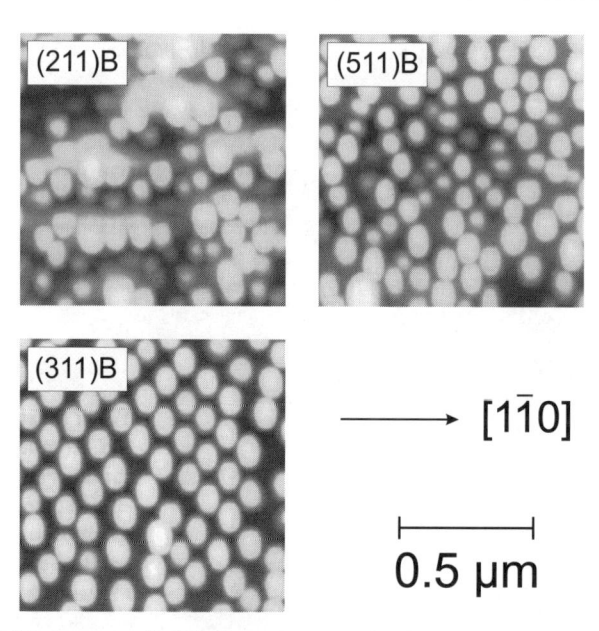

Fig. 4.17 AFM images of AlGaAs microcrystals formed by nominal 3.5 nm thick In$_{0.4}$Ga$_{0.6}$As deposition on GaAs (211)B, (311)B, and (511)B substrates. (Reproduced by permission of American Institute of Physics from Nötzel et al., 1994c)

4.4 OTHER MATERIAL SYSTEMS

The formation of ordered arrays of islands in strained heteroepitaxy is a general phenomenon observed in a large number of material systems. In the following results are compiled for a number of heterostructures. Similar results to those on InGaAs/AlGaAs have been observed for other III–V compounds such as InAlAs/AlGaAs on GaAs (Leon et al., 1995), InAs on InP (Ponchet et al., 1995; Fafard et al., 1996; Taskinen et al., 1997; Li et al., 1998b), and InP/GaAs (Sopanen et al., 1995; Marchand et al., 1997). InAs/GaP islands were found to nucleate in the Volmer–Weber growth mode (Leon et al., 1998). In the following sections results for InP/GaInP on GaAs(001) (Section 4.4.1), Sb-containing compounds on GaAs and InP (Section 4.4.2), Ge on Si (Section 4.4.3), group III–V on Si (Section 4.4.4), II–VI compounds (Section 4.4.5) and group III nitrides (Section 4.4.6) are compiled.

4.4.1 InP ON GaInP/GaAs

InP islands have been grown on InGaP/GaAs (001) using MOCVD (Carlsson et al., 1994; Georgsson et al., 1995). A 4 ML thick GaP cap was grown on top of the InGaP layer before InP growth was started in order to increase the density of fully developed islands and decrease the density of tiny InP islands (Georgsson et al.,

GROWTH OF SELF-ORGANIZED QDS 81

1995). Upon 4 ML InP deposition at 580 °C pyramidal islands develop whose size and shape have been investigated in detail using HREM (Georgsson *et al.*, 1995) (Fig. 4.18). The average width is found to be 40–50 nm, the average length is 55–65 nm. The side facets are low index planes including {001}, {110} and {111} (Fig. 4.18). Using MBE at a deposition temperature of 580 °C, InP dots were

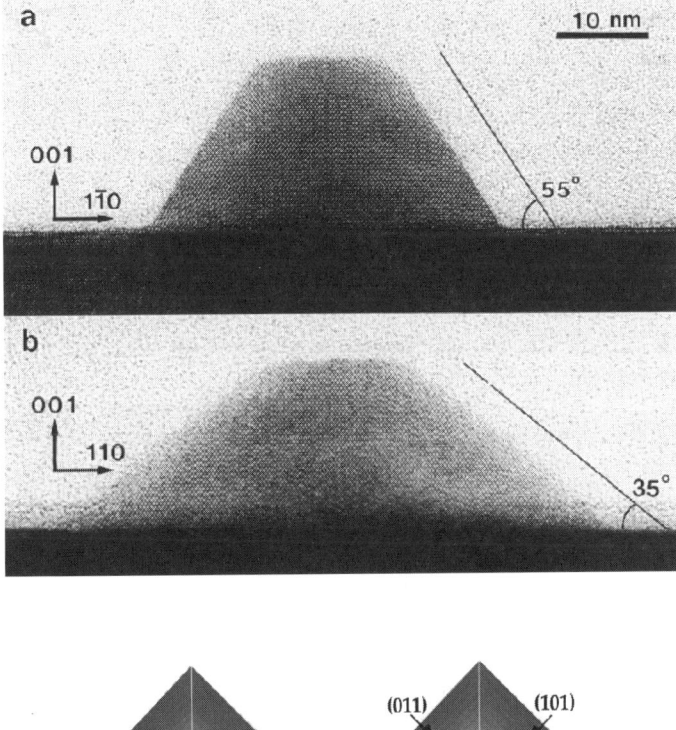

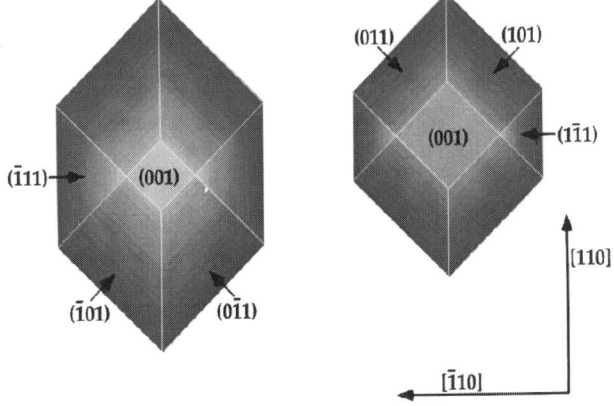

Fig. 4.18 Cross-sectional HREM of uncapped InP/InGaP SK islands on GaAs (001) along the [110] (a) and [-110] (b) directions together with a schematic model of the island shape. (Reproduced by permission of the American Institute of Physics from Georgsson *et al.*, 1995)

fabricated on GaInP/GaAs (001) by Kurtenbach *et al.* (1995). A diameter of 30–50 nm and a height of about 5 nm were found. The luminescence transition of such islands is close to 1.7 eV.

4.4.2 Sb COMPOUNDS

It is expected that the incorporation of antimony compounds will lead to usable QD luminescence in the wavelength range well above 1.1 µm on GaAs substrates. The growth of pseudomorphic GaSb films on GaAs using metal-organic vapor deposition was reported by Chidley *et al.* (1989); above a critical thickness of 1.5 nm large (> 100 nm) elastically relaxed islands were observed. Formation of coherent dots in molecular beam epitaxy (MBE) was first reported by Hatami *et al.* (1995) and evidenced by transmission electron microscopy (TEM) images (Fig. 4.19a). The in-plane projection of the quantum dot shape appears rectangular with rounded edges. Some additional contrast can be seen in TEM images due to the two-dimensional wetting layer resolved in between the GaSb dots. A statistical analysis of the size and shape of about 10^2 dots is shown in Fig. 4.19b. The density of the dots is approximately 4×10^{10} cm^{-2}. Most of the dots have a slightly asymmetric base shape. The dots exhibit an average lateral size $\bar{L} = \sqrt{ab}$ of 22 nm with a standard deviation σ_L of 4.2 nm, a and b being the axes of the dots in the $\langle 100 \rangle$ and $\langle 010 \rangle$ directions, respectively (Figure 4.19b). The relative standard deviation $\sigma_L/\bar{L}$ is 19%. The asymmetry $\lambda = (a - b)/(a + b)$ has an average value of $|\bar{\lambda}| \approx 0.07$ (Fig. 4.19c).

Growth of GaSb/GaAs QDs was also reported by Thibado et al. (1996); in scanning tunneling microscopy (STM) a dot height of 9 nm was found. Reported lateral QD sizes are 22 nm (Hatami *et al.*, 1995), 28 nm (Bennett *et al.*, 1996) and 40–60 nm (Glaser *et al.*, 1996) for GaSb, 50 nm diameter for InSb and 56 nm diameter for AlSb (Bennett *et al.*, 1996) quantum dots on GaAs. InSb dots were also fabricated on InP substrates (Ferrer *et al.*, 1996; Uztmeier *et al.*, 1996). Ferrer *et al.* (1996) observed a typical dot size of 15 nm. Part of the mismatch between dot material and substrate was found here to be accommodated by misfit dislocations. We note that (In,Ga,Al)Sb/GaAs heterostructures exhibit a type-II (staggered) band line-up featuring recombination from spatially indirect excitons and are thus electronically fundamentally different from their well-known As-containing type-I counterparts (see Section 6.8).

4.4.3 Ge ON Si

Ge islands in Si are particular interesting, given the prominence of Si technology for electronic device technology. Optoelectronic applications are also of interest, e.g. the extension of the spectral range of Si-based detectors toward larger wavelengths and the design of novel light emitting devices.

Formation of Ge or Ge-rich islands upon deposition on Si(001) has been found in a number of works (Eaglesham and Cerullo, 1990; Mo *et al.*, 1990; Hansson *et al.*,

GROWTH OF SELF-ORGANIZED QDS 83

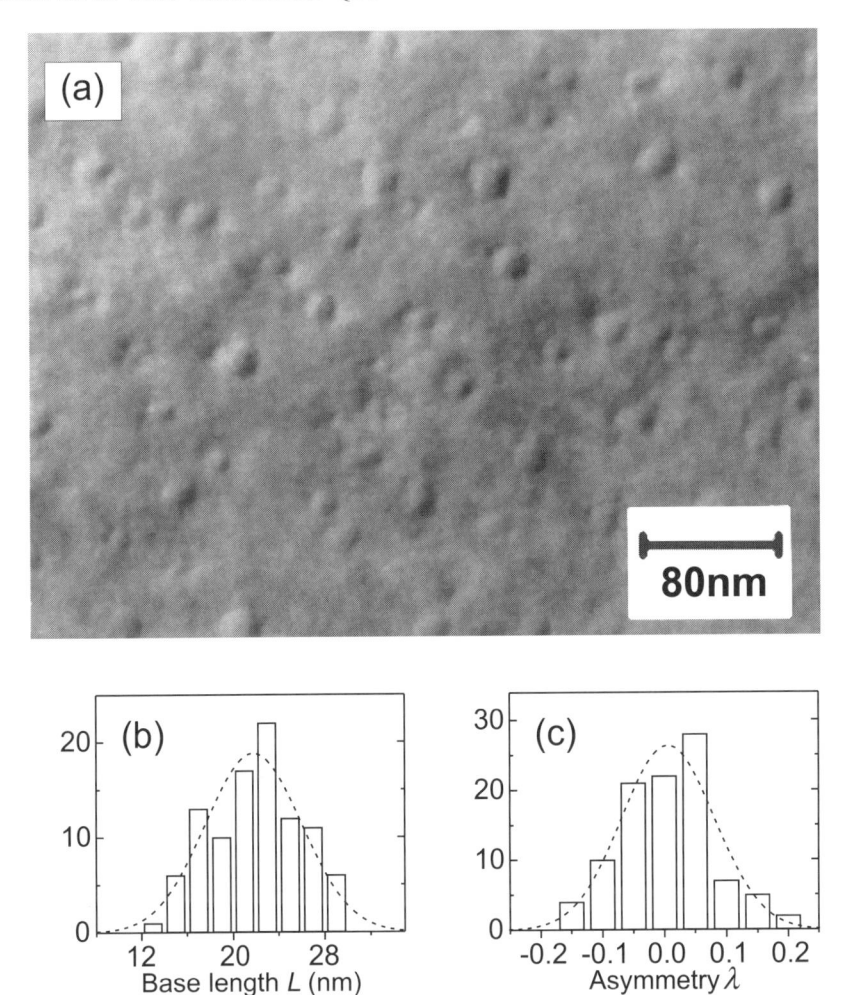

Fig. 4.19 (a) Plan-view weak-beam TEM image of GaSb/GaAs quantum dots grown by MBE. Part of the contrast is due to strain fields. (b) Histogram of island size distribution. (c) Histogram of island asymmetry. (After Hatami et al., 1995)

1994; Schittenhelm et al., 1995; Abstreiter et al., 1996; Kamath et al., 1997a, 1997b; Brunner et al., 1998). Highly strained Ge films on Si surfaces were shown to decompose to coherently strained Ge islands with rectangular base and facets along the [001] and [010] directions (Mo et al., 1990). The elongated Ge islands were found to be oriented along one of these directions with equal probability. However, no ordering in size and aspect ratio was discussed at that time. Abstreiter et al. (1996) reported Ge-rich islands ordered in size which had been grown using MBE. For deposition at 670 °C an area density of 10^9 cm^{-2} was achieved; the lateral dot size was typically 150 nm and the height 15 nm.

4.4.4 III–V ON Si

The integration III–V optoelectronic devices with Si electronics has attracted much interest (e.g. see Kaminishi, 1987; Razeghi et al., 1988; Grundmann et al., 1991b; Yamada et al., 1997). New efforts are now triggered by the possibilities opened up by quantum dots.

Due to fast carrier capture into dots and resulting immobilization the detrimental role of nonradiative defects is reduced. This has been demonstrated in a comparative study of InGaAs/GaAs quantum wells and InAs/GaAs QDs grown on GaAs substrate and on top of highly dislocated (10^7 cm^{-2}) GaAs/Si layers (Lacombe et al., 1997). Island quality for both cases is very similar; on GaAs/Si the islands appear to be a little bit smaller (9 nm versus 11 nm) with a slightly higher density (1.4 versus 1.1×10^{11} cm^{-2}). While the luminescence efficiency of the quantum well drops by a factor of 6 on the GaAs/Si substrate due to diffusion of carriers toward the dislocations, the intensity of the QD luminescence is not changed.

Another promising approach towards integration of III–V material and Si is the direct growth of III–V quantum dots in a crystalline Si matrix. Island growth of III–V on Si has been investigated for many years, e.g. GaAs/Si (Biegelsen et al., 1987; Hull and Fischer-Colbrie, 1987) or GaP/Si (Soga et al., 1991). The growth apparently occurs in the Volmer–Weber mode since in between the islands no III–V material is found. For a low coverage of 1.0 nm GaAs deposition the islands were found to be coherent (Hull and Fischer-Colbrie, 1987). Defects such as stacking faults and dislocations are typically generated when the islands grow or coalesce.

A high-density ensemble (5×10^{11} cm^{-2}) of ~4 nm high InAs dots on Si (001) has been developed in MBE upon deposition of 5.5 ML InAs (Cirlin et al., 1998, Tsatsul'nikov et al., 1998). Initially a two-dimensional layer forms (SK growth). An AFM image of the array of islands is shown in Fig. 4.20. The base sides of the islands appear to be along [100] and [110], with lateral dimensions of about 12 and 20 nm, respectively. Such islands covered by Si show intense luminescence at 1.28 µm at 50 K.

4.4.5 II–VI COMPOUNDS

The formation of islands in strained heteroepitaxy has been reported for a number of II–VI material systems. This material system is particularly challenging since due to the high dielectric constant the bulk exciton size is only a few nanometers. Therefore quantum dots exhibiting significant confinement effects must be very small.

$Zn_{1-x}Cd_xSe$/ZnSe quantum wells were found to exhibit three-dimensional growth and dot formation and zero-dimensionally confined states for sufficiently high strain (Cd content $x > 0.3$) (Lowisch et al., 1996) (see also Section 6.5). Stranski–Krastanow CdSe dots (Flack et al., 1996) can be used as stressors for underlying ZnCdSe quantum wells (Nikitin et al., 1997). CdSe islands on ZnSe are also reported by Hommel et al., (1997). ZnSe dots are reported by Arita et al., (1997) to be typically 27 nm high and 200 nm wide. CdSe dots on GaAs (110) (Ko et

GROWTH OF SELF-ORGANIZED QDS

Fig. 4.20 AFM image of InAs islands on Si(001) grown by MBE. (After Cirlin *et al.*, 1998)

al., 1997) were found to be 10 nm high and 47 nm wide. Three-dimensional nucleation of ZnCdSe on ZnSe (110) has been reported by Zhang *et al.* (1997a). ZnSe dots on GaAs (110) were reported to be 3 nm high and about 50 nm wide (Zhang *et al.* 1997a). Stacks of flat 5 nm wide (Cd,Zn)Se islands in a ZnSe matrix have been reported by Strassburg *et al.* (1998). Volmer–Weber growth of ZnSe QDs during MOCVD on ZnS/Si and ZnS/GaAs layers was reported by Harris Liao *et al.*, (1997)

4.4.6 GROUP III NITRIDES

Quantum dot formation has been also observed in the (In,Al,Ga)N system. Narukawa *et al.* (1997) reported that InGaN layers decompose into regions of laterally increased and decreased In concentration. The regions of higher In content represent localizing potentials for carriers from which optical recombination is observed.

Stranski–Krastanow growth of GaN on $Al_xGa_{1-x}N$ surfaces has been reported by Tanaka *et al.* (1996). Typical GaN dots grown on $Al_{0.2}Ga_{0.8}As$ using MOCVD exhibited an average width of ~ 40 nm and a height of ~ 6 nm, as determined from AFM images. The dot density was $3 \times 10^9 \, cm^{-2}$. The GaN dot structures were only observed when the AlGaN surface was exposed to tetraethylsilane (TESi) prior to GaN deposition, which is attributed to an 'antisurfactant' effect, enabling 3D growth (Tanaka *et al.*, 1996). Daudin *et al.* (1997) fabricated 10 nm wide and 2 nm high GaN dots on AlN. Vertical stacks of GaN QDs with AlN barriers (for stacking of QDs see Section 4.5) were reported by Feuillet *et al.* (1997) (Fig. 4.21) grown by MOCVD at 710 °C. The growth of InGaN QDs on AlGaN was reported by

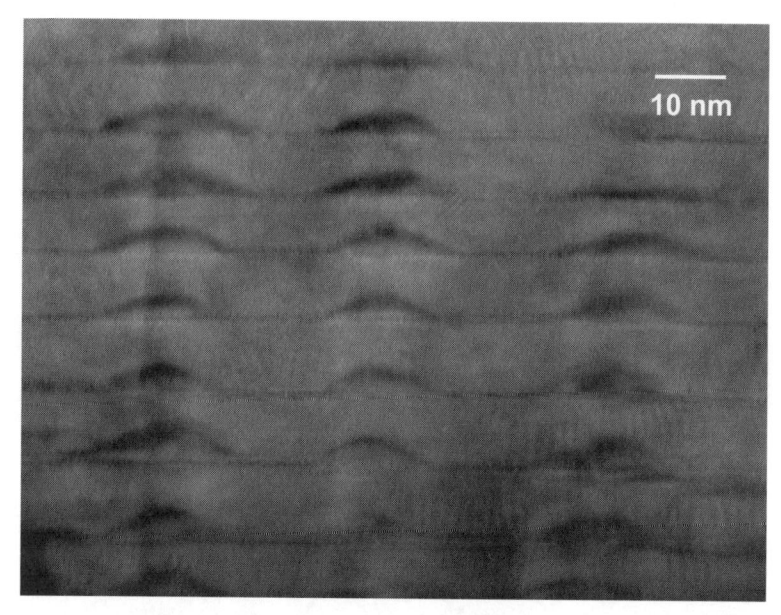

Fig. 4.21 Multiple stack of GaN/AlN dots grown by MBE. (After Feuillet *et al.*, 1997)

Hirayama *et al.* (1997). A high-density ($3 \times 10^{10}\,\text{cm}^{-2}$) ensemble of 5 nm high and 10 nm wide dots was found in AFM images.

4.5 VERTICAL STACKING OF QUANTUM DOTS

Vertical stacking of layers containing quantum dots is important for most device applications in order to increase the filling factor of quantum dots in a given sample. Already Goldstein *et al.* (1985) had observed that islands arrange vertically on top of each other in subsequent layers (Fig. 2.5). A theoretical discussion of such ordering phenomena is given in Section 3.6. Depending on the spacer (barrier) thickness and growth temperature rather different morphologies arise.

For thin barriers, which do not entirely cover the first layer of quantum dots, island shape transformation has been observed (Ledentsov *et al.*, 1996d). In Fig. 4.22 the cross-sectional (a) and plan-view (b) TEM images of InAs vertically coupled quantum dots (VECODs) formed by three-cycle InAs/GaAs deposition are shown. The average thickness of InAs deposited in each cycle equals 0.55 nm. Each GaAs deposition cycle corresponds to an average thickness of 1.5 nm. As can be seen in Fig. 4.22b, the average lateral size of the islands increases from bottom to top from ≈ 11 nm to $\approx 17 \pm 1$ nm. The islands have a square base with main axes along the [100] and [010] directions. The histogram of the nearest-neighbor dot orientation (Fig. 4.22c and d) proves that the dots are arranged in a 2D square lattice with primitive lattice vectors along the same directions. This ordering is clearly

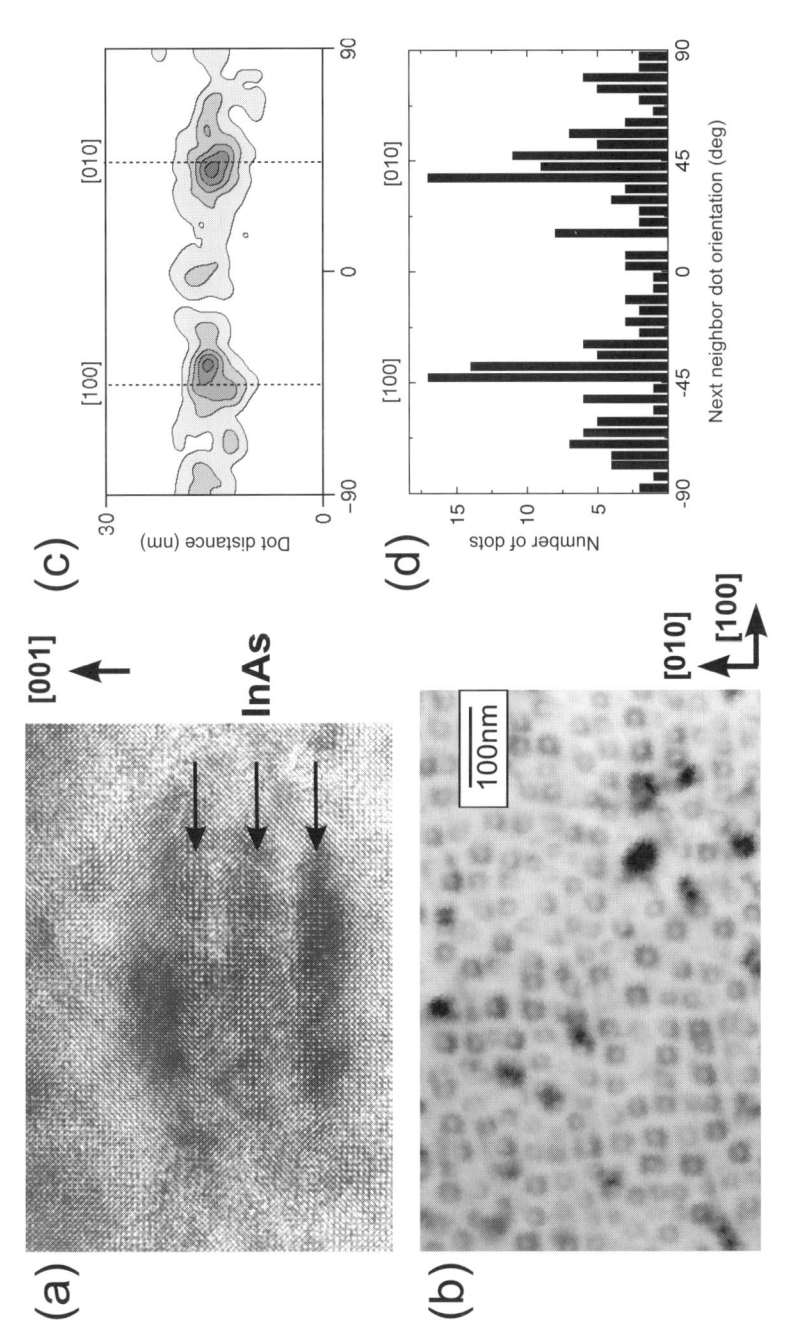

Fig. 4.22 Vertically coupled InAs quantum dots (VECODs) in a GaAs matrix: (a) high-resolution electron microscopy [010] cross-sectional image formed by nine beams, defocus is 60 nm; note the different spot density in the InAs and GaAs regions; (b) bright-field plan-view transmission electron microscopy (TEM) micrograph under [001] zone axis illumination; (c) 2D histogram of next two neighbors center-to-center distance and direction for the TEM of part (b); (d) projection of part (c) on to the angular axis. Maxima in [100] and [010] directions prove the VECOD arrangement in a 2D square lattice. (After Ledentsov et al., 1996d)

observed in all parts of the TEM image and is found to agree perfectly with theory (Section 3.6.5). The resulting arrangement of InAs insertions can be considered as an artificial three-dimensional semiconductor crystal. Stacking of closely spaced islands has been also reported by Solomon *et al.* (1996), Darhuber *et al.* (1997a) and Sugiyama *et al.* (1997).

Vertically correlated stacking of well-separated InAs islands in a GaAs matrix hass been reported by Goldstein *et al.* (1985), Xie *et al.* (1995) and Sugiyama *et al.* (1996) for MBE growth and by Heinrichsdorff *et al.* (1997c, 1997d) for MOCVD

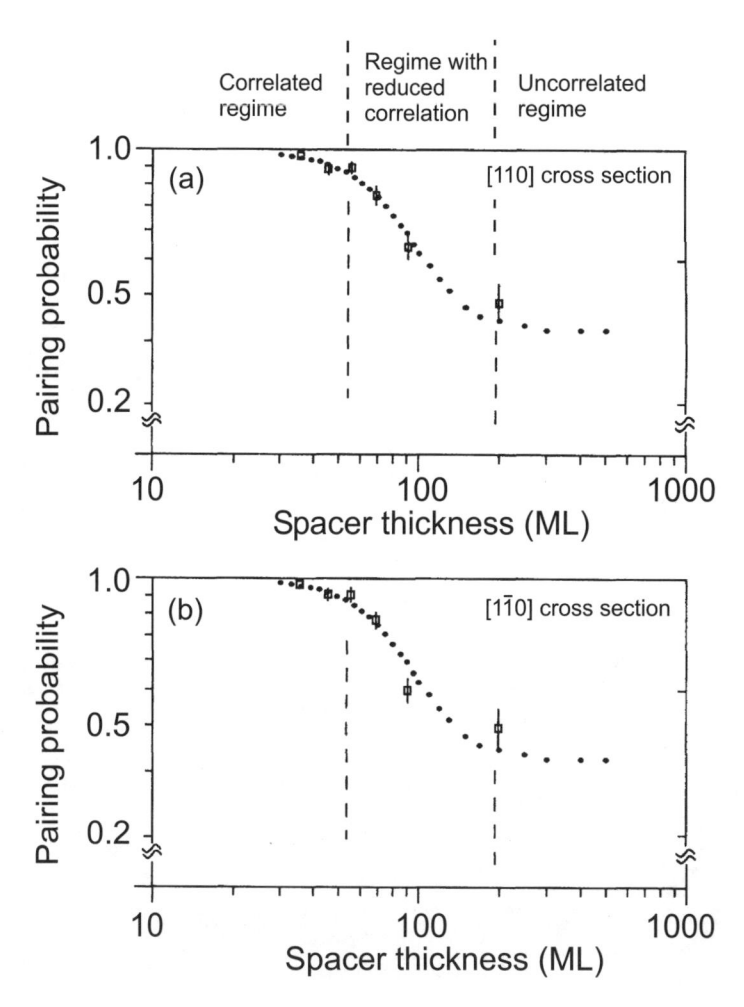

Fig. 4.23 Experimentally observed pairing probability in MBE grown stacks of InAs/GaAs quantum dots as a function of the spacer layer thickness. Data are taken from (a) (110) and (b) (1-10) cross-sectional TEM images. The filled circles are fitted to the data from the theory of correlated island formation under strain fields. (After Xie *et al.*, 1995)

GROWTH OF SELF-ORGANIZED QDS

growth. Vertical correlation has been observed also in the Ge/Si system (Xie *et al.*, 1995; Rahmati *et al.*, 1996; Darhuber *et al.*, 1997b). With increasing separation layer thickness the vertical correlation of InAs islands is lost (Fig. 4.23) (Xie *et al.*, 1995) because the surface strain field due to the underlying dots becomes too weak to influence growth kinetics (Section 3.7.1). The loss of correlation with increasing barrier thickness has also been observed for Ge/Si dots (Rahmati *et al.*, 1996). Merging of islands of different initial size is found to be the dominant mechanism leading to a uniform size distribution for SiGe/Si multilayer QDs (Mateeva *et al.*, 1997). Cross-sectional STM experiments have revealed dissolution of the wetting layer and the presence of indium between the dot columns (Wu *et al.*, 1997) for fivefold stacks (3 ml InAs, 5.6 nm GaAs) grown at 480 °C by MBE.

Up to 25 stacks of separated InGaAs/GaAs dot layers of excellent crystallographic perfection have been fabricated using MBE at the optimum growth temperature of 485 °C (Fig. 4.24a). The importance of the growth temperature can be seen in Fig. 4.24b where a cross-sectional TEM image of a twentyfold stack grown at 520 °C is shown. The initial dot size is much larger and subsequent interfaces become strongly corrugated.

On a well-ordered three-dimensional quasi-crystal of islands (Fig. 4.24a) X-ray diffraction experiments have been performed (Darhuber *et al.*, 1997a) that average over much larger surface areas than TEM images. Reciprocal space maps (RSMs) have been recorded around the (004) and several asymmetric reflections. The (004) RSM is found to exhibit a sharp coherent peak and some diffuse scattering around it, but no pronounced signatures of the dots. In Fig. 4.25a the RSM around the (404)

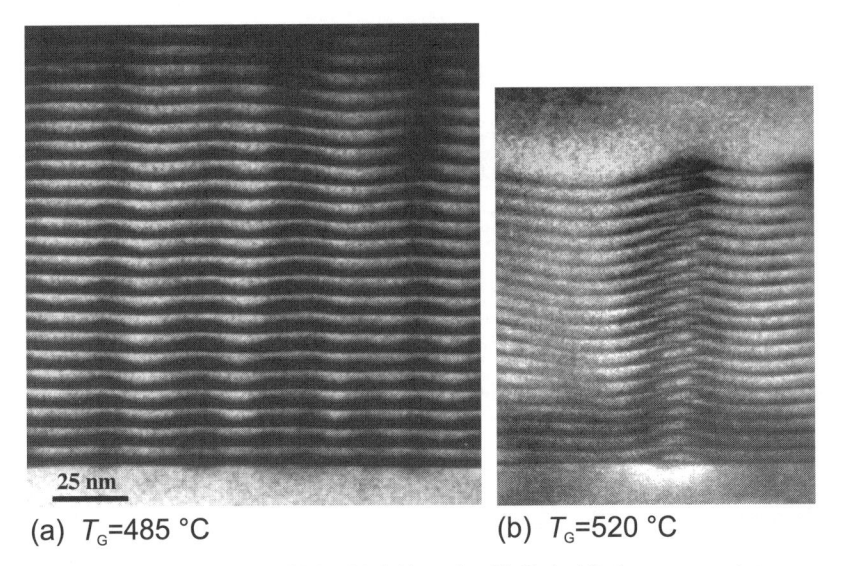

(a) T_G=485 °C   (b) T_G=520 °C

Fig. 4.24 TEM micrographs of (a) a 25-fold stack of InGaAs/GaAs quantum dots grown at 485 °C and (b) a 20-fold stack grown at 520 °C using MBE

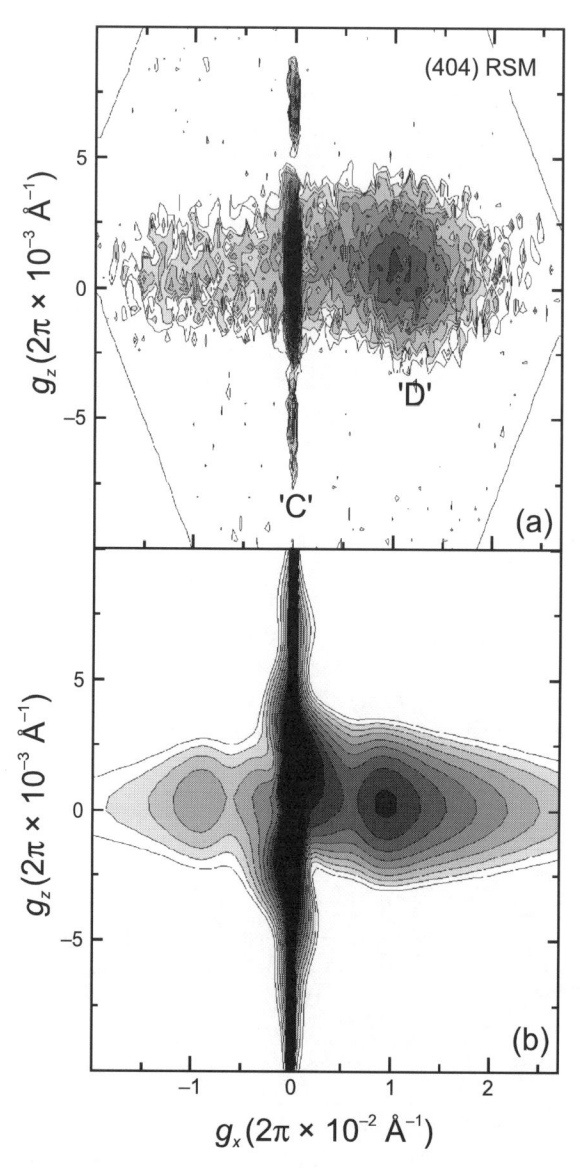

Fig. 4.25 (a) Measured and (b) calculated reciprocal space map around the (404) reflection. 'C' labels the coherent signal from the multiple QW (WL), 'D' the diffuse signal from the dots. (After Darhuber *et al.*, 1997a)

GROWTH OF SELF-ORGANIZED QDS 91

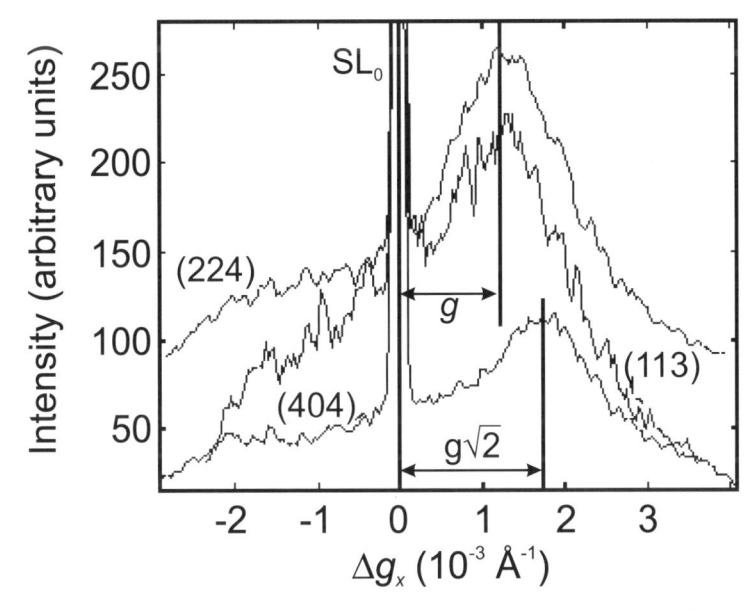

Fig. 4.26 Integration of the (404), (224), and (113) reciprocal space maps in the g_z-direction. The peak separation differs by a factor $\sqrt{2}$ in the (404) and (224) directions. (After Darhuber et al., 1997a)

reflection is shown. The broad diffuse satellite peak (labeled 'D' in Fig. 4.25a) is ascribed to lateral ordering along the [100] direction. The far field of the strain distribution around the quantum dots contributes to the scattering intensity. A theoretical diffraction pattern is shown in Fig. 4.25b. A similar diffuse diffraction peak is found in the (224) RSM, but with a separation from the coherent peak that is *decreased* by 2. The same peak separation has been found in the (113) RSM. The projections of the (404), (224), and (113) RSM on the g_x axis are shown in Fig. 4.26. The change in scattering vector by 2 is consistent with a lateral ordering of the QDs in a square lattice with vectors along [100] and [010]. The dot distance in the [100] direction is determined to be ~ 55 nm in agreement with TEM experiments. We note that the signal from the short-range strain field is expected at higher-order satellites.

The vertical alignment of stacked quantum dots also occurs during MOCVD growth of multiple layers of quantum dots (Ledentsov et al., 1996a; Heinrichsdorff et al., 1997b). A prerequisite for the successful stacking of quantum dots is that the first layer does not contain any defects (see Fig. 4.16). In Fig. 4.27 the plan-view and cross-sectional TEM images of a threefold stack are shown (Heinrichsdorff et al., 1997b). For the present spacer layer thickness of 4 nm vertical alignment of dots is obvious. The dot size increases in higher layers. The corrugation of the AlGaAs layer is found only in the [110] direction. For larger separation layer thicknesses ~ 18 nm the vertical correlation and the variation of dot size are lost.

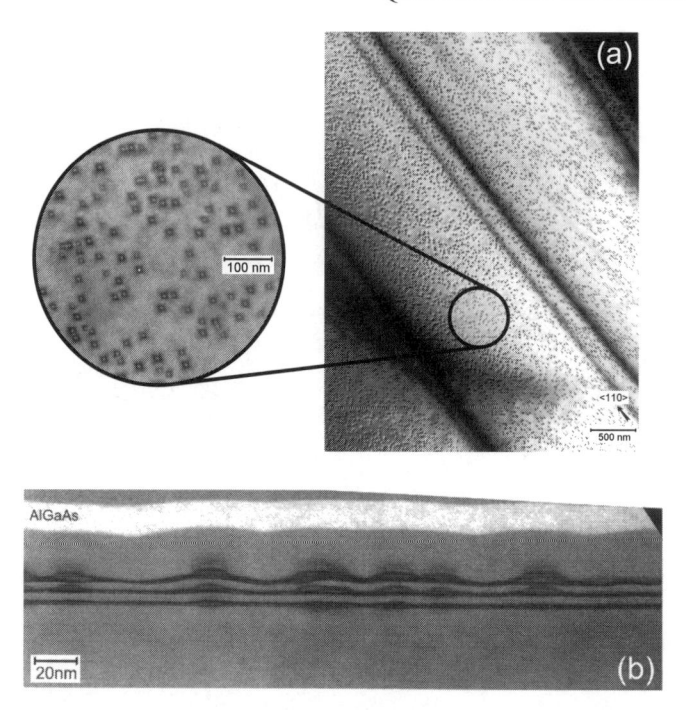

Fig. 4.27 (a) Plan-view and (b) cross-sectional TEM images of a threefold stack of InAs/GaAs quantum dots grown by MOCVD. (After Heinrichsdorff et al., 1997b)

4.6 ARTIFICIAL ALIGNMENT OF QUANTUM DOTS

Various attempts have been undertaken towards improving site control for self-organized quantum dots, which is of importance for fabrication of devices incorporating QDs of homogeneous size. The basic concepts introduced are: ordering of dots along multiatomic steps (Kitamura et al., 1995; Leon et al., 1997), undulations caused by misfit dislocations (Häusler et al., 1996; Xie et al., 1997), and growth on corrugated (Mui et al., 1995; Jeppesen et al., 1996; Seifert et al., 1996b; Tsui et al., 1997; Konkar et al., 1998) or masked surfaces (Kamins et al., 1997).

Kitamura et al. (1995) demonstrated the alignment of quantum dots along surface steps for (001) GaAs substrates misoriented by 2° toward [010], [110], and [1-10]. Due to step bunching approximately 2 nm high steps were formed. Figure 4.28 visualizes the alignment of InGaAs quantum dots with a diameter below 20 nm along multiatomic surface steps for the [010] misoriented surface.

In the vicinity of a dislocation the local strain is changed. Depending on the sign of the amount of change of strain in- or out-diffusion of strained material is induced. Häusler et al. (1996) reported alignment of InP dots along ⟨110⟩ on the misfit dislocations created in a $In_{0.61}Ga_{0.39}P$ buffer on GaAs (001) grown by MBE. Ge dots were found to nucleate on the intersection of [110] and [1-10] misfit dislocations (Fig. 4.29) (Xie et al., 1997). The dislocation network was formed at the interface of

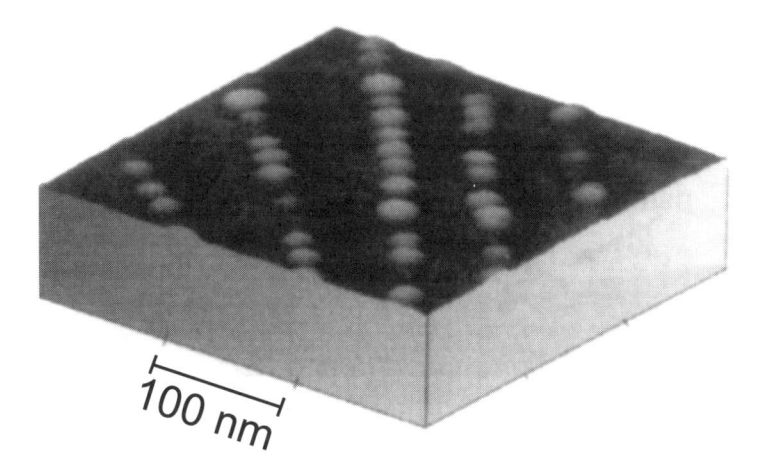

GROWTH OF SELF-ORGANIZED QDS 93

Fig. 4.28 AFM image (300 nm × 300 nm) of InGaAs quantum dots aligned at multiatomic steps on GaAs (001) surface misoriented by 2° toward [010]. (Reproduced by permission of American Institute of Physics from Kitamura *et al.*, 1995)

the Si substrate and a SiGe buffer layer. Ge dots were deposited on top of a thin Si cap. The average dislocation distance is 100 nm, but dislocation spacing varies in a random fashion. The typical Ge dot size is 200 nm. The spacing of rows and columns decorated with Ge dots is about 1 μm, much larger than the *average* dislocation distance, and corresponds to groups of dislocations that are closely spaced (Xie *et al.*, 1997).

On corrugated substrates islands tend to nucleate at characteristic sites of the structure, such as edges, side walls or trenches. Mui *et al.* (1995) reported the alignment of InAs islands grown using MBE along ridges that had been fabricated using wet chemical etching. Rather large pitch (1 μm) islands are found to nucleate on the side walls for ridges along [110] and are found on the (001) top of the mesas or at the foot of the mesa for ridges along [1-10]. In short (0.28 μm) pitch V-grooves islands are only found at the bottom and the side wall of the grooves. Alignment of InAs islands grown using chemical beam epitaxy in chains at the bottom of trenches and preferred nucleation in cylindrical holes was reported in Jeppesen *et al.* (1996). The alignment of InP dots grown using MOCVD on patterned GaAs/GaInP surfaces was reported in Seifert *et al.* (1996b). The use of compliant substrates might present another alternative for ordered growth of QDs (Ejeckam *et al.*, 1997).

Kamath *et al.* (1997a, 1997b) reported the preferential location of Ge islands on top of Si (001) ridges with {110} side walls grown on stripe-like openings in an SiO_2 mask (Fig. 4.30b). For fairly wide ridges (670 nm up to 1.7 μm) several rows of Ge islands are formed on the mesa top along the edges. Towards the center the islands have random positions. On narrow ridges (450 nm) one row of Ge islands is found to sit along each edge (Fig. 4.30a) (Kamath *et al.*, 1997a, 1997b). In these two rows, having a distance of 300 nm, the Ge islands are almost perfectly periodic with an average distance of about 85 nm.

94 QUANTUM DOT HETEROSTRUCTURES

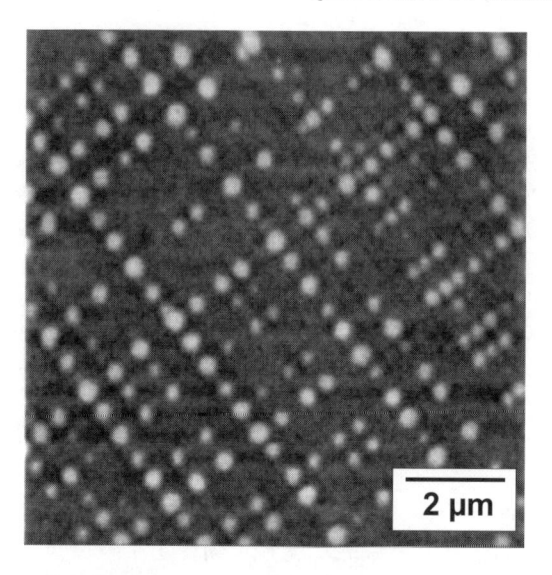

Fig. 4.29 AFM image (10 μm × 10 μm) of Ge islands on the Si layer above a misfit dislocation network. (Reproduced by permission of American Institute of Physics from Xie *et al.*, 1997)

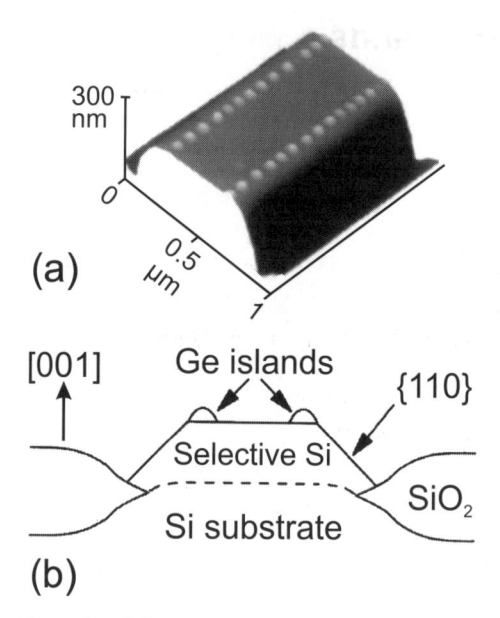

Fig. 4.30 (a) Three-dimensional display of an AFM image of a 450 nm wide Si ridge along ⟨100⟩. Ge islands line up along the edges of the mesa. (b) The corresponding schematic cross section. (Reproduced by permission of American Institute of Physics from Kamins and Williams (1997))

5 Modeling of Ideal and Real Quantum Dots

Elastic, electronic and optical properties of ideal and real quantum dots are theoretically derived in this chapter. Semiconductor lasers with QD layers as active medium are modeled. The main focus lies on self-organized strained quantum dots. The results presented in Sections 5.1 to 5.4 are, however, of general validity for quantum dots independent of the way they have been created.

Strain caused by the differences of the lattice constants of dot and substrate (barrier) materials is decisive for both the self-organization mechanisms and the electro-optical properties. Therefore the strain distribution in and around dots and the impact of strain on band structure and phonon spectrum is elucidated first in Section 5.1. Electronic levels in arbitrary three-dimensional potentials are calculated in Section 5.2. Coulomb interaction and its interplay with three-dimensional localization are considered in Section 5.3. Besides excitons, states including more than two particles are covered, such as biexcitons and multielectron systems. Optical transitions between electronic levels both of single quantum dots and of dot ensembles are treated in Section 5.4 and will present a basis for understanding and deriving information from experiments. A theory of carrier statistics in QDs is presented in Section 5.5. The impact of external fields is illustrated in Section 5.6. Calculations of phonon spectra and their role in energy relaxation are described in Section 5.7. A theory of lasers with quantum dots as the active medium is outlined in Section 5.8.

Some topics, also some not covered by the present theoretical chapter, e.g. third-order nonlinearities, are discussed in detail in the book of Bányai and Koch (1993). The Coulomb blockade and related electronic effects are treated in much greater detail in Grabert and Horner (1991), Fukuyama and Ando (1992) and Grabert and Devoret (1992).

5.1 STRAIN DISTRIBUTION

The strain distribution in solids can be treated in the continuum mechanical (CM) approximation (Saada, 1974), in the framework of the valence force field (VFF) model (Musgrave and Pople, 1962; Keating, 1966; Martin, 1970; Kane, 1985), or using density functional techniques (Scheffler et al., 1985). Bernard and Zunger (1994) have shown that for InAs/GaAs quantum wells on (001) surfaces, featuring a mismatch of about 7%, down to a thickness of one monolayer all three theoretical models result in virtually identical strain distributions. In the following we will mainly discuss continuum theoretical approaches since most insight of the nature of strain can be derived from them. In Section 5.1.3 we will discuss the VFF model and compared it to the CM approximation. Substrate bending will be generally neglected since the strained layers are assumed to be much thinner than the substrate (Chaudhari, 1969).

The strain distribution of a three-dimensional dot of arbitrary shape and elastic properties in a matrix cannot be calculated in closed form but has to be solved numerically with finite element or finite difference methods. Simplification to isotropic materials allows a solution for spheres (Section 5.1.2.1). An extension to arbitrarily shaped dots can be made for the case of identical elastic properties of dot and matrix (Section 5.1.2.2).

5.1.1 STRESS–STRAIN RELATIONS

The relations between the stress components σ_{ij} and the strains ε_{ij} for arbitrary materials are

$$\sigma_{kl} = C_{klmn}\varepsilon_{mn}$$

and (5.1)

$$\varepsilon_{kl} = S_{klmn}\sigma_{mn}$$

where C_{klmn} are the compliances and S_{klmn} the stiffness coefficients. For materials with cubic symmetry only three of the 81 components are independent. These are $C_{11} = C_{1111} = C_{2222} = C_{3333}$, $C_{12} = C_{1122} = C_{2233}$, etc., and $C_{44} = C_{1212} = C_{2323}$, etc., and accordingly S_{11}, S_{12} and S_{44}. Other components are zero. For many semiconductors Poisson ratio (Eq. 5.3) is in the vicinity of 1/3.

In isotropic materials there exist only two independent elastic coefficients and the anisotropy constant $C_0 = 2C_{44} - C_{11} + C_{12}$ is zero. The stress–strain relation can then be written in terms of Young's modulus E and Poisson's ratio v only:

$$E = \frac{1}{S_{11}} = \frac{(C_{11} - C_{12})(C_{11} + 2C_{12})}{C_{11} + C_{12}} \quad (5.2)$$

$$v = -ES_{12} = \frac{C_{12}}{C_{11} + C_{12}} \quad (5.3)$$

Thus

$$\sigma_{ij} = \frac{E}{1+v}\varepsilon_{ij} + \frac{vE}{(1+v)(1-2v)}\delta_{ij}\varepsilon_{nn}, \quad \varepsilon_{ij} = \frac{1}{E}[(1+v)\sigma_{ij} - v\delta_{ij}\sigma_{nn}] \quad (5.4)$$

5.1.2 STRAIN IN DOTS

5.1.2.1 Spherical dot, isotropic materials

The case of an elliptical dot *and* isotropic materials was treated in Eshelby (1957). In order to obtain some general insight about solutions of three-dimensional strain problems it is more lucid to treat a spherically symmetric problem. A solution for the simple case of a spherical (homogeneous) dot of a lattice constant different to that of the matrix with different (isotropic) elastic properties was given in Grundmann et al. (1995c). The dot has a radius ρ_0 and volume $V_0 = (4\pi/3)\rho_0^3$; the relative lattice mismatch is called ε_0.

Spherical coordinates (ρ, θ, ϕ) are introduced. The solution is based on the radial displacement for a spherical shell with inner and outer radii ρ_i and ρ_a, subjected to inner and outer pressures P_i and P_a, given by (Saada, 1974).

$$u_\rho(\rho) = \frac{\rho(1+\nu)}{E}\left[P_i \frac{\frac{1-2\nu}{1+\nu} + \frac{1}{2}\left(\frac{\rho_a}{\rho}\right)^3}{\left(\frac{\rho_a}{\rho_i}\right)^3 - 1} + P_a \frac{\frac{1-2\nu}{1+\nu} + \frac{1}{2}\left(\frac{\rho_i}{\rho}\right)^3}{\left(\frac{\rho_i}{\rho_a}\right)^3 - 1}\right] \quad (5.5)$$

When a sphere (quantum dot) and a hollow matrix are brought into contact, a procedure known as *shrink fit*, the mismatch of the lattice constants of the inner and outer materials, imposes the following condition for the radial displacement u_ρ:

$$(u_\rho^{in} - u_\rho^{out})|_{\rho=\rho_0} = \varepsilon_0 \rho_0 \quad (5.6)$$

The contact pressure P depends on the elastic constants of the inner and outer material (the ambient pressure on the system is neglected):

$$P = -\varepsilon_0 \frac{1}{\frac{1+\nu^{out}}{2E^{out}} + \frac{1-2\nu^{in}}{E^{in}}}, \quad P' = -\varepsilon_0 \frac{2E}{3(1-\nu)} \quad (5.7)$$

The prime denotes, and also in the following, the case where the elastic constants in the inner and outer materials are identical. This assumption allows algebraic simplifications of the following formulas. The strains are (Grundmann et al., 1995c)

$$\varepsilon_{\rho\rho}^{in} = \frac{2}{3}\varepsilon_0 \frac{1-2\nu}{1-\nu} = \varepsilon_{\theta\theta}^{in} = \varepsilon_{\phi\phi}^{in}$$
$$\varepsilon_{\rho\rho}^{out} = \frac{2}{3}\varepsilon_0 \frac{1+\nu}{1-\nu}\left(\frac{\rho_0}{\rho}\right)^3 = -2\varepsilon_{\theta\theta}^{out} = -2\varepsilon_{\phi\phi}^{out} \quad (5.8)$$

In the case of $\nu = 1/3$, we find $\varepsilon_{\rho\rho}^{in} = \frac{1}{3}\varepsilon_0$ and $\varepsilon_{\rho\rho}^{out} = \frac{4}{3}\varepsilon_0(\rho_0/\rho)^3$. The radial displacements are

$$u_\rho^{in} = \frac{2}{3}\varepsilon_0 \frac{1-2\nu}{1-\nu}\rho \quad \text{and} \quad u_\rho^{out} = -\frac{1}{3}\varepsilon_0 \frac{1+\nu}{1-\nu}\rho_0^3 \frac{1}{\rho^2} \quad (5.9)$$

We note that, starting with Eq. (5.5), results can be obtained in a similar fashion for spherical dots with concentric caps of finite thickness, as outlined in Rockenberger et al. (1998).

A comparison of the strain distribution for a pseudomorphic sphere (dot), a cylinder (quantum wire) of radius r, and a slab (quantum well) shows that around a slab the barrier is completely unstrained, for the cylinder the strain in the barrier decays like r^{-2} and for a sphere the strain decays like ρ^{-3}. The distribution of the distortion energy between the inner and outer materials is fundamentally different for the three geometries. The total strain energy E_0 of the inner material *and* the barrier per unit volume of the inner material of all three structures are identical and given by

$$E_0 = \frac{E\varepsilon_0^2}{1-\nu} \qquad (5.10)$$

As depicted in Fig. 5.1 (for $\nu = 1/3$ and identical elastic constants of the inner and outer materials), the energy of dilatation $E_s = E_0/3$, due to the isotropic part I of the strain:

$$I = \varepsilon_{xx} + \varepsilon_{yy} + \varepsilon_{zz} \qquad (5.11)$$

$$I = 3\varepsilon_0 \left[1 + \frac{1+\nu^{\text{out}}}{1-2\nu^{\text{in}}} \frac{E^{\text{in}}}{2E^{\text{out}}} \right]^{-1} \rightarrow 2\varepsilon_0 \frac{1-2\nu}{1-\nu} \qquad (5.12)$$

is also identical and completely stored in the inner material for all three cases. This is also true for the case of anisotropic material parameters (Eshelby, 1957).

In the case of a slab the energy of distortion (due to the anisotropic component of the strain) is fully stored in the slab. For a cylinder, the barrier becomes strained and stores three-quarters of the energy of distortion. For the spherical geometry the entire energy of distortion resides in the outer material. The strain in a planar geometry (slab) does not depend on the elastic constants of the barrier material, while it does depend on them in the case of a cylinder and a sphere.

These fundamental differences become important when in realistic structures dots coexist with a quantum well (wetting layer) and exhibit different hydrostatic strain. For $\nu^{\text{in}} = \nu^{\text{out}} = 1/3$ the hydrostatic part I of the strain in the quantum well is given

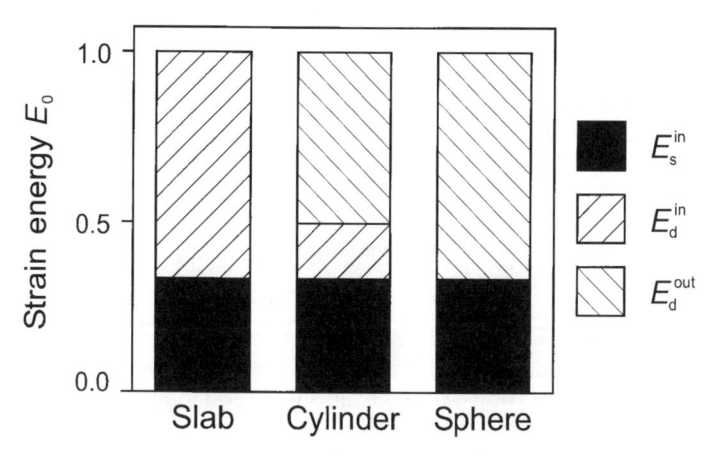

Fig. 5.1 Energy of dilatation (solid black) and energy of distortion in the inner (hatched) and outer (cross-hatched) materials (with identical isotropic elastic constants and $\nu = \frac{1}{3}$ for three different geometries. E_0 is the total strain energy according to Eq. (5.10)

MODELING OF QUANTUM DOTS

by $I_{\text{slab}} = \varepsilon_0$, but for the sphere by $I_{\text{sph}} = 3\varepsilon_0(2E^{\text{in}}/E^{\text{out}} + 1)^{-1}$. This difference reflects the intuitive result that a stiffer barrier ($E^{\text{out}} > E^{\text{in}}$) leads to a larger hydrostatic strain in the dot.

For arbitrary distributions of $\varepsilon_0(\rho)$ a numerical solution of a one-dimensional differential equation resulting, for example, from the equilibrium condition for the strain energy density U, $\partial U/\partial \rho = 0$, can be found. In Fig. 5.2 the strain distribution for spherical geometry is shown for $v = 1/3$ for the case of a homogeneous sphere.

5.1.2.2 Arbitrary shape, isotropic material

From the solution for a sphere a solution for an arbitrarily shaped dot of volume V can be obtained. Dividing the displacements (5.9) by V_0, we obtain the displacement per unit volume of the dot. From the displacements we can derive the stress σ_{ij}^0 per unit volume, where the x_i denote the coordinates:

$$\sigma_{ii}^0 = \frac{1}{4\pi} \frac{E\varepsilon_0}{1-v} \frac{2x_i^2 - x_j^2 - x_k^2}{\rho^5} \quad \text{and} \quad \sigma_{ij}^0 = \frac{3}{2} \frac{1}{4\pi} \frac{E\varepsilon_0}{1-v} \frac{x_i x_j}{\rho^5} \quad (5.13)$$

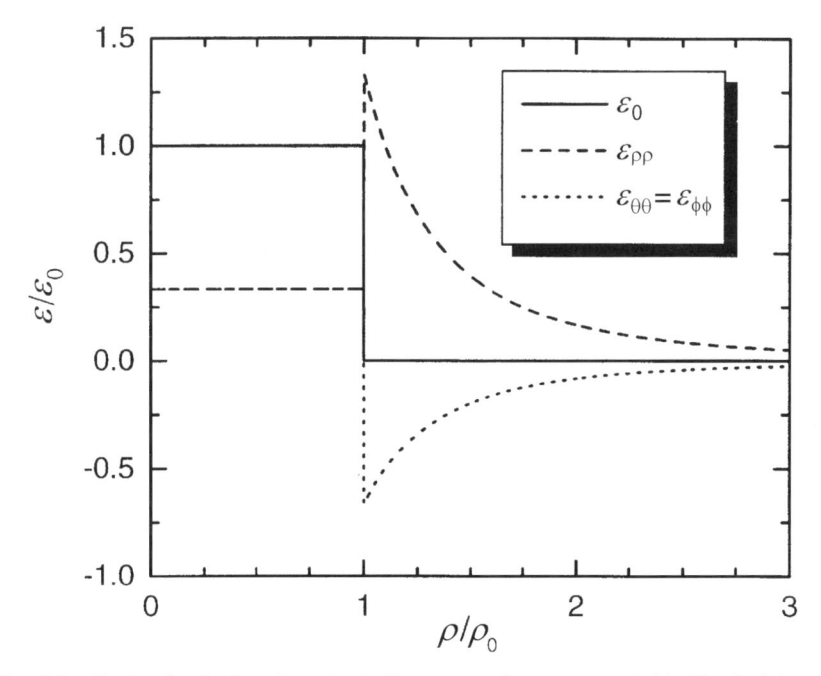

Fig. 5.2 Strain distribution for spherically symmetric geometry (with identical isotropic elastic constants and $v = \frac{1}{3}$ throughout the structure). Within the sphere of radius ρ_0 the lattice mismatch ε_0 is constant

with $i \neq j \neq k$. Due to the linear superposition of stresses, the stress distribution σ_{ij}^V for an arbitrary dot of volume V can be obtained by integrating over V:

$$\sigma_{ij}^V(\mathbf{r}_0) = \iiint_V \sigma_{ij}^0(\mathbf{r} - \mathbf{r}_0) d^3\mathbf{r} \qquad (5.14)$$

From the stress the strain can be calculated from Eq. (5.1). This scheme can only be applied when the material constants of the dot and the matrix are identical. Otherwise the solution for the infinitesimal 'δ-dot' depends on the position within the dot. As a generalization of Eq. (5.14), ε_0 can be taken as $\varepsilon_0(\mathbf{r})$ in Eq. (5.13).

In the case that ε_0 is constant within V, the volume integral can be readily transformed into an integral over the surface ∂V using Gauss' theorem. With the 'potentials' A_{ij} we fulfill div $A_{ij} = \sigma_{ij}$:

$$A_{ii} = -\frac{1}{4\pi} \frac{E\varepsilon_0}{1-v} \frac{x_i e_i}{\rho^3} \quad \text{and} \quad A_{ij} = -\frac{1}{2}\frac{1}{4\pi} \frac{E\varepsilon_0}{1-v} \frac{x_i e_j + x_j e_i}{\rho^3} \qquad (5.15)$$

for $i \neq j$, $\mathbf{e}_i$ being the unit vector in the i direction.

Special care must be taken at the singularity $\mathbf{r} = \mathbf{r}_0$ if $\mathbf{r}_0$ lies within V, because the stress within the 'δ inclusion' is not singular (in contrast to the electrostatic analog of a δ charge). Thus we find

$$\sigma_{ij}^V(\mathbf{r}_0) = \oiint_{\partial V} A_{ij} \, d\mathbf{S} + \delta_{ij} \frac{E\varepsilon_0}{1-v} \iiint_V \delta(\mathbf{r} - \mathbf{r}_0) d^3\mathbf{r} \qquad (5.16)$$

A similar scheme has been independently developed in Downes *et al.* (1997).

5.1.2.3 Pyramid, isotropic material

For self-organized growth quantum dots of pyramidal shape are observed (Bimberg *et al.*, 1995). Using Eq. (5.16), numerical integration over the facets of such a dot can be readily achieved. First a pyramid with a 45° facet angle, i.e. $b = 2h$, b being the base width and h being the height, is considered (Fig. 5.3a) assuming $v = 1/3$. In Fig. 5.3a the strains ε_{xx}, ε_{yy}, and ε_{zz} are shown along the z direction for a line through the pyramid top. Additionally the isotropic (hydrostatic) part I of the strain (Eq. 5.11) and the biaxial part

$$B^2 = (\varepsilon_{xx} - \varepsilon_{yy})^2 + (\varepsilon_{yy} - \varepsilon_{zz})^2 + (\varepsilon_{zz} - \varepsilon_{xx})^2 \qquad (5.17)$$

are depicted in Fig. 5.3b. At the tip of the pyramid the strain components diverge within the continuum theory. The hydrostatic part of the strain is confined to the inner material. The anisotropic part is shared between the dot and the barrier. In Fig. 5.3c the tetragonal strain

$$T = 2\varepsilon_{zz} - \varepsilon_{xx} - \varepsilon_{yy} \qquad (5.18)$$

is shown. For $\varepsilon_0 < 0$ (as is the case for InAs in GaAs) the strain has a biaxial compressive character for $T < 0$; for $T > 0$ the strain is biaxial tensile. In Fig. 5.4 the

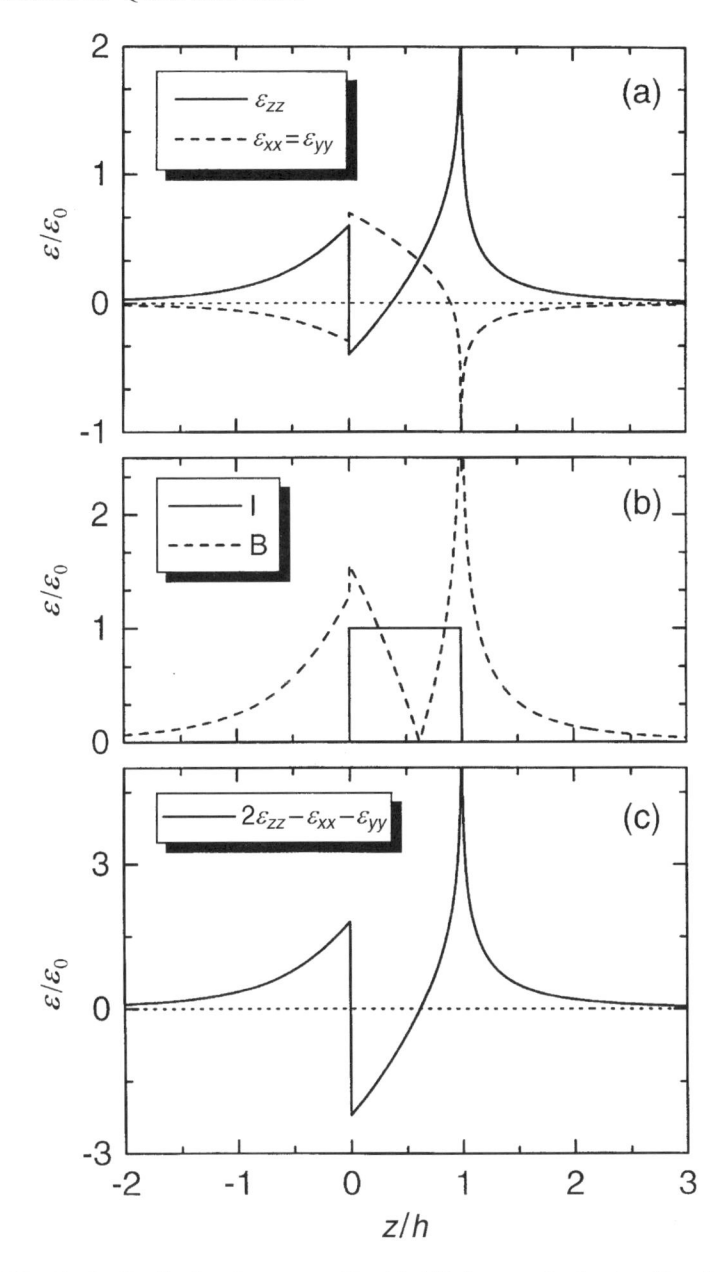

Fig. 5.3 Strain distribution for a pyramid with 45° facet angle, i.e. $b = 2h$, where b is the base width and h the height, along the line through the pyramid top (line A in Fig. 5.7). Identical isotropic elastic constants and $v = \frac{1}{3}$ are taken throughout the structure: (a) displays the strain components ε_{zz} and $\varepsilon_{xx} = \varepsilon_{yy}$, the latter relation imposed by symmetry only for this specific cut; (b) shows the hydrostatic part I and anisotropic part B of the strain defined in Eqs. (5.11) and (5.17). In (c) the tetragonal strain T as defined in Eq. (5.18) is depicted

strain distribution in the (*x,z*) plane through the pyramid center is shown. For the present case of $v = 1/3$, the iso-strain lines go continuously through the side facets for ε_{xx} and ε_{zz}; for ε_{yy} and ε_{xz} they have a discontinuity of ε_0. Symmetry requires that ε_{xy} and ε_{yz} are zero in this plane.

In Fig. 5.5 the strain for a truncated pyramid (with a flat top) with a 45° facet angle and $h = \frac{3}{4}(b/2)$ is shown. No divergence exists in this case.

5.1.2.4 Pyramid, realistic materials

Numerical methods like finite element (FE) or finite difference (FD) schemes are required if realistic elastic material properties, i.e. anisotropy and different elastic constants for different parts of the structure, are to be accounted for. For finite element method (FEM) calculations of uncovered GeSi islands on Si (Christiansen *et*

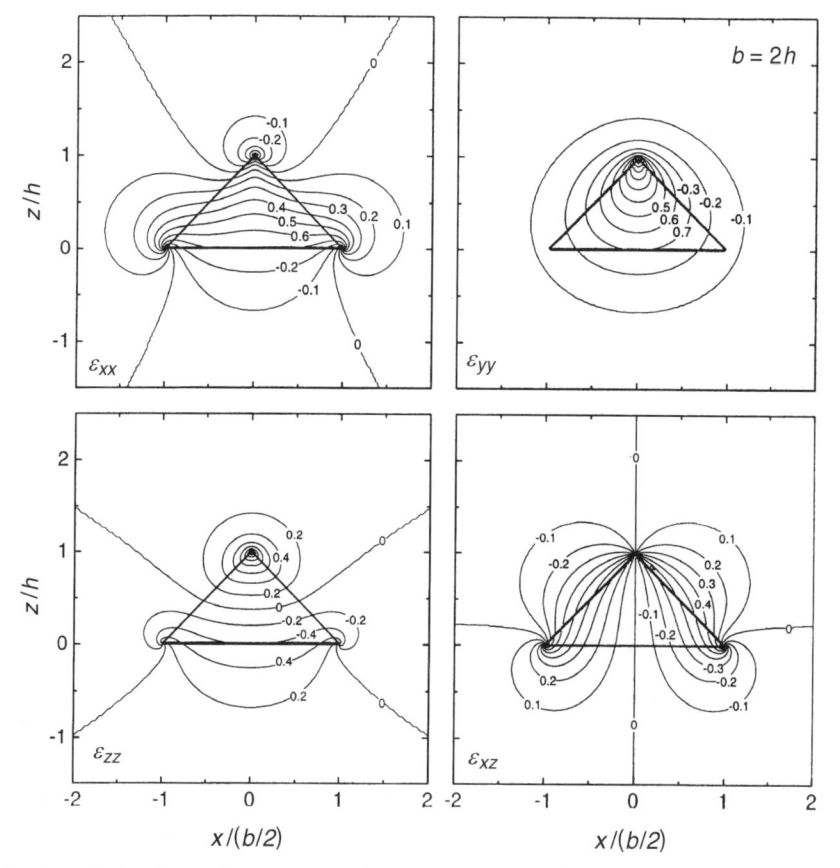

Fig. 5.4 Strain distribution for a pyramid with a 45° facet angle in the (*xz*) plane through the pyramid top. Identical isotropic elastic constants and $v = \frac{1}{3}$ are taken throughout the structure. ε_{xx}, ε_{yy}, ε_{zz} and ε_{xz} are shown; due to symmetry in this plane, ε_{xy} and ε_{yz} are zero

MODELING OF QUANTUM DOTS

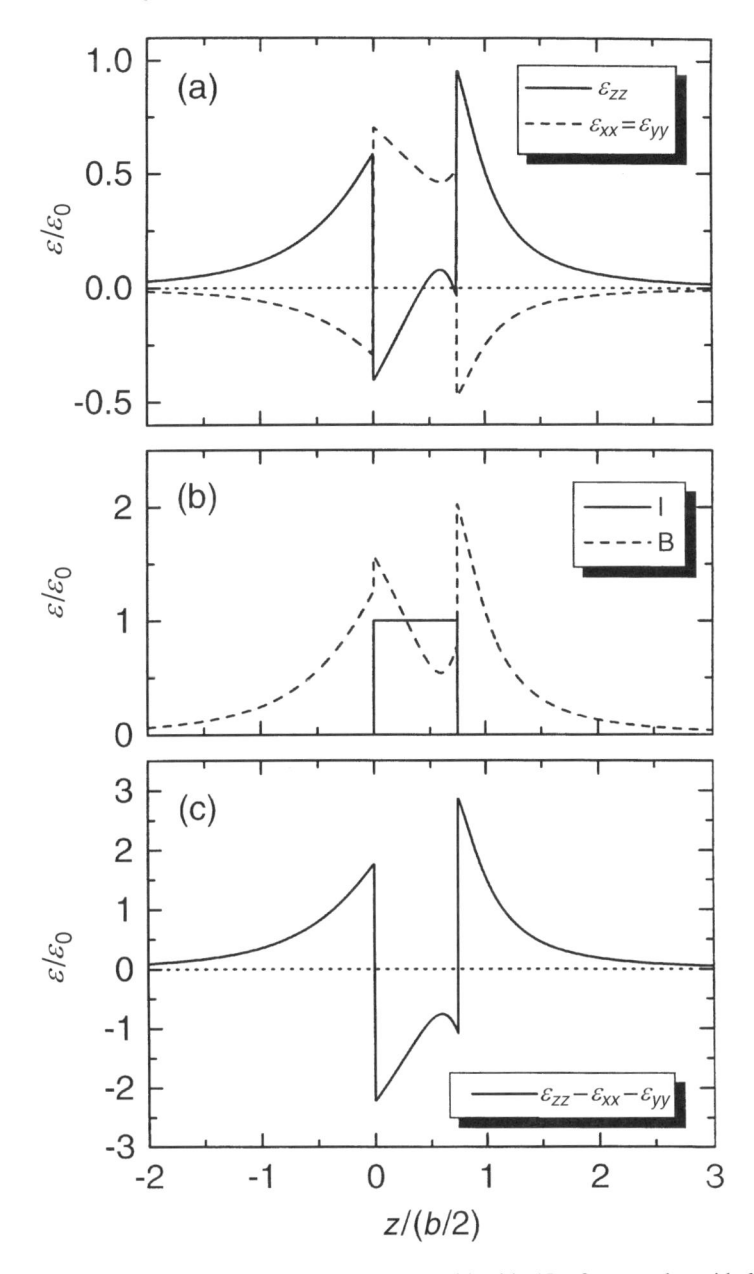

Fig. 5.5 Strain distribution for a truncated pyramid with 45° facet angle, with $h = 3b/8$, where b is the base width and h the height, along the line through the pyramid top (line A in Fig. 5.7). Identical isotropic elastic constants and $v = \frac{1}{3}$ are taken throughout the structure: (a) displays the strain components ε_{zz} and $\varepsilon_{xx} = \varepsilon_{yy}$, the latter relation imposed by symmetry only for this specific cut; (b) shows the hydrostatic part I and anisotropic part B of the strain, respectively. In (c) the tetragonal strain T is depicted

al., 1994) isotropic elastic properties have been assumed. Their results show that truncated pyramidally shaped strained islands impose a considerable strain in the substrate. In the substrate at approximately 3.5 times the island height a *z* displacement of one-tenth of the maximum displacement is found. The *x* and *y* components of the displacement reach about two-thirds of the island height into the substrate.

Self-organized InP quantum dots with the shape of flat discs have been used by Sopanen *et al.* (1995) as stressors (Kash *et al.*, 1991) on top of a GaAs/InGaAs/GaAs heterostructure. The strain field induced by these stressors has been simulated with the FEM (Tulkki and Heinämäki, 1995). The stressor induces a tensile strain extending into the quantum well, so underneath it the conduction and valence bands are lowered. We note that for stressors the presence of a surface is essential. If they were embedded in an infinite medium the dilatation (hydrostatic strain) outside the stressor would vanish and hence not affect the conduction band at all.

The strain distribution in buried InAs pyramids on a thin InAs wetting layer in GaAs can be calculated in a finite difference scheme (Grundmann *et al.*, 1995c). Numerically the total strain energy at about 2×10^6 voxels is minimized, using a conjugate gradient method. At interfaces, the condition of a continuous stress tensor yields the proper boundary conditions. In order to avoid oscillatory solutions arising when symmetric difference quotients are used, the energies from the eight possible combinations of forward and backward differences in the three directions have been averaged by Grundmann *et al.* (1995c). The InAs pyramid considered has base sides along the [100] and [010] directions and {101} side facets. It resides on a thin planar wetting layer.

The strain distribution for a 6 nm high pyramid on a 1 ML thick wetting layer is depicted in Fig. 5.6, where the solid line denotes ε_{zz}, the dashed line ε_{xx} and the dotted–dashed line ε_{yy}. In Fig. 5.6a strain components along the linescan intersecting the wetting layer far from the dot are shown. In Fig. 5.6b the intersection goes through the top of the pyramid (line A in Fig. 5.7). In both cases symmetry imposes $\varepsilon_{xx} = \varepsilon_{yy}$. In Fig. 5.7c the hydrostatic component *I* and the biaxial component *B* along line A are visualized. In the wetting layer (Fig. 5.6a) the strain is biaxial and entirely confined to InAs. Compressive (negative) interfacial strain causes an expansion along the *z* direction (ε_{zz} is positive), known from quantum wells as tetragonal distortion. The wetting layer is affected by the QD only in its vicinity, within a distance of about half of the pyramid's base length. The strain distribution in and around the pyramid is qualitatively similar to the case of pyramids of isotropic material, as discussed in Section 5.1.2.3.

The hydrostatic strain in the wetting layer and the pyramid itself are different, since in the latter case it depends on the elastic properties of the barrier material as discussed above. The pyramid suffers the larger hydrostatic compression because GaAs is stiffer than InAs. This situation is typical since usually the barrier material with the larger bandgap has the larger Young's modulus. This is true for a sequence of III–V semiconductors (GaP, AlAs, GaAs, InP, AlSb, GaSb, InAs, InSb) and also

MODELING OF QUANTUM DOTS 105

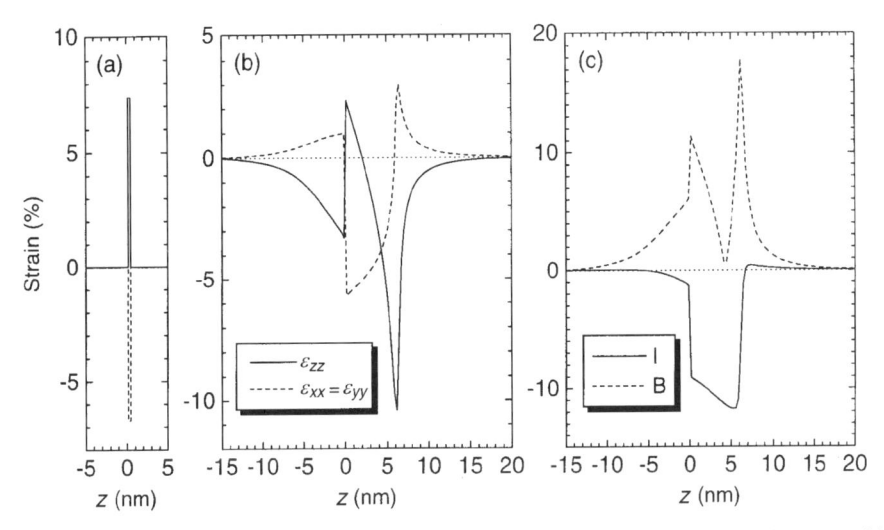

Fig. 5.6 Strain distribution in and around a pyramidal InAs quantum dot (height 6 nm with {101} facets) on a 1 ML thick InAs wetting layer surrounded by GaAs for linescans in the [001] direction: (a) through the wetting layer far away from the dot; (b) and (c) along line A of Fig. 5.7. The solid line in (a) and (b) denotes ε_{zz}, the dashed line ε_{xx}. In (c) the solid (dashed) line represents the hydrostatic (biaxial) strain component

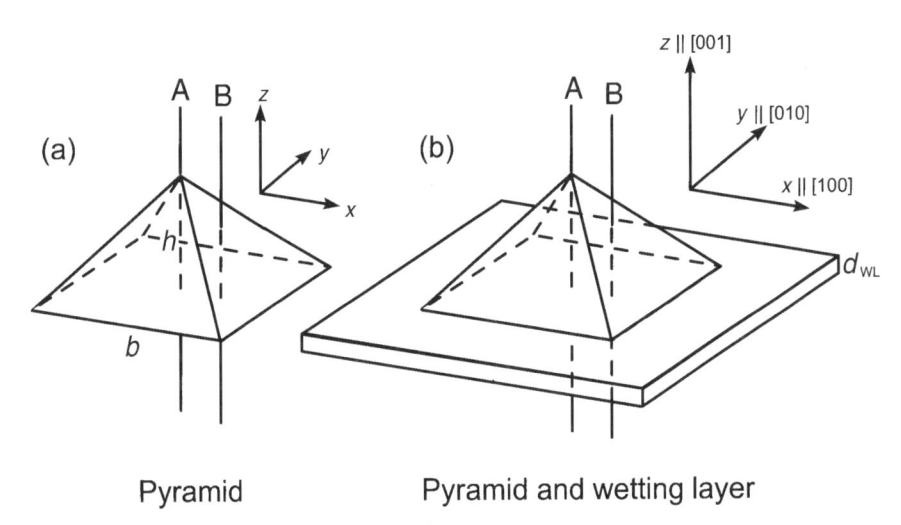

Fig. 5.7 Schematic geometry of (a) pyramid and (b) a pyramid on the wetting layer quantum well. Lines A and B denote particular cuts through the structure. The pyramid height is h, base width is b, wetting layer thickness d_{WL}

for the pair (Si, Ge). However, the two covalent crystals may not be ranged directly in the III–V series. A comparison with results of the atomistic valence force field model can be found in Section 5.1.3.

5.1.2.5 Induced surface strain

After coverage by a thin layer, which in the following is assumed to be planar, the buried dot induces a surface strain field. If the lattice constant of the material of the dot is larger than that of the surrounding matrix the surface lattice constant directly above the dot will be increased, while further away it becomes smaller. This variation of the surface lattice constant acts as the driving force for the vertical alignment of islands (see Section 3.7).

The strain on the (x, y) surface at $z = 0$ above a spherical dot at $x, y = 0$ and $z = -h$ can be obtained from the superposition of stresses of that sphere in an infinite medium and an image sphere† at $z = +h$. Using identical isotropic elastic properties for the dot and the (semi-infinite) matrix, the change of surface unit cell area is given by

$$\frac{\delta A}{A} = \varepsilon_{xx} + \varepsilon_{yy} = \frac{2}{3}\varepsilon_0 \left(\frac{\rho_0}{\sqrt{r^2 + h^2}}\right)^3 \left(1 - \frac{3h^2}{r^2 + h^2}\right) \quad (5.19)$$

where $r = \sqrt{x^2 + y^2}$ is the distance from the center of the dot. For $r < r_0$, $r_0 = \sqrt{2}h$, the surface unit cell is enlarged (for $\varepsilon_0 < 0$), for $r > r_0$ it is shrunk, and for $r \to \infty$ it is unchanged. We note that r_0, the position of the change in sign, depends on h, but is independent of the size ρ_0 of the underlying dot. The maximum change in surface area (at $r = 0$),

$$\left.\frac{\delta A}{A}\right|_{max} = -\frac{4}{3}\varepsilon_0 \frac{\rho_0^3}{h^3} \quad (5.20)$$

is reached directly above the dot.

The change of surface unit cell area above an InAs/GaAs pyramid was calculated by Grundmann et al. (1996d) and is shown in Fig. 5.8. The maximum change directly above the pyramid is compared in Fig. 5.8a with the result for an isotropic sphere (Eq. 5.20) of equal volume positioned at the center of gravity of the pyramid. The d^{-3} decay is obvious, and such a sphere turns out to be a good approximation. We conclude that the strain on top of the barrier is rather weakly dependent on the detailed shape of the dot. In Fig. 5.8 the radial distribution of the change in surface unit cell area is depicted for a surface 5 nm above the pyramid's top. For this calculation periodic in-plane boundary conditions at ± 16 nm have been used (Grundmann et al., 1995c).

The induced surface strain is predicted to lead to self-organization in the vertical direction (Ledentsov et al., 1996d; Tersoff et al., 1996). Close-by islands in the

---

† The concept of image islands is problematic: It ensures that σ_{zz} at the surface is zero. However, the tangential stresses, σ_{xz} and σ_{yz} are not zero.

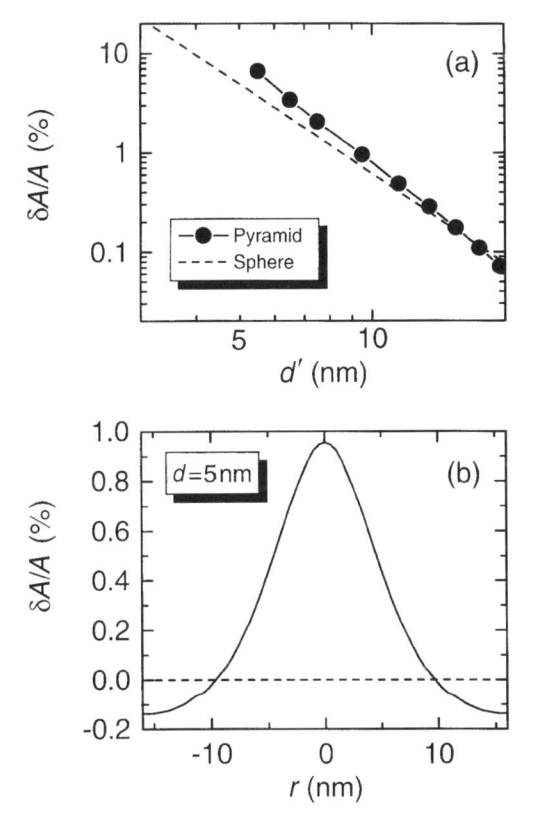

Fig. 5.8 (a) Maximum change of surface unit cell area above an InAs/GaAs pyramid ($b = 12$ nm) (circles, line is guide to the eye). The distance d' is measured from the center of gravity of the pyramid (at $\frac{1}{4}$ the height). The dashed line is the analytical result of Eq. (5.20) for a sphere of equal volume with the same center of mass. (b) Radial dependence of $\delta A/A$ for a surface 5 nm above the pyramid top ($d' = 9.5$ nm) for periodic boundary conditions

buried layer will result in a joint strain field at the surface, leading to the nucleation of only one island in the next layer. Islands that are sufficiently far apart allow the development of an additional island in between. Thus with increasing number of layers the ensemble is expected to become more homogeneous (Fig. 5.9).

5.1.3 VALENCE FORCE FIELD MODEL

Sufficiently small quantum dots can be also directly modeled using the atomistic VFF model (Musgrave and Pople, 1962; Keating, 1966; Martin, 1970; Kane, 1985). The elastic energy of each atom is written in terms of the positions of its neighbor atoms (and second next neighbors) and then added up for all atoms. In the version of

108   MODELING OF IDEAL AND REAL QUANTUM DOTS

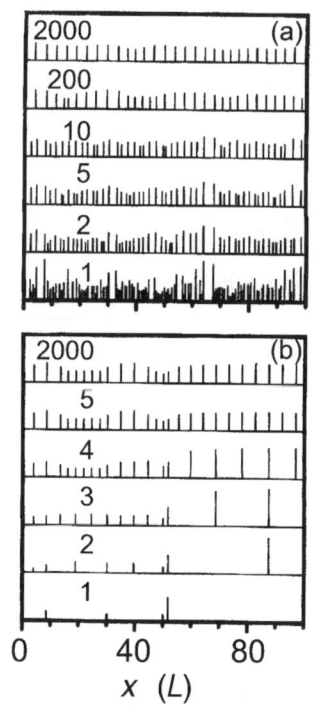

Fig. 5.9 Calculated island positions and sizes (height of lines represents island volume relative to average for the particular layer) in selected successive layers of vertically stacked quantum dots starting with (a) closely spaced and (b) widely spaced islands. (After Tersoff *et al.*, 1996)

Keating (1966), the elastic energy U of the crystal is written as a sum over all atoms i and given as

$$U = U_\alpha + U_\beta \qquad (5.21)$$

where

$$U_\alpha = \frac{1}{4}\sum_i \left[\frac{\alpha_{ij}}{4} \sum_j \frac{(\mathbf{r}_{ij} \cdot \mathbf{r}_{ij} - 3d_{ij}^2)^2}{d_{ij}^2} \right]$$

$$U_\beta = \frac{1}{4}\sum_i \left[\sum_j \sum_{k \neq j} \frac{\beta_{ijk}}{2} \frac{(\mathbf{r}_{ij} \cdot \mathbf{r}_{ik} + d_{ij}d_{ik})^2}{d_{ij}d_{ik}} \right]$$

where the sums over j and k run over the four tetrahedrally arranged neighbor atoms, $\mathbf{r}_{ij}$ denotes the vector from the ith atom towards its jth neighbor, and $4d_{ij}$ is the lattice constant of the binary (or elementary) i–j constituent. U_α in Eq. (5.21) is nonzero when the bond *length* is changed from the strain-free state and is thus called a 'bond-stretching' interaction. U_β in Eq. (5.21) is non-zero when the *angle* between bonds is altered and is thus called a 'bond-bending' interaction.

By comparison with the cubic strain tensor it follows that α and β can be expressed in terms of C_{11} and C_{12} (Keating, 1966):

$$\alpha = d(C_{11} + 3C_{12}), \qquad \beta = d(C_{11} - C_{12}) \qquad (5.22)$$

In a ternary compound or across a heterointerface the β parameter is geometrically averaged if the j and k atoms are not identical: $\beta_{ijk} = \sqrt{\beta_{ij}\beta_{ik}}$ (Podgorny *et al.*, 1985). However, since the model works with two parameters, C_{44} is no longer an independent elastic modulus but is fixed to the value

$$C_{44} = \frac{\alpha\beta}{(\alpha+\beta)d}, \qquad \text{i.e. } \kappa = \frac{2C_{44}(C_{11}+C_{12})}{(C_{11}-C_{12})(C_{11}+3C_{12})} \equiv 1 \qquad (5.23)$$

This relation is approximately fulfilled for a number of semiconductors as shown in Table 5.1.

The potential of Eq. (5.21) is not harmonic. However, as remarked by Kane (1985), anharmonic effects due to the higher-order terms have not been shown to be satisfactorily treated by the Keating model. In particular, the nonlinear elastic moduli C_{ijk} (McSkimin and Andreatch, 1964, 1967) do not enter the theory. Therefore, a linearized version of U_α and U_β in Eq. (5.21) was proposed by Kane (1985); in addition, second-neighbor bond stretching, contiguous bond bending and the MSBN interaction (McMurry *et al.*, 1967; Solbrig, 1971) have been included. The eight parameters of that model have to be determined from fits to experimental phonon dispersion curves. It has been applied by Cusack *et al.* (1996) to InAs/GaAs pyramids. The difference between strain distributions in pyramids of the nonlinear Keating VFF model and CM has been discussed by Pryor *et al.* (1998).

The differences between the strain distribution in a pyramid calculated within the continuum strain theory and the linearized VFF model are shown in Fig. 5.10. Two different continuum mechanical models are compared, one (CM) with the regular value for C_{44}, one (CM(C_{44}^{VFF})) with the value implied in the two-parameter Keating model according to Eq. (5.23). The main modifications between VFF and CM models are (M. Grundmann, 1998, unpublished):

(a) At the tip of the pyramid the strain components ε_{ii} diverge in continuum theory. The VFF model yields a well-defined finite value. A numerical solution of the CM model (with a voxel size of one atom in finite differences), however, creates a similar value.
(b) At interfaces the VFF and CM models differ on the atoms directly adjacent to the interface.

Table 5.1

	Si	Ge	GaAs	InAs	InP
κ	0.99	1.07	1.13	1.22	1.20

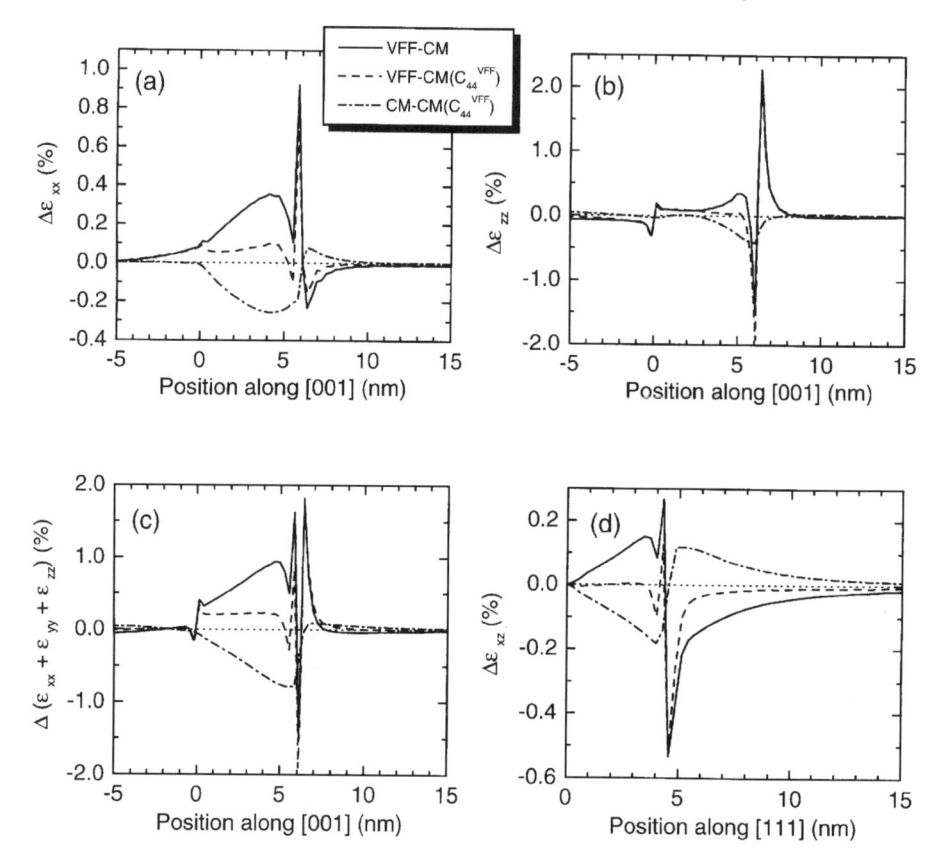

Fig. 5.10 Comparison of strain distributions of a 6 nm (20 ML) high InAs pyramid bound by {101} facets on 1 ML thick InAs WL within a GaAs matrix. The differences between the (linearized) valence force field (VFF) model, continuum mechanics (CM, Fig. 5.6) and continuum mechanics with the value for C_{44} used in VFF (CM(C_{44}^{VFF})) are compared. (a) ε_{xx}, (b) ε_{zz}, (c) hydrostatic strain along [001] through the center of the pyramid, (d) shear component ε_{xz} along the [101] direction, originating at the bottom center of the pyramid. Solid (dashed) represents the difference between the VFF and CM(CM(C_{44}^{VFF})) models. Dash–dotted line shows the difference between the CM and CM(C_{44}^{VFF}) models

(c) In the volume, differences between VFF and CM are mainly due to the incorrect value of C_{44} in VFF; they disappear if VFF is compared to CM(C_{44}^{VFF}).

(d) The symmetry of the continuum elastic strain tensor in the (001) plane is C_{4v}. The tetrahedral configuration of atoms in the VFF model leads to a C_{2v} symmetry, i.e. strain components are different along the [110] and [1–10] directions. This is visualized in Fig. 5.11, where the difference of the strain component ε_{zz} along the [110] and [1–10] directions is plotted; for CM (with any value for C_{44}) this difference is *a priori* zero.

MODELING OF QUANTUM DOTS   111

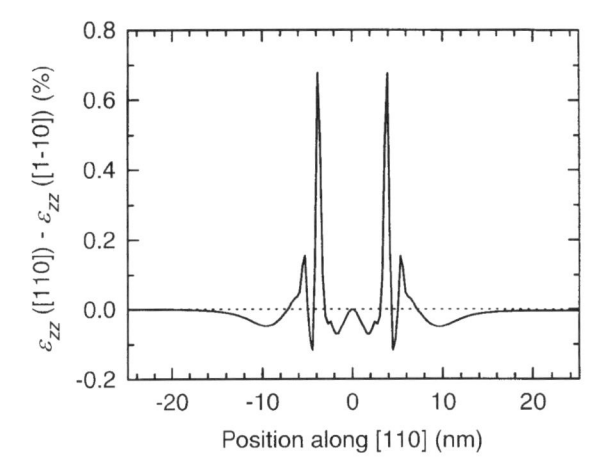

Fig. 5.11 Difference of ε_{zz} along the [110] and [1-10] directions through the center (position 0) of the InAs/GaAs pyramid

5.1.4 IMPACT ON BAND STRUCTURE

The band structure of a material is altered by the presence of strain, which changes the lattice constant and reduces the symmetry of the crystal. Strain modifies energy gaps and lifts degeneracies. For a discussion of electronic properties a decomposition into isotropic (hydrostatic) and anisotropic parts is suitable again.

For zincblende-type semiconductors the effects of strain can be treated within the $k \cdot p$ perturbation theory (Bahder, 1990; Zhang, 1994b; Enders et al., 1995), originally developed by Kane (1957). In the eight-band model the conduction band, the heavy and light holes valence bands, and the spin–orbit interaction split-off valence band are considered for both spin directions. A group theoretical treatment has been worked out by Bir and Pikus (1974). The case of homogeneous strain has been reviewed by Pollak (1990). Strained pyramids actually represent a case of inhomogeneous strain. The inclusion of terms containing the gradient of the strain into the Hamiltonian has been treated (Zhang, 1994b). However, this Hamiltonian has not yet been applied to actual structures. The general form of the strain-induced modification of the band structure is described by a product of strain components times a deformation potential. The values of typical deformation potentials are between 1 and 10 eV; they determine the amount of band structure modification and enter the calculations as basic material properties.

At the Γ point ($\mathbf{k} = 0$) the conduction bands are decoupled from the valence bands (when additionally the small coupling of the conduction band to shear deformations due to the lack of inversion symmetry in zincblende material is neglected). The conduction band is modified by the hydrostatic shift

$$\delta E_H^c = a_c(\varepsilon_{xx} + \varepsilon_{yy} + \varepsilon_{zz}) = a_c I \quad (5.24)$$

where a_c is the conduction band hydrostatic deformation potential. The effect on the hole levels depends largely on the symmetry of the strain and can be obtained from the remaining 6 × 6 Hamiltonian. A simple case is biaxial strain in the (001) plane: the heavy and light hole bands exhibit a shift

$$\delta E_H^v = a_v(\varepsilon_{xx} + \varepsilon_{yy} + \varepsilon_{zz}) = a_v I \quad (5.25)$$

and a splitting

$$\delta E_S^{v(001)} = b(2\varepsilon_{zz} - \varepsilon_{xx} - \varepsilon_{yy}) = bT \quad (5.26)$$

where a_v is the hydrostatic deformation potential of the valence band and b a shear deformation potential. The change of the bandgap due to hydrostatic strain scales with the total hydrostatic deformation potential $a = a_c + a_v$. The sign of $\delta E_S^{v(001)}$ is such that for compressive in-plane strain the $|\frac{3}{2},\frac{3}{2}\rangle$ (heavy hole) band becomes the top valence band. The splitting is modified in second order by the inclusion of a finite split-off energy. For biaxial strain in the (111) plane the splitting is given by

$$\delta E_S^{v(111)} = 2\sqrt{3}d|\varepsilon_{xy}| \quad (5.27)$$

where d is another shear deformation potential.

A theoretical analysis of the impact of strain on the masses in tetrahedral semiconductors was carried out by Aspnes (1978), performing a third-order expansion of the Pikus–Bir Hamiltonian (Bir and Pikus, 1974). The value of the conduction band mass m_c^* in the presence of hydrostatic strain I is given by

$$\frac{m_0}{m_c^*} = 1 + \frac{2p_{cv}^2}{3m_0}\left\{\frac{2}{E_g}\left[1 - \frac{I}{3}\left(2 + \frac{3a}{E_g}\right)\right] + \frac{1}{E_g + \Delta_0}\left[1 - \frac{I}{3}\left(2 + \frac{3a}{E_g + \Delta_0}\right)\right]\right\} \quad (5.28)$$

where

$$m_c^* \approx m_c^{0*}/\left[1 - \frac{I}{3}\left(2 + \frac{3a}{E_g + \Delta_0/3}\right)\right]$$

where m_0 is the free electron mass and m_c^{0*} denotes the conduction band mass in the strain-free case; p_{cv} is the momentum matrix element between the conduction band and the valence band states. For a number of III–V and II–VI materials p_{cv}^2/m_0 has a value of around 10 eV (Ehrenreich, 1961; Cardona, 1963). Δ_0 denotes the split-off energy. Compression leads to an increase of the electron mass.

Aspnes and Cardona (1978) have given expressions for tetragonal (uniaxial stress along $\langle 100 \rangle$) and shear strains (uniaxial stress along $\langle 111 \rangle$). In these cases, the mass becomes anisotropic and has different values parallel and perpendicular to the stress axis.

5.1.5 IMPACT ON PHONON SPECTRUM

In diamond-type materials the optical phonons at $\mathbf{k}=0$ are triple degenerate while for zincblende-type material an LO phonon and twofold degenerate TO phonons exist. The strain impacts these vibrational modes due to a change in the spring constant. The shift of frequencies of the optical phonons is given by the eigenvalues $\lambda = \Omega^2 - \omega_0^2$, $\Omega \approx \omega_0 + \lambda/(2\omega_0)$ of the matrix (Cerdeira et al., 1972):

$$\begin{pmatrix} P\varepsilon_{xx} + q(\varepsilon_{yy} + \varepsilon_{zz}) & 2r\varepsilon_{xy} & 2r\varepsilon_{xz} \\ 2r\varepsilon_{xy} & p\varepsilon_{yy} + q(\varepsilon_{zz} + \varepsilon_{xx}) & 2r\varepsilon_{yz} \\ 2r\varepsilon_{xz} & 2r\varepsilon_{yz} & p\varepsilon_{zz} + q(\varepsilon_{xx} + \varepsilon_{yy}) \end{pmatrix} \quad (5.29)$$

where $p = K_{11}\omega_0^2$, $q = K_{12}\omega_0^2$, $r = K_{44}\omega_0^2$, and K_{ij} are the dimensionless phonon deformation potentials (Jusserand and Cardona, 1989) and ω_0 is the phonon frequency in the strain-free case.

The hydrostatic component of the stress leads to a shift $\Delta\Omega_H$ of the phonon frequencies. A uniaxial stress X along [001] or [111] lifts the cubic symmetry and splits ($\Delta\Omega$) the triplet in diamond-type material into a singlet (Ω_s) and a doublet (Ω_d). For zincblende-type material one of the doublet components is the LO phonon (with a different ω_0); in this case a different set of K_{ij} exists for the TO and LO phonons. For strain in the [001] or [111] direction the phonon frequencies are (Cerdeira et al., 1972)

$$\text{LO phonon} : \Omega_d = \omega_{LO} + \Delta\Omega_H - \tfrac{1}{3}\Delta\Omega$$

$$\text{TO phonon} : \begin{cases} \Omega_d = \omega_{TO} + \Delta\Omega_H - \tfrac{1}{3}\Delta\Omega \\ \Omega_s = \omega_{TO} + \Delta\Omega_H + \tfrac{2}{3}\Delta\Omega \end{cases} \quad (5.30)$$

with

$$\Delta\Omega_H = \frac{X}{\omega_0}\frac{p+2q}{6}(S_{11} + 2S_{12})$$

$$\Delta\Omega = \Omega_s - \Omega_d = \frac{X}{2\omega_0}\begin{cases} (p-q)(S_{11} - S_{12}), & X\|[001] \\ rS_{44}, & X\|[111] \end{cases} \quad (5.31)$$

For a biaxially strained quantum well with (001) interfaces one obtains

$$\Delta\Omega_H = \frac{1}{2\omega_0}(p+2q)I \quad \text{and} \quad \Delta\Omega = \frac{1}{2\omega_0}\frac{(p-q)}{2}T \quad (5.32)$$

The above results are valid in the case of homogeneous strain. For slowly varying strain the phonon spectrum can be calculated as the average over the entire structure. In heterostructures the elastic properties of the constituents are usually different. This leads to novel effects of phonon quantization like interface modes (Sood et al., 1985) and zone folding (Colvard et al., 1980) when the structural dimensions are of the order of only a few lattice constants.

5.1.6 PIEZOELECTRIC EFFECTS

Due to the lack of inversion symmetry the zincblende semiconductors are piezoelectrically active. Shear strain leads to a piezoelectric polarization **P** with

$$P_i = e_{ijk}\varepsilon_{jk} \tag{5.33}$$

e_{ijk} being the piezoelectric tensor, which has for zincblende material only nonzero components if $i \neq j \neq k$. The only nonzero component, $e_{123} = e_{14}$, is also called the piezoelectric module and has been experimentally determined for a number of zincblende materials. Piezoelectric effects are well known for quantum wells (Caridi et al., 1990) and superlattices (Smith, 1986; Mailhiot and Smith, 1987) on (111) substrates.

The polarization induces a fixed charge ρ_P given by

$$\rho_P = -\text{div } \mathbf{P} \tag{5.34}$$

which gives rise to the piezoelectric potential V_P, where the effects due to the difference of dielectric constants are neglected:

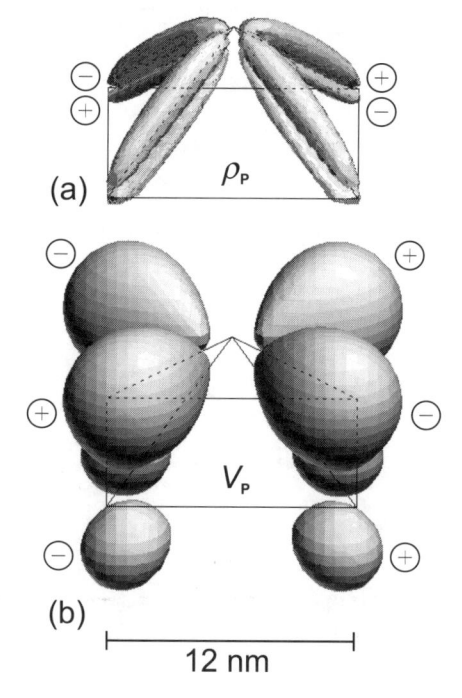

$$V_P(\mathbf{r}_0) = \frac{1}{4\pi\varepsilon_0\varepsilon_r}\iiint \frac{\rho_P(\mathbf{r})}{|\mathbf{r} - \mathbf{r}_0|} d^3\mathbf{r} \tag{5.35}$$

Fig. 5.12 (a) Piezoelectric charge density due to shear strains in a 12 nm base width pyramidal quantum dot (isosurfaces for a volume charge density $|\rho_P| = 0.3 \, \text{e/nm}^3$). The front left edge is {112}A type. (b) Resulting piezoelectric potential (isosurface for $|V_P| = 30 \, \text{meV}$)

Since the polarization is connected with the shear strain, effects are mainly expected in the ⟨111⟩ directions of quantum dot heterostructures. The charge distribution and the resulting piezoelectric potential for a strained InAs pyramid in GaAs (Grundmann et al., 1995c) are depicted in Fig. 5.12. The charge is mainly concentrated along the {112}-like pyramid edges and has opposite sign for adjacent edges. The resulting potential V_P has a quadrupole-like character and reduces the C_{4v} symmetry of the potential of the pyramid to C_{2v}.

5.2 QUANTUM CONFINEMENT

For the calculation of electronic states in quantum dots several schemes have been used with different levels of sophistication. The simplest models are the effective mass particle-in-a-sphere calculations for infinite barriers, refined by finite barriers and different masses for the inner and outer material. Those will be treated in Section 5.2.2. For the large class of close to spherical quantum dots (e.g. II–VI dots in a glass matrix, which are discussed in detail by Hennberger et al., 1992, and Woggon, 1997) this treatment is mostly already sufficient. For quantum dots of a different shape the potential is nonseparable and the Schrödinger equation has usually to be solved numerically. A semianalytical approach has been given by Marzin and Bastard (1994) for conical quantum dots. For spherical dots a multiband envelope-function theory has been presented by Sercel and Vahala (Sercel and Vahala, 1990a; Vahala and Sercel, 1990).

Other more atomistic theoretical approaches are tight-binding calculations (e.g. see Lippens and Lannoo, 1990) and the pseudopotential method (Wang and Zunger, 1994). The latter has been carried out for Si (Wang and Zunger, 1994) and II–VI (CdSe) (Wang and Zunger, 1996a) quantum dots with 10^3 atoms and has been extended to systems with several 10^5 atoms using a pseudopotential-based multiband $k \cdot p$ method (Wang and Zunger, 1996b).

The presence of Coulomb interaction between the localized carriers modifies the single-particle picture depending on the strength of the confinement in comparison to the electrostatic energy and will be treated in the next section, Section 5.3.

5.2.1 PARTICLE IN A HARMONIC POTENTIAL

The solutions for the d-dimensional harmonic oscillator, i.e. the eigenenergies for the Hamiltonian

$$H = \frac{\mathbf{p}^2}{2m} + \sum_{i=1}^{d} \frac{1}{2} m \omega_0^2 r_i^2, \qquad d = 1, 2 \text{ and } 3 \qquad (5.36)$$

are given by

$$E_n = \left(n + \frac{d}{2}\right)\hbar\omega_0, \qquad n = 0, 1, 2, \ldots \qquad (5.37)$$

The energy levels of one- and two-hole states in GaAs/Al$_{0.3}$Ga$_{0.7}$As QDs with parabolic in-plane confinement have been calculated by Pedersen and Chang (1996) on the basis on a multi-band effective mass theory. It was found that for sufficiently large confinement potential both the single hole and the two-hole ground states are changed from primarily heavy-hole character to light-hole character.

More detailed treatments can be found in most quantum mechanics textbooks.

5.2.2 PARTICLE IN A SPHERE

The problem of a particle in a centrosymmetric finite potential well with different masses m_1 in the dot and m_2 in the barrier has been treated in Tran Thoai et al. (1990a). The Hamiltonian and the potential are given by

$$H\Psi(\mathbf{r}) = \left\{\nabla \frac{\hbar^2}{2m^*}\nabla + V(\mathbf{r})\right\}\Psi(\mathbf{r}) \quad \text{and} \quad V(r) = \begin{cases} -V_0, & r \leq R_0 \\ 0, & r > 0 \end{cases} \quad (5.38)$$

The wavefunction can be separated in radial and angular components $\Psi(\mathbf{r}) = R_{nlm}(r)Y_{lm}(\vartheta, \phi)$, where Y_{lm} are the spherical harmonic functions. For the ground state ($n = 1$) the angular momentum l is zero and the solution for the wavefunction (being regular at $r = 0$) is given by

$$R(r) = \begin{cases} \dfrac{\sin(kr)}{kr}, & r \leq R_0 \\ \dfrac{\sin(kR_0)}{kR_0} \exp[-\kappa(r - R_0)], & r > R_0 \end{cases} \quad (5.39)$$

where

$$k^2 = \frac{2m_1(V_0 + E)}{\hbar^2} \quad \text{and} \quad \kappa^2 = \frac{2m_2(-E)}{\hbar^2}$$

From the boundary conditions that both $R(r)$ and $(1/m)[\partial R(r)/\partial r]$ are continuous across the interface at $r = R_0$, the transcendent equation

$$kR_0 \cot(kR_0) = 1 - \frac{m_1}{m_2}(1 + \kappa R_0) \quad (5.40)$$

is obtained. From this formula the energy of the single particle ground state in a spherical quantum dot can be determined. For a given radius the potential needs a certain strength $V_{0,\min}$ to confine at least one bound state; this condition can be written as

$$V_{0,\min} < \frac{\pi^2 \hbar^2}{8m^* R_0^2} \quad (5.41)$$

for $m_1 = m_2 = m^*$. For a general angular momentum l, the wavefunctions are given by spherical Bessel functions j_l in the dot and spherical Hankel functions h_l in the barrier. Tran Thoai et al. (1990a) gave the transcendent equation for the energy of the first excited level:

$$kR_0 \cot(kR_0) = 1 + \frac{k^2 R_0^2}{\dfrac{m_1}{m_2}\dfrac{2 + 2\kappa R_0 + \kappa^2 R_0^2}{1 + \kappa R_0} - 2} \tag{5.42}$$

In the case of infinite barriers ($V_0 = \infty$), the wavefunction vanishes outside the dot and is given by (normalized)

$$R_{nlm}(r) = \sqrt{\frac{2}{R_0^3}} \frac{j_l(k_{nl} r)}{j_{l+1}(k_{nl} R + 0)} \tag{5.43}$$

where k_{nl} is the nth zero of the Bessel function j_l, e.g. $k_{n0} = n\pi$. With two-digit precision the lowest levels are determined from the following:

k_{nl}	$l=0$	$l=1$	$l=2$	$l=3$	$l=4$	$l=5$
$n=0$	3.14	4.49	5.76	6.99	8.18	9.36
$n=1$	6.28	7.73	9.10	10.42		
$n=2$	9.42					

The $(2l+1)$ degenerate energy levels E_{nl} are ($V_0 = \infty$, $m = m_1$)

$$E_{nl} = \frac{\hbar^2}{2m} \frac{k_{ln}^2}{R_0^2} \tag{5.44}$$

The 1s, 1p, and 1d states have smaller eigenenergies than the 2s state. For a cubic dot of side length a_0 and infinite potential barriers one finds the levels $E_{n_x n_y n_z}$:

$$E_{n_x n_y n_z} = \frac{\pi^2 \hbar^2}{2m} \frac{n_x^2 + n_y^2 + n_z^2}{a_0^2}, \qquad n_x, n_y, n_z = 1, 2, \ldots \tag{5.45}$$

For a sphere the separation between the ground and the first excited state is $E_1 - E_0 \approx E_0$; for a cube and a two-dimensional harmonic oscillator it is exactly E_0. For a three-dimensional harmonic oscillator this quantity is $E_1 - E_0 = \frac{2}{3} E_0$.

The valence band structure near the Γ point has been described by the Luttinger Hamiltonian (Luttinger, 1956) for spherical quantum dots with infinite (Xia, 1989) and finite (Sercel and Vahala, 1990a; Vahala and Sercel, 1990) confinement potentials. The split-off bands have been additionally taken into account in Ekimov et al. (1993). In the spherical approximation it is assumed that $\gamma_2 = \gamma_3$, and the Hamiltonian for the hole quantization can be reduced to the form (Baldereschi and Lipari, 1973)

$$H = \frac{\gamma_1}{2m_0} \left[p_h^2 - \frac{\mu}{9}(p_h^{(2)} J^{(2)}) \right] + V_h(\mathbf{r}_h) \tag{5.46}$$

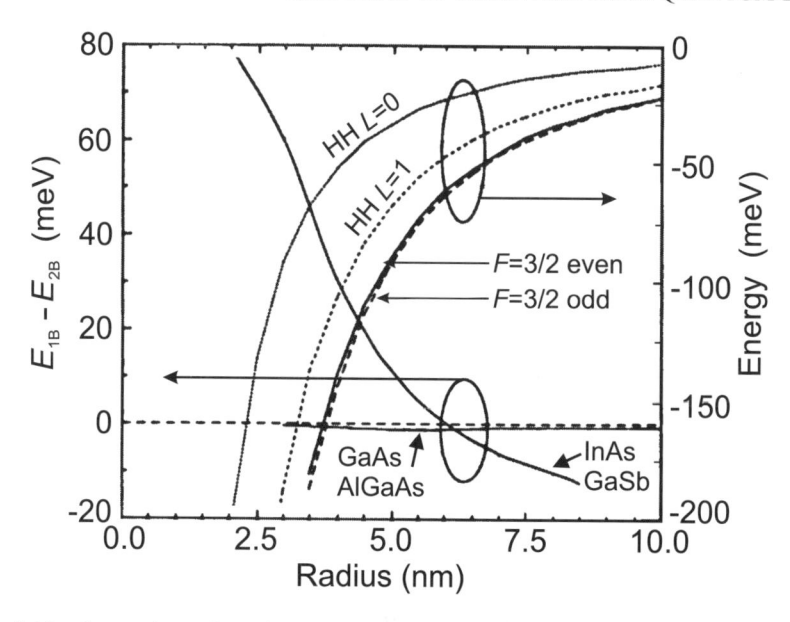

Fig. 5.13 Comparison of one-band effective mass and multiband calculations for spherical quantum dots. Left axis: difference between the ground state energies for type-I GaAs/AlGaAs and type-II GaSb/GaAs quantum dots. Right axis: confinement energy for the two lowest energy valence band states in the GaAs quantum dot. (After Vahala and Sercel, 1990)

which is irreducible under the full rotation group. Here m_0 denotes the free electron mass, $\mu = 2\gamma_2/\gamma_1$, and γ_1, $\gamma_2 = \gamma_3$ are the Luttinger parameters, and $p_h^{(2)}$ and $J^{(2)}$ are the spherical rank 2 tensors for the momentum operator p_h and the angular momentum operator $J = 3/2$ (Baldereschi and Lipari, 1973). The Hamiltonian for the electrons remains of the form in Eq. (5.38). In Fig. 5.13 one-band effective mass theory is compared with calculations taking into account the valence band coupling between heavy and light holes. The most pronounced differences are found for the type-II system (see Section 5.3.2) InAs/GaSb in the regime of strong hole confinement (Vahala and Sercel, 1990).

5.2.3 PARTICLE IN A CONE

For the case of InAs islands and an InAs wetting layer on a GaAs substrate a semianalytical approach was carried out by Marzin and Bastard (1994). The shape of pyramids with {104} side facets, as derived from AFM experiments (Moison et al., 1994), is approximated by 12° facet angle cones. For this rather flat geometry they assume the strain state of the InAs island to be the same, namely biaxial, as in the 2D layer. In the calculation the same electron mass, $m_e = 0.067$ (GaAs value), is used for the dot, quantum well, and barrier. The confinement energies of the first two electron levels are shown as a function of the cone base radius in Fig. 5.14 for two different

MODELING OF QUANTUM DOTS

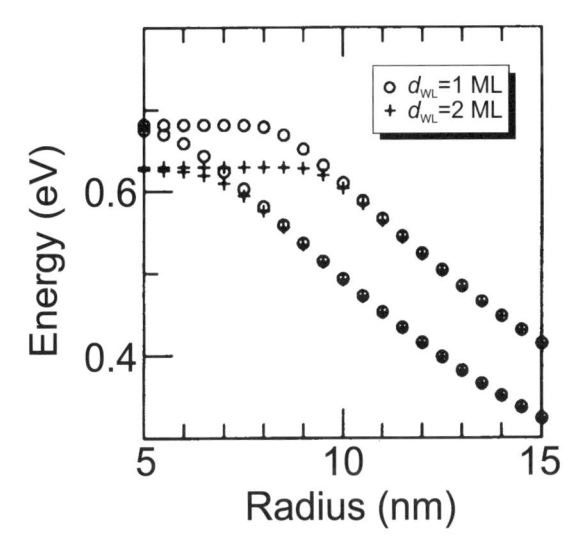

Fig. 5.14 Confinement energies of the first two electron levels as a function of cone base radius for two wetting layer thicknesses: 1 ML (○) and 2 ML (+). (After Marzin *et al.*, 1994)

wetting layer thicknesses d equal to one and two monolayers. As expected for small radii, the levels converge toward the ground state of the respective InAs quantum well. For large radii the quantum dot states practically do not depend on the wetting layer thickness because the states are well confined and do not overlap with the 2D layer. The energy separation between the first and second electron level (if a second one exists) is around 100 meV for a cone base radius up to 15 nm. The transition energies for the first two electrons to heavy hole transitions and the first electron to light hole transitions are shown in Fig. 5.15a. In Fig. 5.15b additionally an indium concentration with a profile

$$x(z) = \frac{a_0}{l} \exp\left(-\frac{z-z_0}{l}\right) \qquad (5.47)$$

has been considered. Such an In concentration distribution leads to an increase of transition energy compared to a homogeneous indium distribution.

5.2.4 PARTICLE IN A PYRAMID

Ruvimov *et al.* (1995) identified self-organized arrays of MBE-grown InAs pyramids with {101} facets (45° facet angle) on a thin InAs wetting layer by comparing cross-sectional HRTEM and image simulation. A similar shape was also found by Oshinowo *et al.* (1994) for MOCVD-grown quantum dots. A numerical solution of the effective mass Schrödinger equation for this geometry was carried out by Grundmann *et al.* (1995c). The equation was discretized on an isotropic cubic cell grid with about 10^6 voxels (lateral resolution ≈ 0.5 nm) by applying a

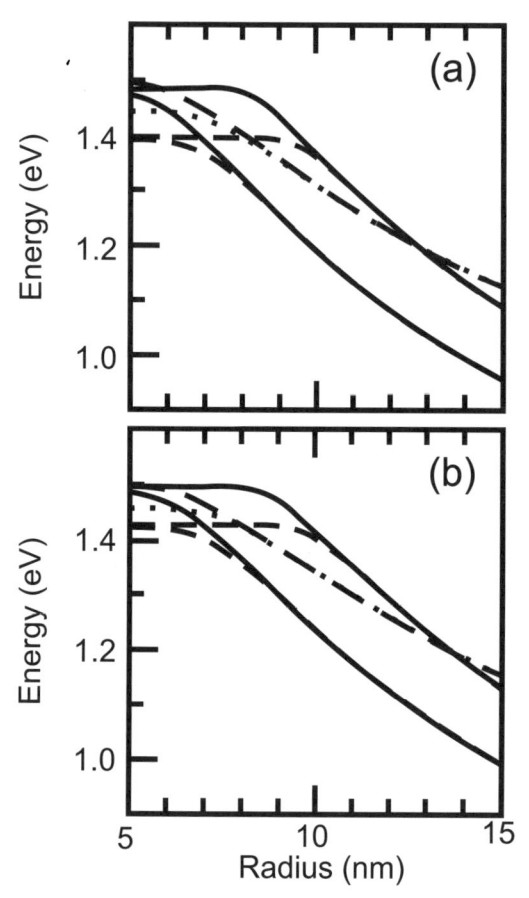

Fig. 5.15 First two heavy hole and first light hole allowed transition energies as a function of the dot base radius (at the lower interface of the wetting layer) (a) without and (b) with inclusion of the effect of indium segregation ($l = 1.1$ nm). The wetting layer thickness is 1 ML (solid lines: heavy hole; long dashed line: light hole) and 2 ML (short dashed lines: heavy hole, dotted line: light hole), respectively. (After Marzin et al., 1994)

symmetrical second-order nonstandard discretization to $(\partial/\partial v)(1/m^*)(\partial/\partial v)$ (harmonic differences: see Li and Kuhn, 1994), employing Dirichlet and Neumann boundary conditions. The resulting matrix eigenvalue problem is solved by a nested iteration generalized block Davidson algorithm (Murray et al., 1992).

The actual quantum dots were embedded in a 14 nm GaAs quantum well with an AlGaAs/GaAs superlattice on top and underneath. For the given geometry the superlattice has practically no impact on the electronic states in the quantum dots. For pyramids smaller than $b = 6$ nm no confined electron state exists for an effective electron mass of $m_e^* = 0.023$. For a larger electron mass such a state exists for smaller sizes (see Section 5.2.8).

MODELING OF QUANTUM DOTS

Choosing a higher bandgap material as the barrier can significantly increase the confinement effects. If the upper barrier is replaced by AlAs or $In_{0.5}Ga_{0.5}P$, the asymmetric potential well of the wetting layer has a 'ground state' above the level for unstrained GaAs for electrons and holes and is no longer bound. The quantum dot ground state, however, remains localized, i.e. the quantized electron level lies below the GaAs conduction band, for $b \geqslant 9$ nm. By comparing luminescence from InAs quantum dot structures, grown on the same GaAs substrate and buffer layers but with a different upper barrier (GaAs or AlAs), the role of the wetting layer for carrier capture and relaxation could be identified and separated from that of the GaAs barrier. When the upper and lower barriers consist of AlAs, the ground state energies are further increased.

Due to the larger hole masses several hole states are localized in the pyramids. Due to the strain-induced potential the hole ground state is squeezed to the bottom of the QD. The degeneracy of the first (*p*-like) excited level is lifted (at zero magnetic field) by the piezoelectric effect. Due to the opposite charge, electron and hole levels of the same symmetry exhibit energy shifts of different sign. Another reason for lifting the degeneracy could be an anisotropy of the quantum dot shape, e.g. a different base width in the [001] and [010] directions. In this case, electron and hole levels would exhibit energy shifts of equal sign. Lifting of the degeneracy by a random potential has been proposed by Tsiper (1996) and will be illustrated in Section 5.4.2. Figure 5.16 illustrates that the expected polarization of luminescence for the different mechanisms is quite different: for the random potential no polarization is expected. In the case of a shape anisotropy, electrons *and* holes will exhibit the energetically lower *p*-like level for polarization parallel to the elongated direction of the dot. The piezoelectric field indicates the lower electron

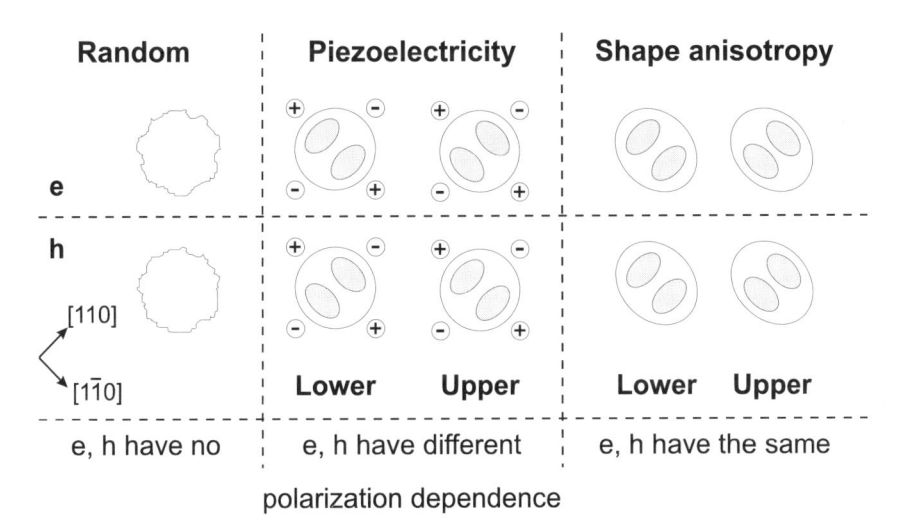

Fig. 5.16 Mechanisms for lifting of the degeneracy of the first excited *p*-like state. Each mechanism leads to a different, characteristic polarization dependence

level for polarization in the [1-10] direction and the lower hole level in the opposite [110] direction.

Cusack *et al.* (1996) calculated the hole levels using the Luttinger Hamiltonian (including four hole bands). Mixing different valence band states leads to a splitting of the first and second excited hole states, even when C_{4v} symmetry is used.

5.2.5 PARTICLE IN A LENS

The quantum dots reported in Leonard *et al.* (1993) are believed to have a lens-like shape, which is described as a part of a sphere with a given base diameter D and a height/diameter ratio of 1:2 (Leonard *et al.*, 1994). Typical dimensions are $D = 20$ nm and height $h = 7$ nm. The ground state wavefunction in such an island has cylindrical symmetry. A characteristic quantity is the in-plane extension $l = \langle r^2 \rangle - \langle r \rangle^2$. The in-plane extension scales essentially with the lens radius. Since confinement in the z direction is much stronger than in-plane, the low-lying excited electron states are in-plane excited states with a well-defined angular momentum. It has been reported in Wojs and Kawrylak (1996) that the exact eigenenergies for the first three electron levels of such dots are well approximated by the states of an infinite (two-dimensional) parabolic potential, also in the presence of a magnetic field (Fig. 5.17).

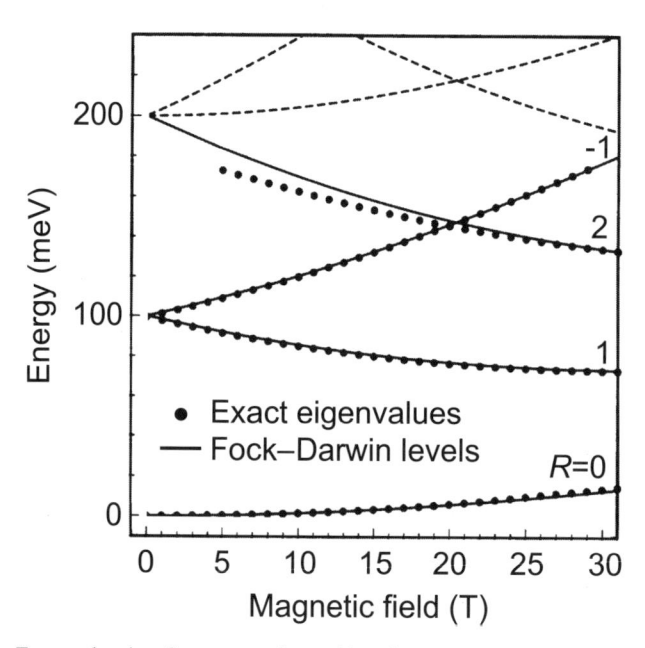

Fig. 5.17 Energy levels of quantum dots with a lens shape (dots) and using a parabolic confinement potential (lines) as a function of the magnetic field. Solid lines correspond to the states included in the basis for diagonalization of the many-body Hamiltonian. (After Wojs and Hawrylak, 1996)

5.2.6 STRESS-INDUCED QUANTUM DOTS

The electronic structure of quantum dots induced by InP self-organized quantum dots acting as stressors for an InGaAs/GaAs quantum well has been calculated by Tulkki and Heinämäki (1995). In Fig. 5.18 the shifts of several transitions in the dots from the quantum well ground state transition are shown as a function of distance d between the quantum well and the surface and the stressor radius R. The levels are essentially equidistant since the strain-induced potential has a prevalent parabolic shape.

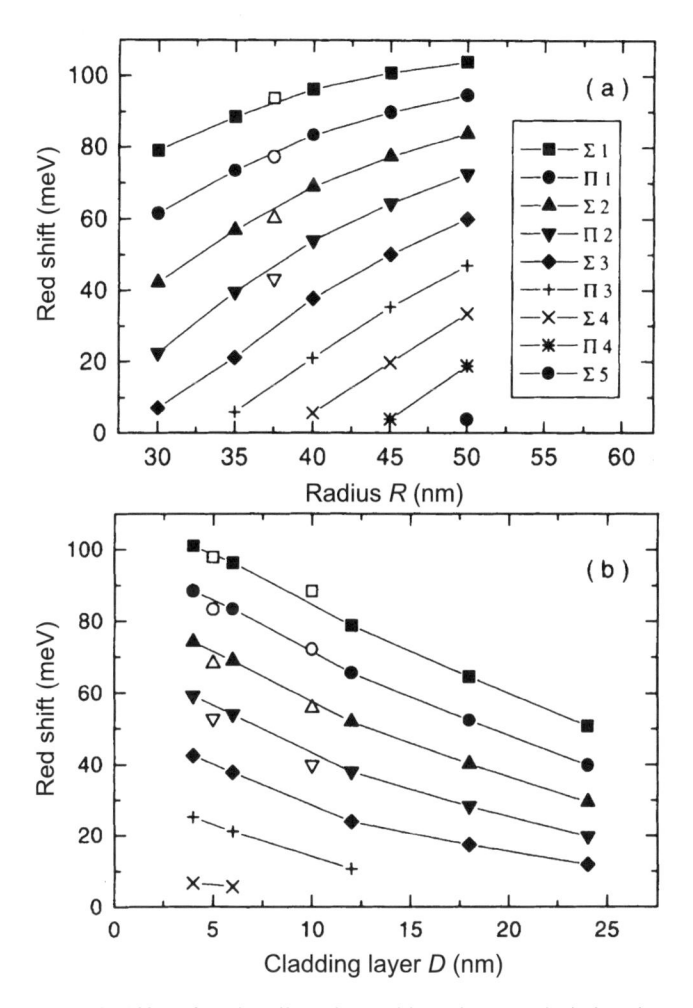

Fig. 5.18 (a) Red shifts of main allowed transitions in a strain-induced quantum dot, measured from the ground state transition in the unpatterned $In_{0.25}Ga_{0.75}As/GaAs$ quantum well as a function of the stressor dot base radius R. The InP stressor dot is assumed to be a flat cone with a top radius of $X = R - 10$ nm. The quantum well thickness is $L_z = 8$ nm and the cap layer thickness is $d = 6$ nm. (b) Red shift as a function of cap layer thickness d for fixed $L_z = 8$ nm, $R = 40$ nm. (After Tulkki and Heinämäki, 1995)

5.2.7 TWOFOLD CLEAVED EDGE OVERGROWTH

Grundmann and Bimberg (1997a) predicted theoretically that electronic quantum dots form at the juncture of three orthogonal quantum wells which could be fabricated with twofold cleaved edge overgrowth (2CEO) (Fig. 5.19) (see also Section 6.6). Already in the single-particle picture electron and hole localization occurs. Additionally, the Coulomb interaction, treated in Grundmann and Bimberg (1997a) within the Hartree approximation (Banyai *et al.*, 1992), leads to further localization with respect to the four connected T-shape quantum wires which act as the effective barriers. Localization energies of up to 10 meV have been predicted for the AlGaAs/GaAs system (Fig. 5.20). Compressive strain, e.g. present in a 2CEO quantum dot made from $In_xGa_{1-x}As$/AlGaAs quantum wells, counteracts and for sufficiently high strain even prohibits formation of localized states (Grundmann *et al.*, 1998).

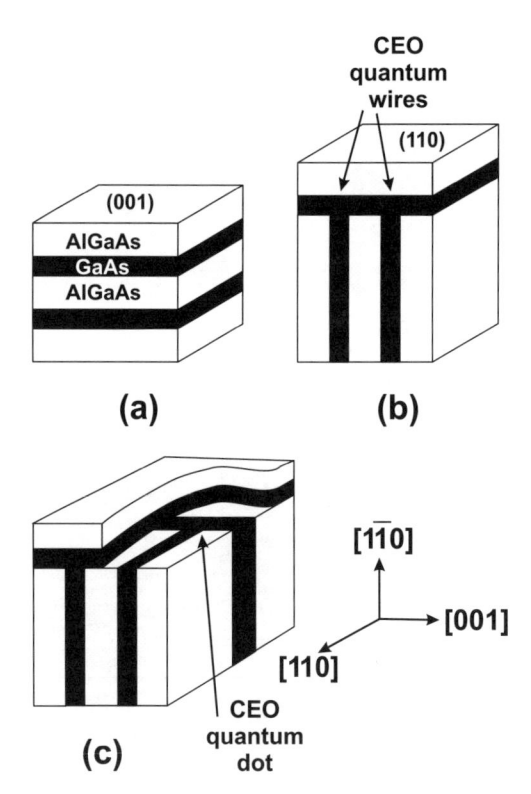

Fig. 5.19 Schematic representation of two-old cleaved edge overgrowth: (a) and (b) describe the standard process used for fabrication of quantum wires. In (c) a second cleave and growth on top of the ($1\bar{1}0$) plane allows the fabrication of quantum dots. (After Grundmann and Bimberg, 1997a)

MODELING OF QUANTUM DOTS 125

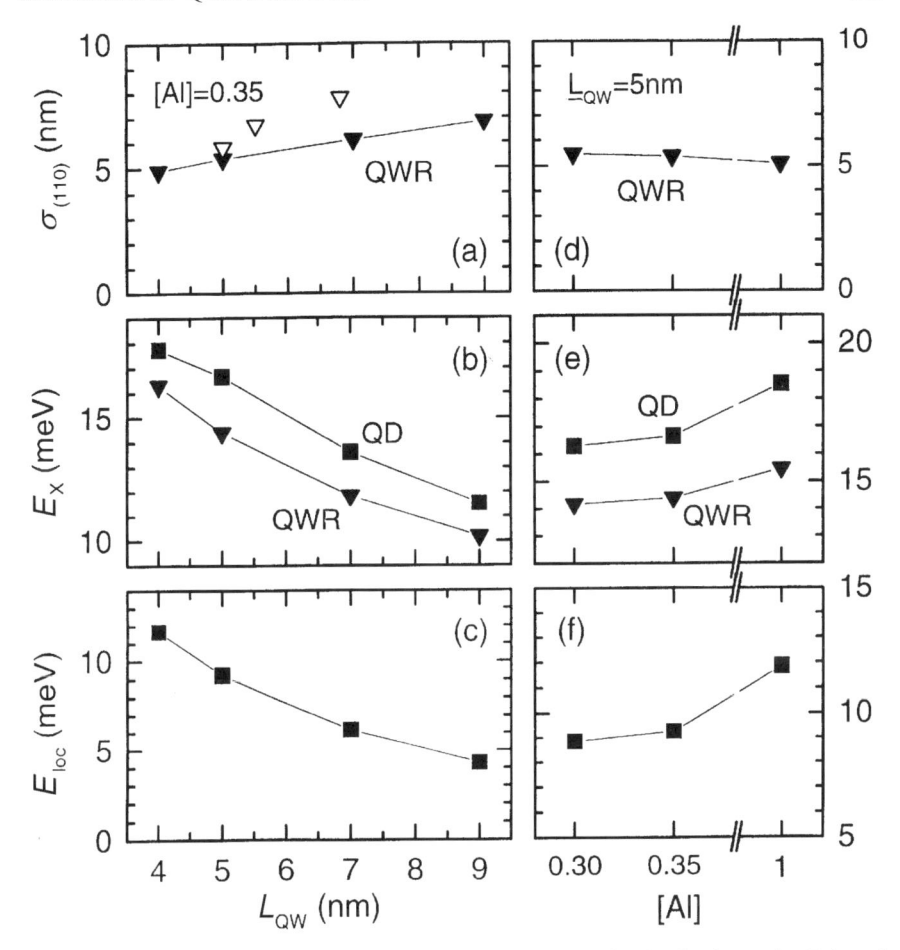

Fig. 5.20 (a), (d) Extension of the exciton wavefunction in the (110) plane of a T-shaped quantum wire. (b), (e) Exciton binding energy for CEO quantum wires and dots. (c), (f) Localization energy for twofold CEO dots. The data are plotted in (a), (b) and (c) as a function of L_{QW} in the GaAs/Al$_{0.35}$Ga$_{0.65}$As system. For (d), (e) and (f) the aluminum content in the barrier is varied for fixed $L_{QW} = 5$ nm. Open triangles in (a) are experimental values for the exciton extension from Someya *et al.* (1995). (After Grundmann and Bimberg, 1997a)

5.2.8 EIGHT-BAND $k \cdot p$ THEORY AND PSEUDOPOTENTIAL METHODS

Refined treatment of electronic states in quantum dots as compared to simple effective mass approximation (EMA) is achieved by application of eight-band k–p theory (Fu *et al.*, 1997; Jiang and Singh, 1998; Pryor, 1998; Stier *et al.*, 1998) and pseudopotential methods (Wang and Zunger, 1994; Fu *et al.*, 1997; Kim *et al.*, 1998a; Zunger, 1998).

Table 5.2

Model	E_1 (meV)	H_1 (meV)	$E_{1,1}$ (eV)
CM, EMA ($m^* = 0.023$)	−151.1	−159.3	1.208
CM, $k \cdot p$	−226.1	−165.5	1.126
CM(C_{44}^{VFF}), $k \cdot p$	−236.6	−172.1	1.109
VFF, $k \cdot p$	−244.5	−175.2	1.098

The eight-band $k \cdot p$ Hamiltonian in the presence of strain has been given, for example, by Pollak (1990) and Zhang (1994b). For InAs/GaAs quantum dots, typical transition energies are around 1.1 eV, while the InAs bulk bandgap lies at 0.4 eV. Strain and confinement effects are large and roughly of the same size. Calculations using coupled valence and conduction bands, treating dispersion and strain effects on the same level simultaneously, will improve the validity of theoretical predictions. Calculations reported in Jiang and Singh (1997) show that the conduction band is nonparabolic and for rather small InAs/GaAs pyramids (base length $b = 11.3$ nm) a second electron state is predicted to exist.

In Table 5.2 results are compared for electron and hole levels in a $b = 13.6$ nm InAs/GaAs pyramid with {101} facets calculated with eight-band $k \cdot p$ theory (Stier et al., 1998). Different strain models discussed in Section 5.1.3, namely the continuum model (CM), the CM(C_{44}^{VFF}) model with the (incorrect) VFF value for

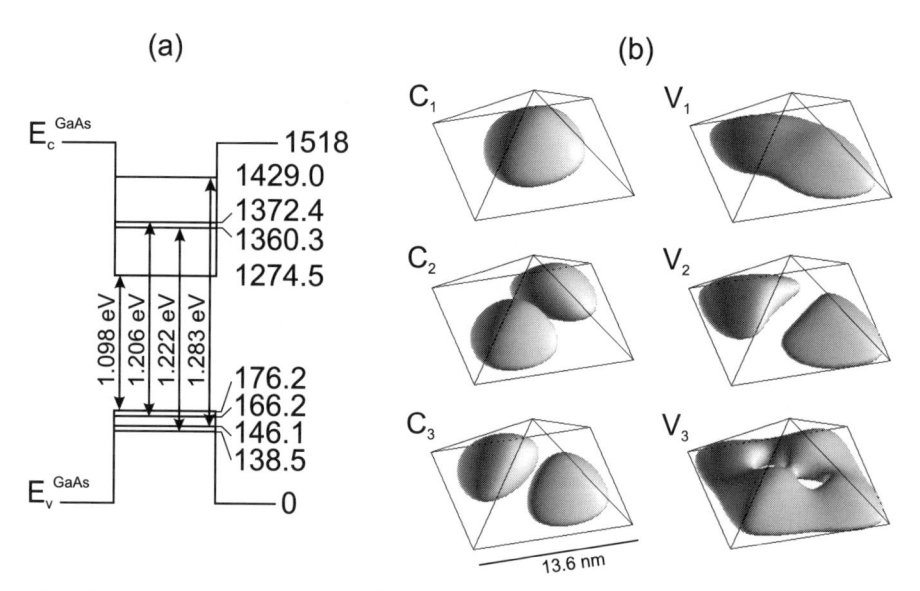

Fig. 5.21 (a) Level structure of $b = 13.6$ nm InAs/GaAs pyramid with {101} facets calculated with eight-band $k \cdot p$ theory and the VFF model; energies are given in millielectronvolts. (b) Wavefunctions of the first conduction (C_1–C_3) and valence (V_1–V_3) band states. (After Stier et al., 1998)

MODELING OF QUANTUM DOTS 127

C_{44}, and the VFF model, are compared. E_1 and H_1 denote the energies of the first quantized electron and hole state, measured from the conduction and valence band edges of unstrained GaAs, respectively. $E_{1,1}$ denotes the energy of the ground state electron-hole recombination (without exciton binding energy). All strain models yield values that are rather close to each other. The level structure and wavefunctions for the first three electron and hole states are shown in Fig. 5.21.

Using a pseudopotential plane-wave approach, transition energies of 1.08 eV (1.00 eV) have been calculated for $b = 9.0$ nm (11.3 nm) InAs/GaAs pyramids with {101} facets. For {113} facets and the same base size transition energies of 1.17 eV (1.10 eV) were found (Kim *et al.*, 1998a). In Fig. 5.22 these results are shown together with the predictions of $k \cdot p$ theory for different pyramid base lengths (Stier *et al.*, 1998).

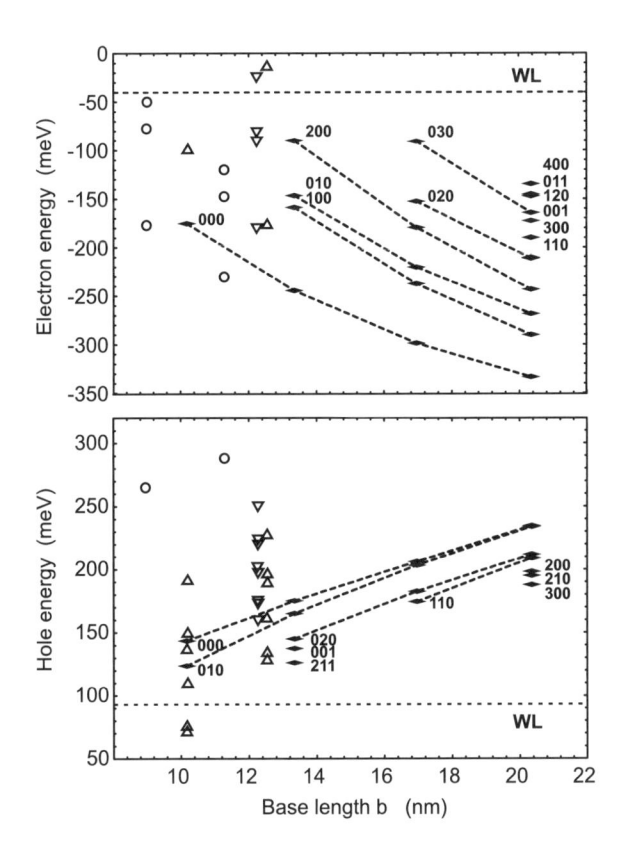

Fig. 5.22 Electron and hole levels for InAa/GaAs pyramids of different base lengths b from eight band $k \cdot p$ theory (strain in the VFF model) (Stier *et al.*, 1998, solid diamond symbols) together with other $k \cdot p$ results (Cusack *et al.*, 1996, six band, open up triangles, Jiang and Singh, 1998, eight band, open down triangles) and pseudopotential calculations (Kim *et al.*, 1998a and Zunger, 1998, open circles)

5.3 COULOMB INTERACTION

Confined charge carriers, electrons and holes interact via the Coulomb interaction. Each pair (i,j) of particles with charge q at position **r** contributes the Coulomb interaction energy W_{ij} to the system:

$$W_{ij}(\mathbf{r}_i, \mathbf{r}_j) = \frac{1}{4\pi\varepsilon_r\varepsilon_0} \frac{q_i q_j}{|\mathbf{r}_i - \mathbf{r}_j|} \tag{5.48}$$

where ε_0 denotes here the vacuum dielectric constant (and not the relative lattice mismatch as used before). For $q_i = q_j$ the particles suffer a repulsion, which increases the total energy of the system; for $q_i = -q_j$ (electron and hole) the energy of the system is lowered and an exciton is formed. The correlation of carrier motion by the Coulomb interaction not only changes the energy spectrum of the system but also alters the oscillator strength of transitions, i.e. their lifetime, which is of particular importance for optoelectronic applications.

For Eq. (5.48) a uniform relative dielectric constant ε_r has been assumed throughout the structure. For a non-uniform dielectric constant the surface polarization and images charges have to be taken into account. This can be done analytically for spherical quantum dots with dielectric constant ε_1 embedded in an infinite medium with dielectric constant ε_2 (Brus, 1984). The correction of W_{ij} for $\varepsilon_2 \neq \varepsilon_1$ is caused by the electric field extending generally much further into the barrier than the wavefunctions. Since in semiconductor heterostructures the barrier material with the larger bandgap has usually the smaller dielectric constant ($\varepsilon_2 < \varepsilon_1$, $\varepsilon = \varepsilon_1/\varepsilon_2 > 1$), the electrostatic interaction is increased by the surface polarization compared to a calculation with the same dielectric constant for the barrier and the dot region. A similar, more extreme situation arises for semiconductor quantum dots in glass or solution, where $1 \approx \varepsilon_2 \ll \varepsilon_1$. The effect of image charges is enhanced for strongly anisotropic structures (the extreme case being a quantum well; see Tran Thoai *et al.*, 1990b), because then a larger fraction of the field lines penetrates the barrier.

In the following we will discuss first the properties of excitons and biexcitons in quantum dots. Then we will review results for multielectron systems in quantum dots. The electron–electron interaction in quantum dots populated with few electrons has drastic consequences for static and dynamic transport properties, the most prominent feature being the so-called Coulomb blockade (Grabert and Horner, 1991).

5.3.1 EXCITONS

When the ground states for electrons and holes in a quantum dot are populated by one carrier each automatically an exciton is formed. In a bulk crystal an exciton can dissociate into a pair of free carriers in the conduction and valence bands, and in principle the exciton binding energy is measurable. The definition of an exciton binding energy in a quantum dot as the difference of the eigenenergies of a

Hamiltonian with and without Coulomb interaction is mainly a theoretical concept since an uncorrelated ground state is nonexistent. The exciton binding energy could be measured in principle if the single-particle energies of electrons and holes are determined independently, e.g. by inter-subband transitions in n- and p-doped quantum dots, respectively.

Three regimes can be defined to compare the importance and role of Coulomb effects with the effects due to the quantization of the kinetic energy (Efros and Efros, 1982):

(a) In the strong confinement regime the Coulomb effects are only a small correction to the dominating quantization of the kinetic energy, i.e. $\Delta E_e \gg E_C$ and $\Delta E_h \gg E_C$, where ΔE_e and ΔE_h are the electron and hole sublevel separations and E_C is the Coulomb interaction energy (Hanamura, 1988). Electron and hole wavefunctions are largely uncorrelated. An example is a small quantum dot, where the dot radius is smaller than the bulk exciton radius (Section 5.3.1.1).

(b) The opposite is true for the weak confinement regime, where the exciton radius is smaller than the dot radius. In this case the electrons and holes form pairs whose center of mass motion is quantized by the confinement potential. ΔE_e and ΔE_h are comparable to or larger than E_C (Section 5.3.1.3).

(c) The intermediate confinement regime is introduced for the case when $\Delta E_e \gg E_C$ and $\Delta E_h \ll E_C$. This situation can arise due to the different masses of electrons and holes. The hole energy is then quantized by the electrostatic potential of the electron orbital (Section 5.3.1.2).

Since the Coulomb energy depends largely on the value of the dielectric constant, dots of the same size can belong to different regimes in different materials. Typically, in III–V compounds, due to the rather small dielectric constants, the bulk exciton radius is > 10 nm, causing a structural quantum dot of similar dimension and sufficiently deep potential to be in the strong confinement regime. However, in II–VI compounds the bulk exciton radius is smaller and typically only a few nanometers, making it much harder to obtain strong confinement in structural quantum dots.

5.3.1.1 Strong confinement regime

For small dot radii, $R_0 \ll a_B$, the kinetic energy due to size quantization is the dominant energy contribution. For both electrons and holes the Coulomb interaction energy is small compared to the separation of ground and excited states. The exciton binding energy E_X can then be calculated using perturbation theory. In first order it is given by the Coulomb interaction matrix element as

$$E_X = \langle 00|W_{eh}|00\rangle = \int\int \Psi_0(\mathbf{r}_e)\Psi_0(\mathbf{r}_h)W_{eh}(\mathbf{r}_e,\mathbf{r}_h)\Psi_0(\mathbf{r}_e)\Psi_0(\mathbf{r}_h) d^3\mathbf{r}_e\, d^3\mathbf{r}_h \quad (5.49)$$

For $R_0 \to 0$ and infinite barriers the exciton binding energy diverges like $1/R_0$. The wavefunctions are to first order not modified by inclusion of the Coulomb interaction. The transition dipole moment between an electron in state m and a hole in state n to the ground state g is given, independently of R_0, by

$$\langle \Psi_{mn}|p|\Psi_g\rangle = p_{cv}\delta_{mn} \qquad (5.50)$$

where p_{cv} is the interband matrix element for the Bloch functions. The δ-function describes the selection rules for the envelope function. If electron and hole masses are different and the confinement potential is finite, this number (wavefunction overlap) is close to 1 for low-lying allowed transitions. The oscillator strength per quantum dot of the transition of an electron and a hole in the state n (Stern, 1963) in the strong confinement regime f_n^{sc} is

$$f_n^{sc} = \frac{2m_0\omega_n}{\hbar}p_{cv}^2 \qquad (5.51)$$

where m_0 denotes the free electron mass and $\hbar\omega_n$ the transition energy. If f_n^{sc} is compared to the exciton oscillator strength f_n^{QW} per unit surface area in a quantum well (Shinada and Sugano, 1966)

$$f_n^{QW} = \frac{2m_0\omega_n}{\hbar}p_{cv}^2 \frac{8}{\pi a_B^2} \qquad (5.52)$$

it follows that only an ensemble of quantum dots with an average center-to-center distance, $d_{av} = a_B\sqrt{\pi}/8 \approx 2a_B/3$, possesses the same oscillator strength per unit surface area. Thus the area coverage with quantum dots has to be rather high to surpass the oscillator strength per unit area of a quantum well (e.g. a_B is 14 nm for GaAs and 50 nm for InAs). The reflectivity of quantum wells and quantum dot arrays has been theoretically compared by Ivchenko and Kavokin (1992) and Ivchenko et al. (1992).

5.3.1.2 Intermediate confinement regime

In this regime $\Delta E_e \gg E_C \gg \Delta E_h$, which can be met if the electron mass is much smaller than the hole mass ($m_e/m_h \ll 1$). The electron is quantized by the confinement potential and the hole moves (adiabatically) in the attractive electrostatic potential build-up by the electron orbital. The hole potential is that of a three-dimensional harmonic oscillator with the respective energy levels.

The oscillator strength f_1 for the $n = 1$ excitonic recombination in a sphere is (with minor approximations):

$$f_1 = \frac{2m_0\omega}{\hbar}p_{cv}(864)^{1/8}\sqrt{\pi}\left(\frac{a_h}{R_0}\right)^{3/8}\exp\left[-\sqrt{\frac{3}{8}\pi}\left(\frac{a_h}{R_0}\right)^{1/2}\right] \qquad (5.53)$$

with $a_h = (4\pi\varepsilon_r\varepsilon_0/e^2)(\hbar^2/m_h)$. The electron hole wavefunction is of the separated form $\psi_e(\mathbf{r}_e)\psi_h(\mathbf{r}_h)$. The transition dipole $\langle p \rangle$ (Efros and Efros, 1982; Hanamura, 1988) is $\langle p \rangle \leq p_{cv}$ and accordingly $f_1 \leq f_1^{sc}$. Within this model the oscillator strength is not enhanced compared to the strong confinement case and the previous conclusion for an enhancement of the oscillator strength (Hanamura, 1988) does not hold. In Efros and Efros (1982) only the (invalid) approximation made in their equation (20) leads to high values for $\langle p \rangle$ larger than p_{cv}.

Variational calculations over a continuous range of dot radii (Takagahara, 1987, 1993; Bryant, 1988; Kayanuma, 1988) taking into account the correlation of carrier motion show a monotonous increase of oscillator strength with increasing dot radius. For dot radii typically larger than $4a_B$ the regime of weak confinement is reached, which will be discussed in the next section.

5.3.1.3 Weak confinement regime

When both ΔE_e and $\Delta E_h \ll E_C$, i.e. $a_e, a_h \ll R_0$, the center of mass motion of the exciton is quantized by the confinement potential and the relative carrier motion is dominated by the Coulomb interaction. The relative carrier motion is described by the function

$$\phi_{1s}(\mathbf{r}_e, \mathbf{r}_h) = \left(\frac{2}{\pi a_B^2}\right)^{1/2} \exp\left(-\frac{|\mathbf{r}_e - \mathbf{r}_h|}{a_B}\right) \tag{5.54}$$

The increase of exciton binding energy E_X due to confinement is marginal and the exciton binding energy is given by the bulk Rydberg value† $E_X^b = E_R$:

$$E_X^b = \frac{\mu}{2\hbar^2}\left(\frac{e^2}{4\pi\varepsilon_r\varepsilon_0}\right)^2 \tag{5.55}$$

where $\mu = m_e m_h/(m_e + m_h)$ is the reduced mass. In the electric dipole approximation $\exp(i\mathbf{q}R_0) \approx 1$, where $\mathbf{q}$ is the wave vector of the radiation field (with typical values such as $\omega_n = 1$ eV and the refractive index $n_r = 3.5$ and $q^{-1} \approx 57$ nm), the exciton oscillator strength per (spherical) quantum dot for the ground state transition is (Hanamura, 1988)

$$f_n = \frac{2m_0\omega_n}{\hbar}p_{cv}^2 \frac{8}{\pi^2}\left(\frac{R_0}{a_B}\right)^3 \frac{1}{n^2}, \quad n = 1, 2, \ldots \tag{5.56}$$

The oscillator strength decreases for higher transitions. It increases with increasing quantum dot size. However, this is true only as long as the electric dipole approximation holds, which breaks down for large R_0. The proper connection with the bulk material oscillator strength for $R_0 \to \infty$ has to be made by properly including the selection rule $\mathbf{K} = \mathbf{q}$ for the wave vectors of the radiation field and the center of mass motion. Such an analysis has been carried out for disk-shaped quantum dots (Sugawara et al., 1995) and will be discussed in the Section 5.3.1.5.

---

† In the definition of the Rydberg energy ε_r is the dielectric constant ε_1 of the dot material.

5.3.1.4 Fine structure

For spherical dots the lowest state $|1s_{1/2}\ 1s_{3/2}\rangle$ is eightfold degenerate in the spherical band approximation (Ekimov et al., 1993). Several mechanisms can lift this degeneracy, like anisotropy of the internal crystal structure or the quantum dot shape (Efros and Rodina, 1993) and the electron–hole exchange interaction, which is proportional to the spatial overlap of the electron and hole wavefunctions (Onodera and Toyozawa, 1967; Takagahara, 1993).

The exchange interaction is sometimes split into a short- and a long-range part. The short-range part is an analog to the singlet–triplet splitting in a two-electron system caused by the Pauli principle. The states of the exciton are described by the total angular momentum $N = F + s$ and its projection N_m, where $F = 3/2$ is the hole momentum with the projections $m_F = \pm 3/2$, $\pm 1/2$ and $s = 1/2$ is the electron spin with $m_s = \pm 1/2$. The resulting states are $|1,1\rangle$ and $|1,-1\rangle$ (also labeled $\pm 1^U$; see Nirmal et al., 1995), $|1,0\rangle$ (0^U), $|2,1\rangle$ and $|2,-1\rangle$ ($\pm 1^L$), $|2,0\rangle$ (0^L), as well as $|2,2\rangle$ and $|2,-2\rangle$ (also labeled ± 2); the superscripts U and L denote the upper and lower states with the same angular momentum projection, respectively.

A detailed discussion of the fine structure and the impact of spin flip dynamics on the spectra from CdS quantum dots in a glass matrix can be found in Nirmal et al. (1995). In Fig. 5.23 the calculated band edge exciton structure in CdSe quantum dots is shown for spherical and elliptical shapes. Radiative recombination from both the ± 2 and 0^L levels, one of them typically being the energetically lowest level, is optically forbidden within the electrical dipole approximation; therefore they are also called dark states.

5.3.1.5 Exciton lifetime in quantum disks

In this section the radiative lifetime of (the optically allowed states of) excitons in quantum dots is discussed. It has already been shown in the previous sections that the oscillator strength depends on the quantum dot size. Thus we expect a similar behavior for the lifetime. Additionally the dependence of the lifetime on temperature is discussed.

Sugawara et al. (1995) presented a theory for quantum disks of fixed height and varying radius which connect continuously the cases of

(a) strong confinement (lifetime is independent of dot size, $\tau \approx 1$ ns),
(b) weak confinement (lifetime decreases proportionally to the volume, $\tau^{-1} \propto V$),
(c) ideal quantum well (lifetime of free excitons, $\tau \approx 10$ ps).

The numerical values chosen by Sugawara et al. (1995) are approximate values for $In_{0.53}Ga_{0.47}As/InP$. Gotoh et al. (1997) calculated exciton radiative lifetimes in rectangular thin GaAs quantum boxes with infinite confinement potential and spatial dimensions between 2D and 0D confinement.

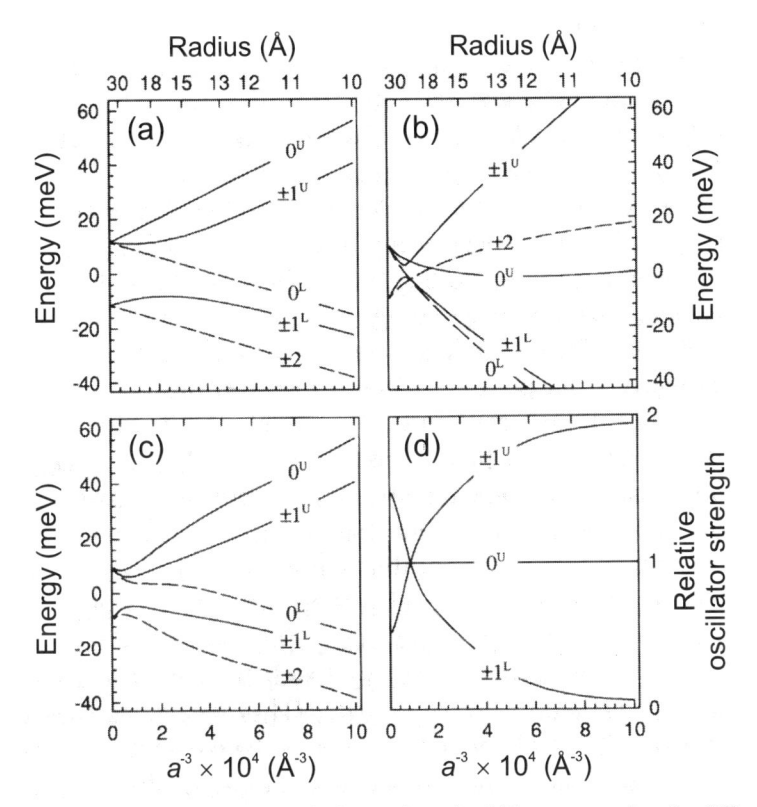

Fig. 5.23 Calculated energies of band edge excitons in CdSe quantum dots for different dot radii a for dots with (a) spherical, (b) elliptical, and (c) size-dependent elliptical shape (from TEM data); for level labels see text. (d) Oscillator strength relative to the 0^U level for the optically active states of (c). (After Nirmal et al., 1995)

For increasingly weaker confinement, the lifetime does not decrease any more because the expression yielding $\tau^{-1} \propto V$ is obtained using the electric dipole approximation, which starts to break down for disk radii of several tens of nanometers. The 2D limit of the quantum well exciton has to be compared with experimental determinations of the lifetime of a true free quantum well exciton (e.g. Deveaud et al., 1991). Low-temperature experiments on quantum wells mostly investigate excitons that are more or less weakly bound to lateral potential fluctuations, which results in radiative lifetimes in the range of several hundreds of picoseconds (Sugawara et al., 1995).

For calculating the radiative lifetime of Wannier excitons in quantum disks, the height of the islands is assumed to be so small that only the first subband is populated (e.g. $L_z = 10$ nm) due to large vertical quantization. The in-plane lateral confinement is modeled as a two-dimensional harmonic oscillator of frequency ω, allowing the relative and center of mass carrier quantization to be separated.

As a result of the center of mass motion we obtain the energy levels E_{kl}^R and wavefunctions $\Psi_{kl}(\mathbf{R})$ of the two-dimensional harmonic oscillator, $k = 0, 1, 2, \ldots$, $l = 0, \pm 1, \ldots, \pm k$. The ground state $\Psi_{00}(\mathbf{R})$ of the center of mass motion is a Gaussian:

$$\Psi_{00}(\mathbf{R}) = \left(\frac{2}{\pi}\right)^{1/2} \frac{1}{\beta} \exp\left(-\frac{R^2}{\beta^2}\right), \qquad \beta = \sqrt{\frac{2\hbar}{M\omega}} \tag{5.57}$$

Since the center of mass motion covers an area $2\pi\beta^2$, $\sqrt{2}\beta = 2\sqrt{\hbar/M\omega}$, is a measure for the disk radius. The selection rule for the in-plane wave vectors $\mathbf{q}$ of the radiation field and $\mathbf{K}$ of the center of mass motion is for all radii $\mathbf{K} = \mathbf{q}$. This supposes that only excitons with a wave vector

$$q < q_c = \frac{\omega_{ex} n_r}{c} \tag{5.58}$$

can recombine, n_r being the index of refraction. For ground state excitons in a disk, the lifetime τ_D for arbitrary disk radius is (Sugawara et al., 1995)

$$\tau_D^{-1} = \frac{e^2 \beta^2}{m_0 \varepsilon_0 c n_r} \int_0^{q_c} \exp\left(-\frac{\beta^2 q^2}{2}\right) [f_\parallel \eta_1(q) + f_\perp \eta_2(q)] q \, dq \tag{5.59}$$

where

$$\eta_1(q) = \left[1 - \left(\frac{q}{q_c}\right)^2\right]^{1/2} + \left[1 - \left(\frac{q}{q_c}\right)^2\right]^{-1/2}$$

$$\eta_2(q) = \left(\frac{q}{q_c}\right)^2 \left[1 - \left(\frac{q}{q_c}\right)^2\right]^{-1/2}$$

$$f_{\parallel/\perp} = \frac{M_{QW,\parallel/\perp}^2(k_\parallel) M_b^2 F^2 |\phi_{10}(0)|^2}{m_0 \hbar \omega_{ex}}$$

$M_{QW,\parallel/\perp}^2$ is the polarization factor in quantum wells for polarization parallel or perpendicular to the quantum well plane (Yamanishi and Suemuno, 1984). For the electron heavy hole transition one obtains for TE polarization:

$$M_{QW,\parallel}^2 = \frac{3}{2} \frac{1 + \cos^2 \vartheta}{2} \tag{5.60}$$

ϑ being 0 for emission perpendicular to the quantum well plane. However, as pointed out in Yan et al. (1990), for spontaneous *emission* the quantum well induced enhancement of *absorption* by 3/2 (due to averaging the electron k vector over a restricted set) has to be omitted from the matrix element. The momentum matrix elements in bulk materials M_b^2 is correctly (Yan et al., 1990)

$$\frac{M_b^2}{m_0} = \frac{1}{3}\frac{p_{cv}^2}{m_0} = \frac{1}{6}\left(\frac{m_0}{m_e^*} - 1\right) \frac{E_g(E_g + \Delta_0)}{E_g + 2\Delta_0/3} \tag{5.61}$$

MODELING OF QUANTUM DOTS

The value used in Sugawara et al. (1995) is a factor of two too small. Thus the numerical results for emission rates have to be corrected by a factor of 4/3.

F^2 is the overlap of the single particle states in the z direction; $F^2 \approx 1$ will be taken as equal to 1 in the following. $\phi_{nm}(r)$ denotes the radial envelope function for the relative motion of the electron and hole; for the quantum well this function is $\phi_{10}(r) = a_B^{-1} \exp(-r/a_B)$. The numerical results for $In_{0.53}Ga_{0.47}As/InP$ disks from Sugawara et al. (1995) are shown in Fig. 5.24 to illustrate the following points.

The equation for the ground state exciton lifetime can be written for the electric dipole approximation, i.e. when $\exp(i\mathbf{qR}) \approx 1$, as

$$\tau_D^{-1} = \frac{e^2 n_r \omega_{ex}^2}{\pi m_0 \varepsilon_0 c^3} \frac{2M_b^2/3}{m_0 \hbar \omega_{ex}} 2\pi\beta^2 |\phi_{10}(0)|^2 \tag{5.62}$$

For $\omega \to \infty$, i.e. $\beta \to 0$, the wavefunction of the relative carrier motion is not affected by the Coulomb attraction and is given by a Gaussian. In this limit of strong confinement the lifetime $\tau_{D,sc}$ is

$$\tau_{D,sc}^{-1} = \frac{e^2 n_r \omega_{ex}^2}{\pi m_0 \varepsilon_0 c^3} \frac{2M_b^2/3}{m_0 \hbar \omega_{ex}} \frac{4m_e m_h}{(m_e + m_h)^2} \tag{5.63}$$

The last factor equals the square of the overlap integral of the single particle wavefunctions and is equal to 1 if $m_e = m_h$. In this case the strong confinement limit yields $\tau \approx 2$ ns. Since M_b is very similar for most III–V materials this value is typical and reached for disk radii smaller than about 5 nm.

As mentioned previously, Eq. (5.62) leads to $\tau_D \to 0$ for $\omega \to 0$, i.e. $\beta \to \infty$. However, the dipole approximation already starts to break down for disk radii larger than about 15 nm and is entirely inapplicable for disk radii larger than about 100 nm. Instead, using the selection rule $\mathbf{K} = \mathbf{q}$, the free exciton lifetime τ_F in a quantum well (for $\beta \to \infty$) is obtained:

$$\tau_F^{-1} = \frac{e^2}{m_0 \varepsilon_0 c n_r} \frac{2M_b^2}{m_0 \hbar \omega_{ex}} |\phi_{10}(0)|^2 \tag{5.64}$$

Essentially the same formula was derived by Andreani et al. (1991). The effect of localization on the exciton lifetime in quantum wells was also treated by Citrin (1993) and Kavokin (1994). For our system the ($L_z = 10$ nm) quantum well free exciton lifetime is $\tau_D \approx 20$ ps. This value is actually reached for disk radii larger than about 150 nm. The lifetime in large disks approaches τ_F like

$$\tau_D^{-1} = \tau_F^{-1}\left[1 - \exp\left(-\frac{\beta^2 q_c^2}{2}\right)\right] \tag{5.65}$$

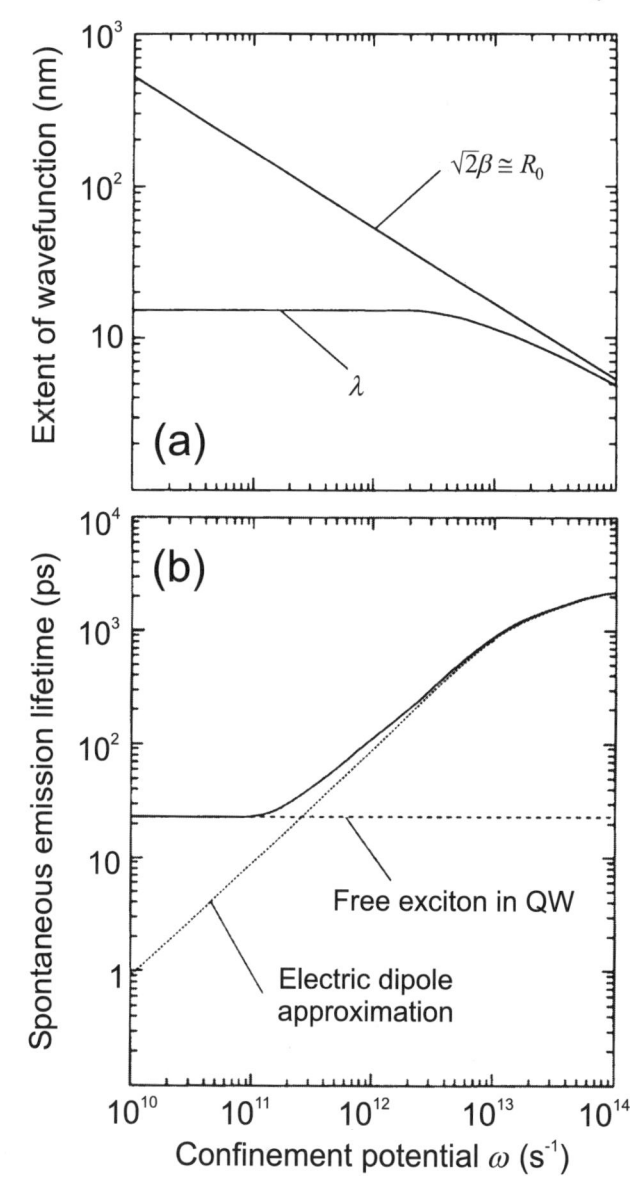

Fig. 5.24 (a) Extent of exciton wavefunction for center of mass motion $\sqrt{2}\beta = 2\sqrt{\hbar/(M\omega)}$ and relative motion $\lambda = \langle \phi_{10}|r|\phi_{10}\rangle$ as a function of lateral confinement frequency ω for $In_{0.53}Ga_{0.47}As/InP$ disks of height $L_z = 10$ nm. (b) Spontaneous emission lifetime of the exciton ground state as a function of ω. The horizontal dashed line is the lifetime of free excitons, the solid line is the lifetime in the electric dipole approximation. (After Sugawara *et al.*, 1995)

MODELING OF QUANTUM DOTS   137

So far, nothing has been said about the temperature dependence. Using the discrete levels E_{kl}^R, the temperature dependence of the lifetime of ground state excitons is written (assuming a Boltzmann distribution) as

$$\tau_D^{-1}(T) = \tau_D^{-1}(0)\left[1 + \sum_{kl} \exp\left(-\frac{E_{kl}^R - E_{00}^R}{k_B T}\right)\right]^{-1} \quad (5.66)$$

For the model system $In_{0.53}Ga_{0.47}As/InP$, the energy separation $\hbar\omega$ in the harmonic oscillator model is equal to $k_B T$ at room temperature for $\beta \approx 6$ nm. This means that quantum dots of this size or smaller (for such sizes they are in the strong confinement limit) and sufficiently deep potential barriers have a smaller radiative lifetime at room temperature than a quantum well of the same material.

The increase of excitonic lifetime in quantum wells with temperature is linear (Feldmann et al., 1987). In contrast, for quantum dots the lifetime is independent of temperature as long quantization effects are strong, i.e. the level separation is larger than $k_B T$. Additionally, the localization energy of the carriers must be large compared to $k_B T$ in order to avoid evaporation of carriers from the dots (see Section 5.3).

5.3.1.6 Energy levels

The ground and first excited state energy of an electron–hole pair in a spherical quantum dot with finite potential barriers has been calculated by Tran Thoai et al. (1990a) using a variational approach (Fig. 5.25). In the limit $R_0 \gg a_B$ the calculation

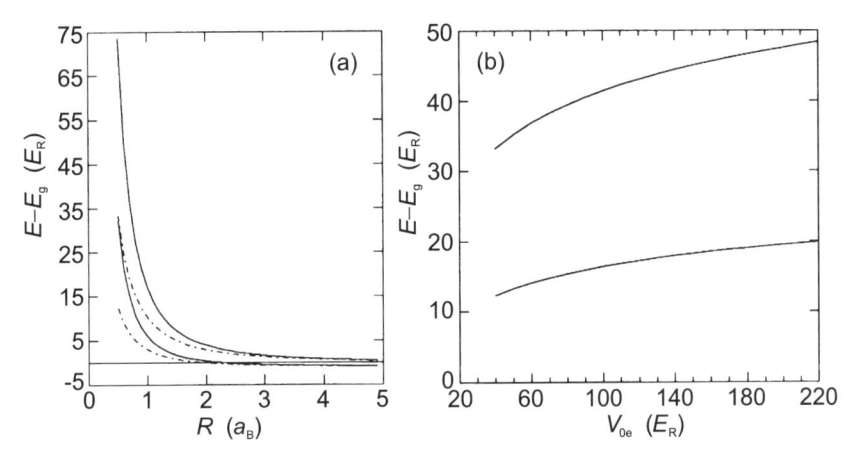

Fig. 5.25 (a) Energies of the two lowest dipole-allowed quantum confined states as a function of sphere radius. Solid lines: infinite confinement potential, dashed lines: $V_0 = 40 E_R$. Other material parameters (CdS) used: $m_e^* = 0.235$, $m_h^* = 1.35$, $E_g = 2.583$ eV, $E_R = 27$ meV, $a_B = 3.0$ nm. (b) Energies of the two lowest quantum confined states as a function of the confinement potential (in units of E_R) for a radius of $a_B/2$. (After Tran Thoai et al., 1990a)

yields the correct values for bulk material. For small radii a significant deviation from the result for infinite barriers occurs for $R_0 < 3a_B$.

The exciton binding energy for GaAs/Al$_x$Ga$_{1-x}$As cylindrical quantum dots has been calculated by Le Goff and Stébé (1992). They find (Fig. 5.26) a maximum energy for cylinder radii comparable to the bulk exciton radius. The radius at maximum energy is only weakly influenced by the height of the cylinder. Exact solutions for excitons in quantum dots with parabolic confinement are reported in Que (1992).

Using matrix diagonalization techniques, where the exciton wavefunction is built up by eigenfunctions of the kinetic energy, numerical results with defined precision can be obtained. The results of Hu et al. (1990) for infinite barriers are shown in Fig. 5.27 for two different ratios of the electron and hole masses. The infinite confinement potential lets the binding energy diverge for $R \rightarrow 0$. Such an approach has been also applied by Gotoh et al. (1996) to rectangular GaAs dots with infinite barriers.

For the InAs/GaAs pyramids observed in experiment the exciton binding energy has been calculated in first-order perturbation theory (Grundmann et al., 1995c) (Fig. 5.28). A value of 20 meV has been obtained for a pyramid base size of 12 nm. This value is large compared to the InAs and GaAs bulk exciton binding energies of about 1 and 4 meV, respectively. In this calculation image the charge effects, which increase the binding energy, have been neglected. With decreasing pyramid size the binding energy shows an increase.

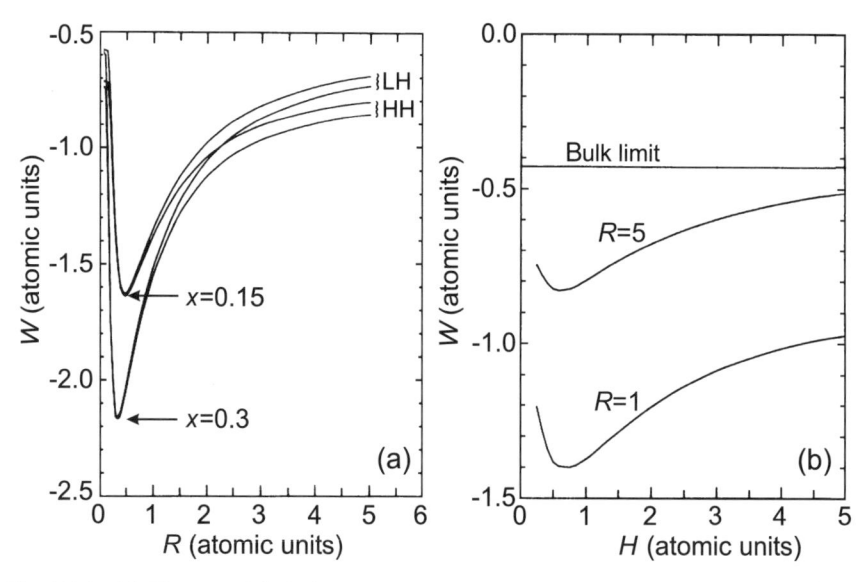

Fig. 5.26 (a) Heavy and light hole exciton binding energies (in units of $2E_R$) in GaAs/Al$_x$Ga$_{1-x}$As disks of different radius (in units of a_B) and fixed height $h = a_B$. (b) Heavy hole exciton binding energy for $x = 0.15$ and varying disk heights for two different radii. (After Goff and Stébé, 1992)

MODELING OF QUANTUM DOTS

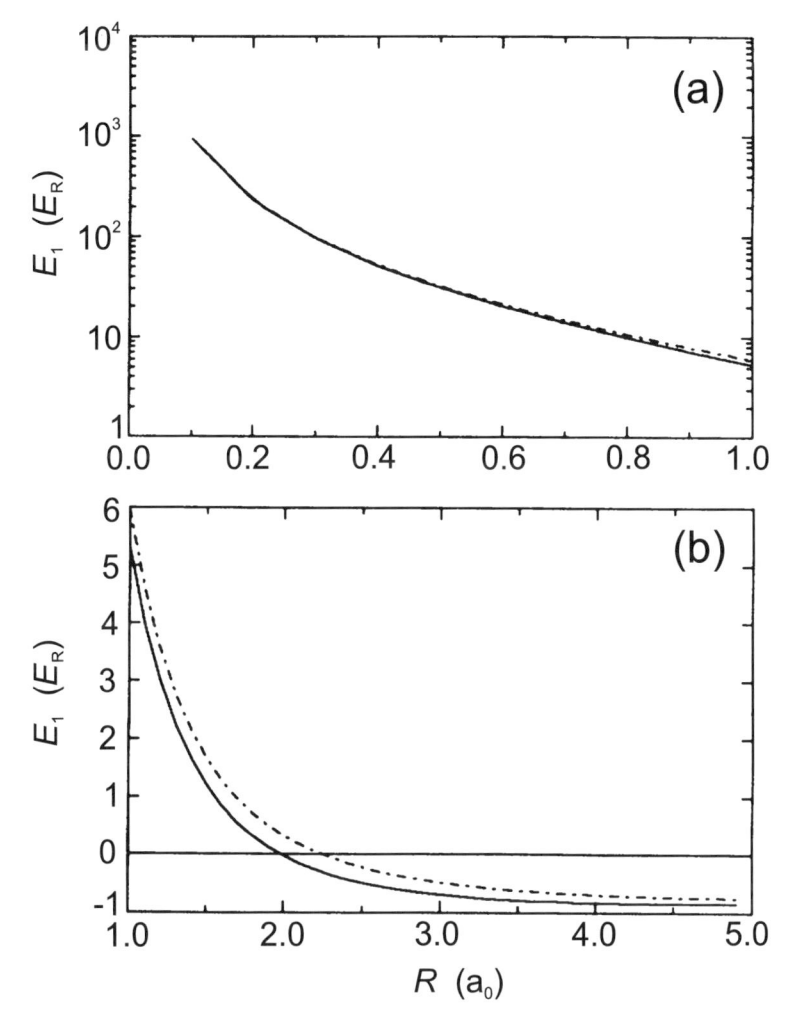

Fig. 5.27 Ground state energy for one electron–hole pair as a function of quantum sphere radius (in units of the Bohr radius) for $\varepsilon_2 = \varepsilon_1$; m_e/m_{hh} is 1 (dashed line) and 0.01 (solid line). (After Hu et al., 1990)

The exciton binding energy for more flat geometries, InAs/GaAs cones and InGaAs/GaAs hemispherical caps, has been calculated by Lelong and Bastard (1996a, 1996b) using a variational approach. For a 12° cone angle the binding energy increases approximately linearly from 25 to 32 meV for a cone base radius ranging from 15 to 7 nm (Fig. 5.29a). In Fig. 5.30 the dependence of the exciton binding energy on In content is shown for hemispherical caps of various radii.

The impact of Coulomb interaction on the spectra of quantum disks with varying disk radii, covering the range from weak to strong confinement, has been calculated by Adolph et al. (1993).

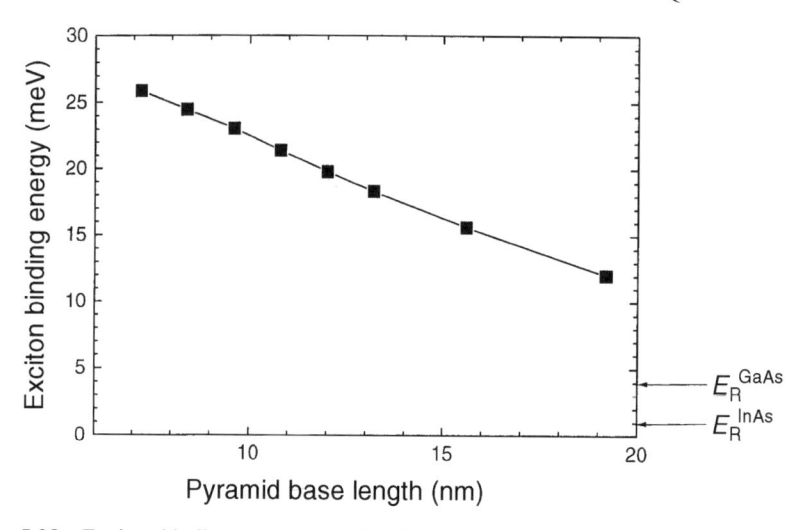

Fig. 5.28 Exciton binding energy as a function of InAs/GaAs pyramid base length. (After Grundmann et al., 1995c)

5.3.1.7 Charged excitons

Lelong and Bastard (1996b) calculated the binding energy of charged excitons, defined for X^+ as the energy difference between the uncorrelated complex formed by two holes and one electron and an exciton plus a single hole in the ground state. For flat InAs/GaAs cones the X^+ binding energy is 2 meV for a radius of 12 nm and drops to zero for a radius of 7 nm (Fig. 5.29). The X^- complex for the same radius range is found to be unbound. For hemispherical InGaAs/GaAs caps the situation is opposite, with the X^- being the bound complex.

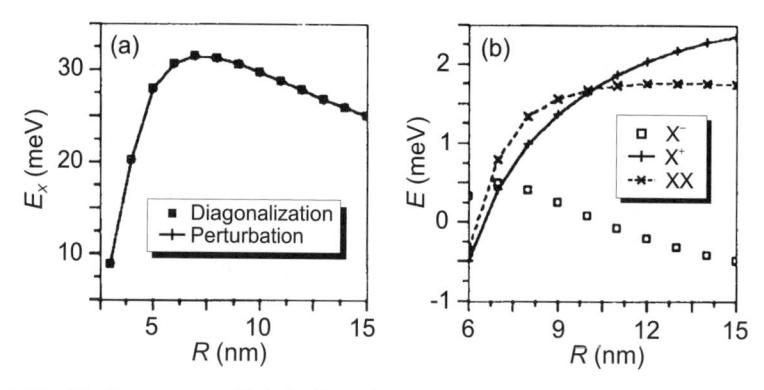

Fig. 5.29 Binding energy of (a) single excitons and (b) exciton complexes versus the basis radius of an InAs/GaAs cone with base angle 12°. (After Lelong and Bastard, 1996b)

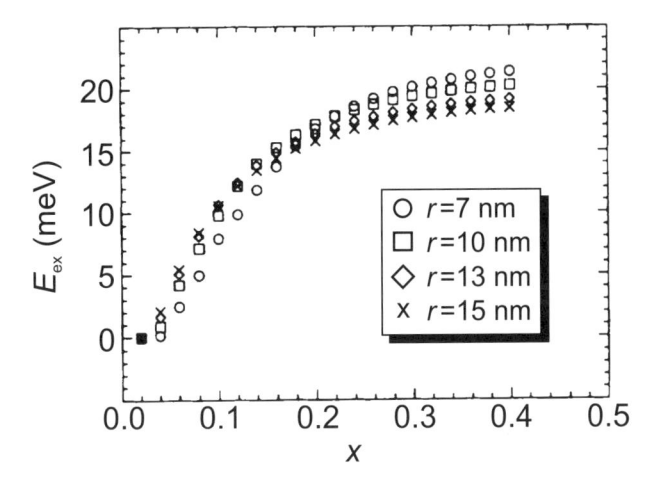

Fig. 5.30 Exciton binding energy in $In_xGa_{1-x}As/GaAs$ hemispherical caps for several values of the cap base radius r. The cap height is fixed to $h = 1.5$ nm, the wetting layer thickness is $d = 0.566$ nm. (After Lelong and Bastard, 1996b)

5.3.1.8 Biexcitons

When the (single-particle) ground states for electrons and holes in a quantum dot are filled by two carriers each, like in a photoluminescence experiment with suitable excitation density, a biexciton is formed instead of two excitons. The biexciton energy E_{XX} differs from the energy of two excitons $2E_X$ by the additional molecular binding energy or biexciton binding energy defined as

$$E_{mol} = 2E_X - E_{XX} \qquad (5.67)$$

In a bulk crystal, the two excitons forming the biexciton can dissociate, i.e. the two excitons can spatially separate so far from each other that the interaction practically vanishes. This is not possible in a quantum dot, where the carriers remain localized in the dot.† The molecular binding energy is not a purely theoretical quantity but of practical interest because it can be experimentally observed, e.g. in recombination spectra or two-photon absorption experiments.

Biexcitons are of special importance for the gain mechanism in quantum dot lasers. The ground state of the dots has to be filled by one electron–hole pair to reach inversion. Stimulated emission occurs between biexciton and exciton energy levels, as will be discussed in Section 5.8.3.3.

E_{mol} vanishes in first-order perturbation theory and has a positive value $E_{mol}^{(2)}$ in second-order perturbation theory (Bányai, 1989), which represents the limit for E_{mol}

---

† A single exciton is not an eigenstate of the four-particle Hamiltonian.

for $R/a_B \to 0$. Upper and lower bounds for $E_{mol}^{(2)}$ are dependent on the ratio of dielectric constants $\varepsilon = \varepsilon_1/\varepsilon_2$ (Bányai, 1989):

$$C_l(\varepsilon) < \frac{E_{mol}}{E_R}\left(2 + \frac{m_e + m_h}{2\mu}\right)^{-1} < C_u(\varepsilon) \tag{5.68}$$

with the following values:

ε	1	4	10
$C_u(\varepsilon)$	0.55	0.79	0.71
$C_l(\varepsilon)$	0.052	0.076	0.104

For III–V compounds $\varepsilon = 1$ is a realistic approximation.

The matrix diagonalization technique delivers values for E_{mol} that are not only in the asymptotic limit (Fig. 5.31) (Hu et al., 1990). Independently of the ratio of electron and hole masses the molecular binding energy is positive and increases with decreasing dot radius. For $R/a_B \leqslant 1$ it takes values around one to two times the bulk exciton binding energy.

Heller et al. (1997) calculated the binding energy of biexcitons bound to island-like interface defects using a variational approach. For a 1 ML deep island with a 20 nm radius in an AlGaAs/GaAs quantum well the typical binding energy is predicted to lie between 1.3 and 1.6 meV for a QW thickness between 2 and 5 nm.

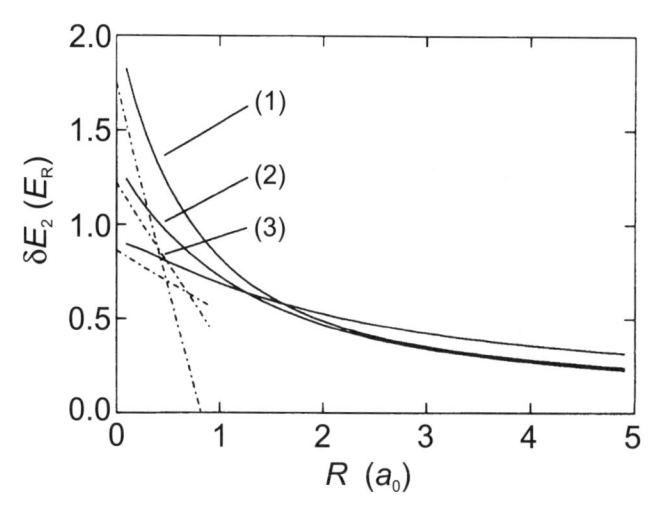

Fig. 5.31 Biexciton binding energy as a function of quantum sphere radius for three different electron to hole mass ratios and $\varepsilon_2 = \varepsilon_1$. Solid lines are the result of numerical matrix diagonalization for $m_e/m_{hh} = 0.1$ (curve 1), 0.2 (curve 2), and 1 (curve 3). The dashed curves are computed using third-order perturbation theory, respectively. (After Hu et al., 1990)

The variational calculations presented in Lelong and Bastard (1996b) for flat InAs/GaAs cones predict biexciton binding energies in the range of 1.5 meV for typical radii larger than 8 nm (Fig. 5.29). This value is comparable to the bulk exciton binding energy of InAs.

Many-body states of up to four excitons have been studied for small quantum dots by Barenco and Dupertuis (1995). The solution is largely analytical. Also non-neutral states, i.e. charged excitons, are discussed in this work.

In the strong confinement limit the biexciton can be viewed as two excitons that can recombine independently. In this case the lifetime τ_{XX} of the biexciton state is given in terms of the exciton lifetime τ_X as

$$\tau_{XX} = \frac{\tau_X}{2} \tag{5.69}$$

If Coulomb correlation plays a significant role, the oscillator strength is increased and the lifetime is shortened.

5.3.2 TYPE-II EXCITONS

In type-II quantum dots one charge carrier is localized inside the dot and the other one in the barrier. The barrier heights are given by V_e and V_h (Fig. 5.32). Due to Coulomb attraction the carrier in the barrier becomes bound to the quantum dot (similar to an 'extended' impurity); the carrier inside the quantum dot modifies its orbital to increase the probability close to the dot boundary. Similar to the situation in type-I dots, a strong and a weak confinement limit exists.

For dot radii much larger than the bulk exciton Bohr radius, the electron and hole are strongly correlated and localized near the dot boundary. For dot radii much smaller than the bulk exciton Bohr radius the 'inside' particle is in its ground state

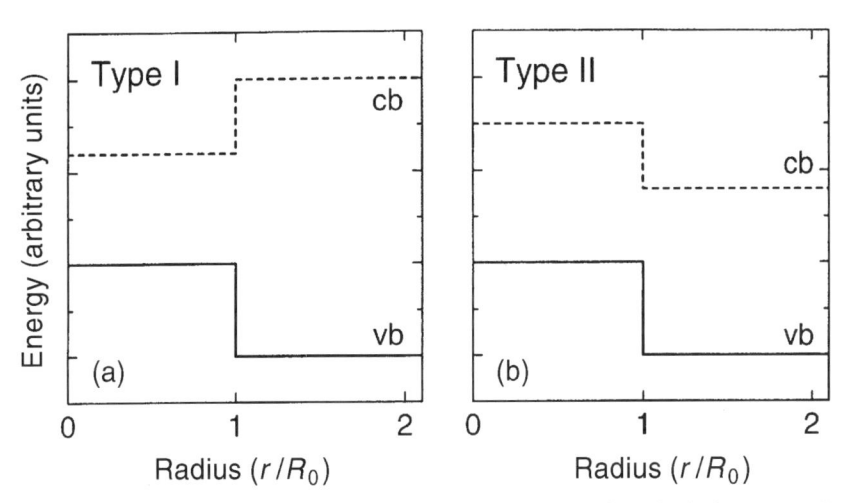

Fig. 5.32 Schematic (a) type-I and (b) type-II band alignment for spherical quantum dot

and hardly affected by the Coulomb interaction. The 'outside' particle sees a potential similar to a bare Coulomb potential. In the limit of the vanishing dot radius the barrier Rydberg energy is recovered.

Calculations for excitons in type-II quantum dots have been presented in Rorison (1993) and Laheld et al. (1995), where spherical quantum dots with finite potential barriers are considered. The previous assumption of infinite barriers (Laheld et al., 1993) leads to a complete spatial separation of electron and hole, i.e. the wavefunction overlap vanishes.

The exciton binding energy as a function of barrier Rydberg energy $\hat{R}$ in the limit of a small dot radius is given by (assuming, without limiting the universality, that the electron is bound in the dot and the hole is localized in the barrier)

$$\hat{E}_X = 1 - 4\hat{R} - 8\hat{R}^2 \ln \hat{R} \qquad (5.70)$$

for $V_e, V_h = \infty$ (Laheld et al., 1993) and

$$\hat{E}_X = 1 - \left(\frac{4}{9} - \frac{2}{3\pi^2}\right)\hat{R}^2 \qquad (5.71)$$

for $V_e = \infty$, $V_h = 0$ (Laheld et al., 1995). The energy is in units of the hole Rydberg energy

$$E_R^h = \frac{m_h}{2\hbar^2}\left(\frac{e^2}{4\pi\varepsilon_r\varepsilon_0}\right)^2$$

and the radius in units of the hole Bohr radius is

$$a_h = \frac{4\pi\varepsilon_r\varepsilon_0}{e^2}\frac{\hbar^2}{m_h}$$

Results for finite band offsets are shown in Fig. 5.33a (Laheld et al., 1995), where identical offsets and effective masses for electrons and holes have been used. The binding energy for infinite offsets is included as a dashed line. It turns out that the binding energy is insensitive to the mass ratio.

The wavefunction overlap is naturally smaller than for a type-I system. In Fig. 5.33b the overlap factor is plotted as a function of the band offset (for $\hat{R} = 3$) and as a function of the radius (for $\hat{V}_e$, $\hat{V}_h = 1$, $\hat{V}_{e/h} = V_{e/h}/\hat{R}$). For a typical band line-up in III–V semiconductor quantum dots, when $\hat{R} \approx 1$ and $\hat{V} \gg 1$, small oscillator strength and long recombination lifetimes are expected.

However, experimental quantum dot geometries may be very different from the idealized case of a sphere in an infinite medium. In a layer of quantum dots a high density of quantum dots can exist in the plane. At moderate carrier density a sheet density of carriers bound in the dots attracts the outside carriers, making the situation much more two-dimensional in character. Additional potential barriers may exist in the growth direction, limiting the extension of the outside particle in that direction (Zimmermann and Bimberg, 1993). Both effects tend to increase the oscillator strength as experimentally confirmed (Hatami et al., 1998).

MODELING OF QUANTUM DOTS

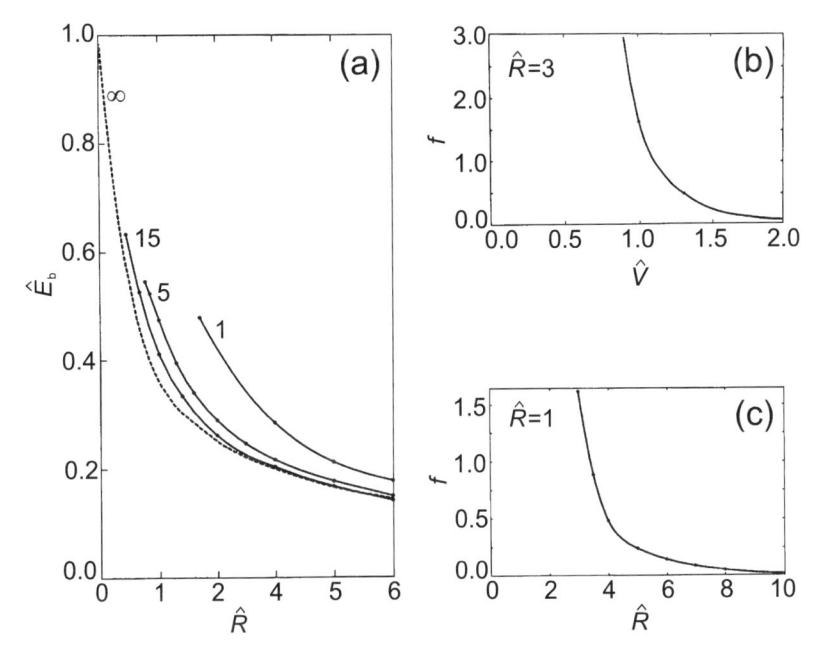

Fig. 5.33 (a) Type-II exciton binding energy (in units of E_R) for equal effective masses and band offsets as a function of quantum sphere radius R (in units of the a_B) for different values of the offset. The dashed line is for $V_e = V_h = \infty$. Wavefunction overlap as a function of confinement potential (b) for $R = 3$ and dot radius (c) for $m_e = m_h$ and $V_e = V_h = 1$. (After Laheld et al., 1995)

In a broken gap configuration, as, for example, is present for InAs/GaSb, self-doping of quantum dots is expected (Sercel and Vahala, 1990b).

5.3.3 ELECTRONICALLY COUPLED QUANTUM DOTS

So far only isolated quantum dots have been discussed. In Section 5.4.2 the properties of quantum dot ensembles will be treated for the case where each quantum dot keeps its properties because they are isolated. Here, the case is considered where quantum dots are so close in real space that electronic coupling arises. The interaction between dots can originate from wavefunction overlap or Coulomb interaction, the latter remaining even for infinite barriers.

The fundamental properties of coupled dots, their spectrum and oscillator strength, are different from those of isolated dots. The fundamental effects due to electronic coupling of two quantum dots have been calculated by Bryant (1992, 1993a, 1993b) for flat islands (laterally confined quantum wells). Without Coulomb interaction, the coupling of two identical quantum dots leads to a splitting of the single particle ground state into a symmetric and an antisymmetric state. The magnitude of the splitting depends on the height and width of the barrier, very similar to the case of double quantum wells. Extensions of this approach can be made for chains (one-dimensional dot superlattice) and arrays (two-dimensional dot superlattice) of dots leading to formation of subbands.

If Coulomb interaction is important, the exciton ground state is the coherent state $|c\rangle_{ex}$ of the excitons in both dots (Bryant, 1993a):

$$|c\rangle_{ex} = \frac{|1\rangle_{ex} + |2\rangle_{ex}}{\sqrt{2}}, \quad |inc\rangle_{ex} = \frac{|1\rangle_{ex} - |2\rangle_{ex}}{\sqrt{2}} \qquad (5.72)$$

The incoherent state $|inc\rangle_{ex}$ has higher energy. The coherent exciton has the sum of the oscillator strengths of the excitons in both dots, while the optical dipole moment of the incoherent exciton vanishes (see also Fig. 5.47).

A tremendous change in excitonic properties can be achieved by applying an electric field along the dot–dot axis, e.g. in a p-i-n structure, which will eventually break the coherence of the excitons and localize electrons and holes in different dots (indirect exciton). These effects are presented in Section 5.6.1.2.

Three-dimensional arrays of disk-like InAs quantum dots in GaAs have been calculated by Li et al. (1996). However, only constant biaxial strain in the InAs has been considered, leaving the GaAs unstrained. With decreasing disk radius and height the wavefunction delocalizes as expected. Heavy and light hole mixing exists at the Γ point.

5.3.4 COULOMB BLOCKADE

The classical electrostatic energy of a quantum dot with capacity C which is capacitively coupled to a gate at a bias voltage V_g is given by

$$E = \frac{Q^2}{2C} - Q\alpha V_g \qquad (5.73)$$

where α is a dimensionless factor relating the gate voltage to the potential of the island and Q is the charge. Mathematically, minimum energy is reached for a charge $Q_{min} = \alpha C V_g$. However, the charge has to be an integer multiple of e, i.e. $Q = Ne$. If V_g has a value, such that $Q_{min}/e = N_{min}$ is an integer, the charge cannot fluctuate as long as the temperature is low enough, i.e.

$$kT \ll \frac{e^2}{2C} \qquad (5.74)$$

Tunneling in or out of the dot is suppressed by the Coulomb barrier $e^2/(2C)$, and the conductance is very low. Analogously, the differential capacitance is small. This effect is called the 'Coulomb blockade' and has been discussed extensively in the literature (e.g. see Grabert and Horner, 1991; Fukuyama and Ando, 1992; Grabert and Devoret, 1992; Kastner, 1993). Peaks in the tunneling current or the capacitance occur when the gate voltage is such that the energies for N and $N+1$ electrons are degenerate, i.e. $N_{\min} = N + \frac{1}{2}$. The expected level spacing is

$$e\alpha \Delta V_g = \frac{e^2}{C} + \Delta \varepsilon_N \tag{5.75}$$

where $\Delta \varepsilon_N$ denotes the change in lateral (kinetic) quantization energy for the added electron; e^2/C will be called the charging energy in the following.

Most research so far has been performed on lithographically defined systems where the lateral quantization energies are small and smaller than the Coulomb charging energy. In this case periodic oscillations are observed, especially for large N. A deviation from periodic oscillations for small N and a characteristic shell structure (at $N = 2,6,12$) consistent with a harmonic oscillator model ($\hbar \omega_0 \approx 3$ meV) has been reported by Tarucha et al. (1996) for ≈ 500 nm large mesas.

In small self-organized dots single-particle level separations can be larger than or similar to the Coulomb charging energy. Classically the capacity for a metal sphere of radius R_0 is given as

$$C_0 = 4\pi \varepsilon_r \varepsilon_0 R_0 \tag{5.76}$$

e.g. $C_0 \approx 6$ aF for $R_0 = 4$ nm in GaAs, resulting in a charging energy of 26 meV.

Quantum-mechanically the charging energy is given (compare with Eq. 5.49) in first-order perturbation theory by

$$E_{21} = \langle 00 | W_{ee} | 00 \rangle = \int\int \Psi_0^2(\mathbf{r}_e^1) W_{eh}(\mathbf{r}_e^1, \mathbf{r}_e^2) \Psi_0^2(\mathbf{r}_e^2) \, d^3\mathbf{r}_e^1 \, d^3\mathbf{r}_e^2 \tag{5.77}$$

where W_{ee} denotes the Coulomb interaction of the two electrons and Ψ_0 the ground state (single-particle) electron wavefunction. The matrix element gives an upper bound for the charging energy since the wavefunctions will rearrange to lower their overlap and the repulsive Coulomb interaction. For lens-shaped InAs/GaAs quantum dots with a radius of 25 nm a charging energy of about 30 meV has been predicted (Wojs and Hawrylak, 1996). The Coulomb charging energy for InAs/GaAs pyramids is shown in Fig. 5.34. It has been calculated, taking into account image charge effects, in the Hartree approximation (Stier et al., 1998). In the present dot system the Coulomb charging energy reaches values equal to $k_B T$ at room temperature.

The impact of electronic coupling of dots on the Coulomb blockade of an array of quantum dots has been analyzed by Stafford and Das Sarma (1994). Three zero-temperature phases were identified depending on the strength of inter-dot tunneling: Coulomb blockade of individual dots (regime of small coupling), collective Coulomb blockade (intermediate coupling where at least separate minibands exist), and the breakdown of Coulomb blockade (strong coupling regime).

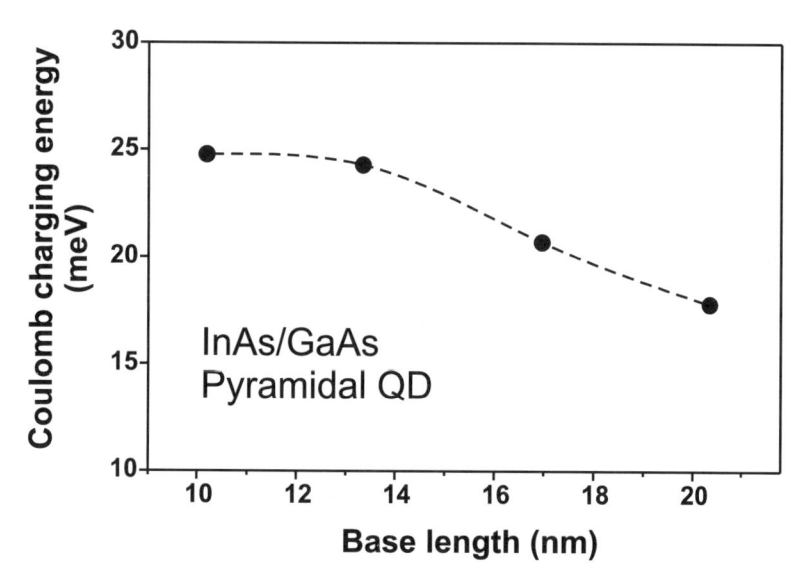

Fig. 5.34 Coulomb charging energy for pyramidal InAs/GaAs QD's of different base length calculated in the Hartree approximation on the basis of eight-band $k \cdot p$ theory.

5.4 OPTICAL TRANSITIONS

The possible (inter-band) optical transitions in a *single* quantum dot consist of a series of δ function lines, whose energy positions depend on the particular three-dimensionally confined levels. In a real quantum dot *ensemble* each individual dot has a slightly different size, shape, strain state, etc.; therefore the energy spectrum varies from dot to dot. This leads to the observation of inhomogeneous broadening of optical spectra. The lifetime of a particular electron and hole (excitonic) state is determined by the oscillator strength of the transition and has been discussed in Section 5.3.1.5. In Section 5.4.3 we discuss (infrared) inter-sublevel transitions in quantum dots which become, in contrast to quantum wells, allowed for light propagation perpendicular to the quantum dot plane.

5.4.1 SINGLE QUANTUM DOTS

The three-dimensional confinement of carriers in a quantum dot leads to a discrete spectrum for electrons and holes. Experimentally small linewidths (FWHM≪1 meV) have been reported for single quantum dots, as summarized in Section 7.2. The small homogeneous linewidth is generally determined by dephasing mechanisms, e.g. the optical recombination itself (lifetime), scattering with phonons, and impurities.

MODELING OF QUANTUM DOTS

The radiative lifetime sets a lower bound for the linewidth Γ:

$$\Gamma \approx \frac{\hbar}{\tau}, \qquad \Gamma(\text{meV}) \approx 0.65822 \frac{1}{\tau(\text{ps})} \tag{5.78}$$

For a typical radiative lifetime $\tau = 300\,\text{ps}$, the broadening is very small, $\Gamma \approx 2.2\,\mu\text{eV}$.

Another consequence of the three-dimensional confinement is the disappearance of the temperature dependence of the linewidth observed for three-, two-, and one-dimensional systems. No continuum of states can be populated, as in the case of a bulk crystal or a quantum well, where the width of the band-to-band recombination spectrum is given by $\text{FWHM} = 1.7954\,kT$ and $\text{FWHM} = \ln(2)\,kT$, respectively. The ionization of excitons by absorption of acoustic phonons to create free electrons and holes is absent; there is no final state for this process (Schmitt-Rink et al., 1987). However, the broadening can be measurable at a sufficiently high temperature when the total lifetime of excitons in the ground state becomes shortened by phonon-activated transitions between different excited and barrier states.

When the ground state of a quantum dot is probed resonantly, absorbed and emitted photon have the same energy, thus making it impossible to separate luminescence and stray light in cw experiments. Experimentally separation is accomplished with time-resolved techniques (Lowisch et al., 1996).

The simplest optical transitions in a (spherical) quantum dot are shown in Fig. 5.35 and described in the following paragraph. An empty quantum dot can absorb

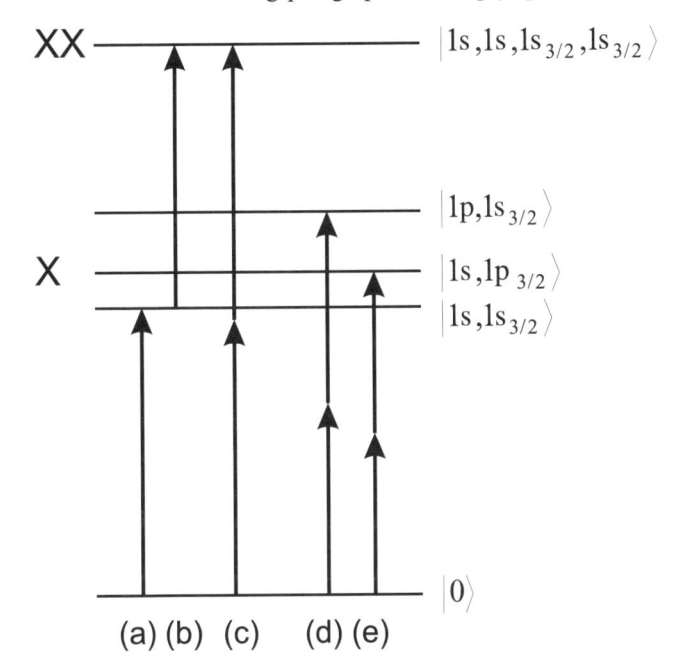

Fig. 5.35 Schematic term scheme for excitons (X) and biexcitons (XX) in spherical quantum dots. $|0\rangle$ denotes the empty quantum dot. The transitions are discussed in the text. (a) and (b) One-photon transitons; (b), (c) and (d) two-photon processes

one photon of energy E_X to create an exciton (a). After being populated with this exciton the quantum dot cannot absorb another photon of this energy E_X but a photon of energy $E_{XX} - E_X$ (b) and is then populated with a biexciton. In reverse, the biexciton state can decay into a photon of energy and an exciton, which subsequently decays into a photon of energy E_X. The electrons and holes making up the biexciton have opposite spins. With a nonlinear two-photon absorption process a biexciton can be created by absorbing two photons of energy $(E_X + E_{XX})/2$ (c). Since two-photon transitions involve changes of the total angular momentum of 0 and 2, the creation of a $|1p, 1s_{3/2}\rangle$ (d) or $|1s, 1p_{3/2}\rangle$ exciton (e) is possible.

5.4.2 QUANTUM DOT ENSEMBLES

In a quantum dot ensemble, as probed by nonspatially resolved experimental techniques, each individual dot has slightly different properties induced by fluctuations in size, shape, strain, etc., leading to a variation of energy levels and subsequently to an inhomogeneous broadening of the ensemble properties, e.g. the spectrum. The impact of size disorder on the spectrum is discussed in Section 5.4.2.1. Typical inhomogeneously broadened linewidths of quantum dot ensemble spectra are 10–100 meV.

The main effect of inhomogeneous broadening results from the distribution of dot sizes (volumes). It is reasonable to assume a Gaussian distribution function P with an average volume V_0 and a standard deviation σ_V. The spectrum of an ensemble of quantum cubes whose side length has a Gaussian distribution has been calculated by Wu et al. (1987). In their spirit we derive here the spectrum for an ensemble of spherical quantum dots with a Gaussian distribution $P(R)$ of radius with an average radius R_0 and a standard deviation $\sigma_R = \langle (R - R_0)^2 \rangle^{-1/2}$:

$$P(R) = \frac{1}{\sqrt{2\pi}\sigma_R} \exp\left[-\frac{(R-R_0)^2}{2\sigma_R^2}\right] \tag{5.79}$$

When infinite barriers are assumed for simplicity, the optical absorption coefficient of a single dot with radius R is given by (neglecting the small homogeneous broadening)

$$\alpha(\hbar\omega, R) = \frac{A}{V} \sum_{nl} (2l+1) \delta\left(\hbar\omega - E_g - \frac{\pi^2\hbar^2}{2\mu} \frac{q_{nl}^2}{R^2}\right) \tag{5.80}$$

where $k_{nl} = \pi q_{nl}$ is the nth zero of the Bessel function j_l (see Eq. 5.44). A collects numerical constants:

$$A = \frac{\pi e^2}{m_0^2 \varepsilon_0 c n_r \omega} \frac{p_{cv}^2}{3}$$

assuming the envelope wavefunction overlaps $F^2 = 1$. Since $E_0 \ll E_g$, A can be taken to be the same for all transitions.

MODELING OF QUANTUM DOTS   151

The ensemble total absorption is given by the superposition, i.e. convolution, of $\alpha(\hbar\omega, R)$ and $P(R)$:

$$\alpha(\hbar\omega) = \frac{A}{\sqrt{2\pi}\sigma_R} \frac{3}{4\pi} \sum_{nl} (2l+1) \int \frac{1}{R^3} \exp\left[-\frac{(R-R_0)^2}{2\sigma_R^2}\right] \delta\left(\hbar\omega - E_g - \frac{\pi^2\hbar^2}{2\mu}\frac{q_{nl}^2}{R^2}\right) dR$$

$$= \frac{B}{\sqrt{2\pi}} \frac{3}{4\pi R_0} \sum_{nl} \frac{2l+1}{\xi q_{nl}^2} \exp\left[-\frac{(q_{nl}/q_0 - 1)^2}{2\xi^2}\right]$$

(5.81)

where the reduced photon energy

$$q_0^2 = \frac{\hbar\omega - E_g}{\pi^2\hbar^2/(2\mu R_0^2)} = \frac{\hbar\omega - E_g}{E_0}$$

and the relative standard deviation $\xi = \sigma_R/R_0$ are introduced. B collects pre-factors which are independent of $\hbar\omega$, n and l with $B = A\mu/(\pi^2\hbar^2)$.

The ensemble spectra for different values of ξ are depicted in Fig. 5.36. They consist (for small ξ) of a series of Gaussian absorption peaks at the peak positions $q_{nl}/q_0 = 1$, which can be written as

$$\hbar\omega_{nl} = E_g + \frac{\pi^2\hbar^2}{2\mu}\frac{q_{nl}^2}{R_0^2}$$

(5.82)

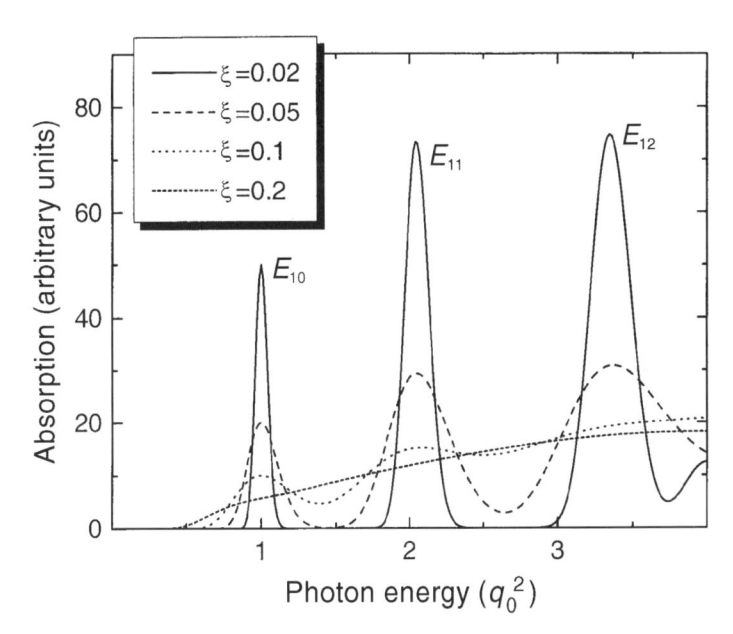

Fig. 5.36 Absorption of quantum sphere ensembles with Gaussian distribution of dot radius with different relative standard deviations ξ

where E_g is the bulk gap energy. Thus the peak positions of the ensemble spectrum are determined solely by the average dot size R_0. When the broadening parameter ξ becomes larger than 0.05, the peaks exhibit a red shift relative to the positions given in Eq. (5.82) and become markedly asymmetric because of the nonlinear dependence of the confinement energy on the dot size. For small values of ξ the absorption peaks are Gaussian with a linewidth FWHM $= 2\sqrt{2 \ln 2}\, \sigma_{nl}$, where σ_{nl} of the (n, l) absorption peaks (generally defined as the one-sided $\exp(-\frac{1}{2})$ decay from the maximum) is given by

$$\sigma_{nl} \approx 2\xi \frac{\pi^2 \hbar^2}{2\mu} \frac{q_{nl}^2}{R_0^2} = 2\xi(\hbar\omega_{nl} - E_g) \qquad (5.83)$$

and increases for higher transitions. For $\xi \to 0$ the linewidth approaches zero, since in this model the homogeneous broadening is neglected. Equation (5.83) also holds true for volume fluctuations of cubic quantum dots. If fluctuations of only one or two sides of the cube are considered, the pre-factor 2 has to be replaced by 2/3 or 4/3, respectively.

In the previous discussion the inhomogeneous broadening of the ground and excited states was assumed to be due to fluctuations of the radius, i.e. volume, of the quantum dot. Another possibility for broadening of the ground state is a variation in bandgap E_g with constant radius, e.g. due to fluctuations of the composition. If the energetic broadening of the ground state is described by a standard deviation σ_{E_g}, the excited states have the same broadening, $\sigma_{nl} = \sigma_{E_g}$, for all n, l. This consideration is easily illustrated for the harmonic oscillator. Including the potential offset E_g, the energy levels E_n of a three-dimensional harmonic potential are given by (compare with Eq. 5.37)

$$E_n = E_g + \left(n + \frac{3}{2}\right)\hbar\omega_0, \qquad n = 0, 1, 2, \ldots \qquad (5.84)$$

A fluctuation of ω_0 (the 'width' of the potential) leads to a broadening $\sigma_{\omega,n} = (n+1)\sigma_{\omega,0}$ proportional to $(n+1)$. In contrast, a fluctuation σ_{E_g} of E_g rigidly shifts up and down the complete level system independently of n. When both effects are present and independent, the inhomogeneous broadening σ_{E_n} of the energy level E_n is given by the convoluted widths:

$$\sigma_{E_n} = \sqrt{\sigma_{E_g}^2 + (n+1)^2 \sigma_{\omega,0}^2} \qquad (5.85)$$

Although excited states suffer larger inhomogeneous broadening, the absorption and subsequently the gain on excited transitions can be higher than for the ground state due to larger degeneracies.

How does the absorption spectrum $\alpha_{E_0}(\hbar\omega)$ of dots with a fixed ground state energy E_0 look above resonance? This question arises, for example, in photoluminescence excitation spectra where the detection wavelength is fixed to a certain energy and the energy of the exciting light is varied. If the quantum dot ensemble has only *one* disorder parameter, like the quantum dot volume, $\alpha_{E_0}(\hbar\omega)$ is a series of sharp lines at the energy positions of the excited levels for the given ground state E_0. If there are at least *two* disorder parameters, e.g. island radius and height, quantum dots with different structural parameters can have the same ground state energy but different excited energy levels. This leads to a 'broadening' of the peaks in $\alpha_{E_0}(\hbar\omega)$. In the following the impact of the disorder on $\alpha_{E_0}(\hbar\omega)$ is modeled (for the case of insufficient spectral resolution to resolve single excited states or in the limit of an infinite number of quantum dots) for a rectangular quantum box with infinite barriers, for which the base sides $a_x = a_y$ and a_z vary independently. Both have Gaussian distributions $P(a_x)$, $P(a_z)$ with average values a_{x0}, a_{z0}, standard deviations σ_x, σ_z, and relative standard deviations ξ_x, ξ_z, respectively.

The energy levels $E_{n_x n_y n_z}$ are given by

$$E_{n_x n_y n_z} = \frac{\pi^2 \hbar^2}{2\mu} \left(\frac{n_x^2 + n_y^2}{a_x^2} + \frac{n_z^2}{a_z^2} \right), \qquad n_x, n_y, n_z = 1, 2, \ldots \quad (5.86)$$

If dots with a ground state energy E_0 are considered, the condition

$$\frac{1}{a_z^2} = \frac{2\mu}{\pi^2 \hbar^2} E_0 - \frac{2}{a_x^2} \quad (5.87)$$

is imposed on the base sides, which are no longer independent of each other. Thus the absorption for a fixed ground state energy can be written as

$$\alpha_{E_0}(\hbar\omega) = A \sum_{n_x n_y n_z} \int \frac{1}{a_x^2 a_z} P(a_x) P(a_z) \delta\left[\hbar\omega - E_g - \frac{\pi^2 \hbar^2}{2\mu} \left(\frac{n_x^2 + n_y^2}{a_x^2} + \frac{n_z^2}{a_z^2} \right) \right] da_x \quad (5.88)$$

For E in the vicinity of

$$E_0 = \frac{\pi^2 \hbar^2}{2\mu} \left(\frac{2}{a_{x_0}^2} + \frac{1}{a_{z_0}^2} \right)$$

the expression reads (using $m^2 = n_x^2 + n_y^2 - 2n_z^2$, $\gamma = a_{x_0}/a_{z_0}$, and B as defined earlier)

$$\alpha_{E_0}(\hbar\omega) = \begin{cases} \dfrac{B}{2\pi} E_0 \sum_{n_x n_y n_z} \dfrac{1}{\xi_x \xi_z |m^2|} \dfrac{1}{1 - 2(x^2 - n_z^2)/m^2} \\ \exp\left\{ -\dfrac{1}{2\xi_x^2} \left(\sqrt{\dfrac{m^2}{(x^2 - n_z^2)(2 + \gamma^2)}} - 1 \right) \right\} \\ \exp\left\{ -\dfrac{1}{2\xi_z^2} \left(\sqrt{\dfrac{m^2 \gamma^2}{[m^2 - 2(x^2 - n_z^2)](2 + \gamma^2)}} - 1 \right)^2 \right\} \end{cases} \quad (5.89)$$

Absorption spectra for a fixed ground state energy are shown in Fig. 5.37 for two different quantum dot shapes as a function of reduced photon energy $(\hbar\omega - E_g)/E_0$. The inhomogeneous broadening parameters have been chosen as $\xi = \xi_x = \xi_z = 0.1$ and $\gamma = 1$ represents a cube. In the case of $\gamma = 2$, the island height is half the lateral base side. In the first case degeneracies let the spectrum appear sharper. In the latter case predominantly the lower lying in-plane levels contribute to the resonances.

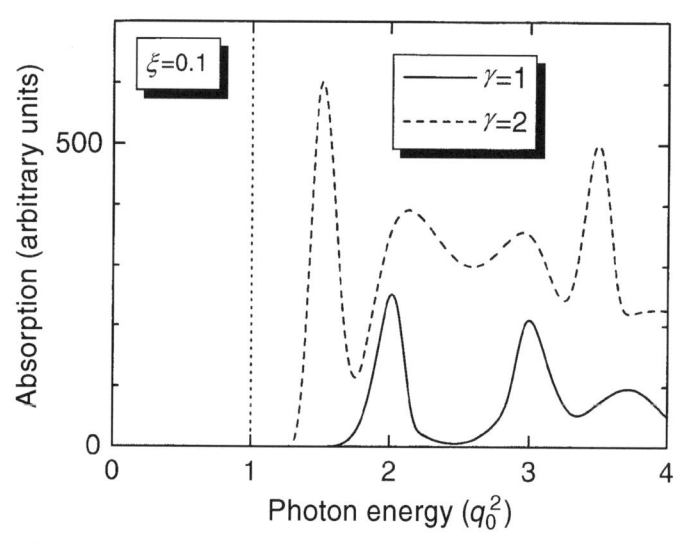

Fig. 5.37 Absorption spectra α_E of ensemble of rectangular quantum dots for fixed ground state energy. The in-plane and vertical extensions, $a_x = a_y$ and a_z, respectively, vary independently with relative standard deviations $\xi_x = \xi_z = \xi = 0.1$. $\gamma = 1$ means cubic geometry, for $\gamma = 2$ the island height is half the lateral base side

A random potential, as, for example, caused by alloy fluctuations in semiconductor alloys, leads to another broadening mechanism. The 'local' composition varies from dot to dot and also inside one dot. The impact of the random potential $V(\mathbf{r})$ with the correlator,

$$\langle V(\mathbf{r})V(\mathbf{r}')\rangle = \gamma\delta(\mathbf{r} - \mathbf{r}') \tag{5.90}$$

is different for nondegenerate and degenerate levels. Without degeneracy the distribution of energy levels E_m is a Gaussian with standard deviation σ_m^2 given by

$$\sigma_m^2 = \gamma \int \psi_m^4(\mathbf{r}) \, d^3\mathbf{r} \tag{5.91}$$

A degenerate level is shifted and split by a random potential. For the simply degenerate first excited state $\psi_{m\pm}(\mathbf{r}) = \psi_m(\mathbf{r})e^{\pm im\phi}$ of an axially symmetric system one finds (Tsiper, 1996) that the eigenenergies measured from the unperturbed level are

$$\delta E_{m\pm} = u \pm \sqrt{x^2 + h^2}, \quad \overline{u^2} = \sigma_m^2 \quad \text{and} \quad \overline{x^2} = \overline{y^2} = \sigma_m^2/2 \tag{5.92}$$

5.4.3 INTER-SUBLEVEL TRANSITIONS

In a two-dimensional quantum well inter-subband transitions are only allowed for light propagation in the quantum well plane because only in this case is the electric field vector parallel to the quantization direction (Bastard, 1988). Thus quantum wells cannot absorb infrared radiation via inter-subband processes for vertical incidence. In

MODELING OF QUANTUM DOTS 155

quantum dots, however, the electronic wavefunctions are quantized in all three dimensions and generally light of all polarization directions can be absorbed.

For the case of pyramidal InAs/GaAs quantum dots we have calculated the inter-sublevel matrix elements from the hole wavefunctions (see Section 5.2.4) and find $\langle \Psi^h_{|100\rangle}|x|\Psi^h_{|000\rangle}\rangle = \langle \Psi^h_{|010\rangle}|y|\Psi^h_{|000\rangle}\rangle = 3.45\,\text{nm}$ and $\langle \Psi^h_{|001\rangle}|z|\Psi^h_{|000\rangle}\rangle = 0.92\,\text{nm}$ for the inter-sublevel transitions between the hole ground state and the first excited hole states. When the piezoelectric field is included, the main axes for in-plane polarization are along the [110] and [1-10] directions. For the pyramidal geometry the transition matrix element for in-plane polarization is larger than that for polarization parallel to the growth direction.

5.5 POPULATION OF LEVELS

5.5.1 THERMAL VERSUS NONTHERMAL DISTRIBUTION

Due to the finite localization energies E_{loc}, i.e. the energy difference between confined level and barrier, carriers can escape from the quantum dot (see Fig. 5.38). This process becomes increasingly important with increasing temperature and decreasing localization energy. In the steady state the average recombination rate in the quantum dots equals the net capture rate:

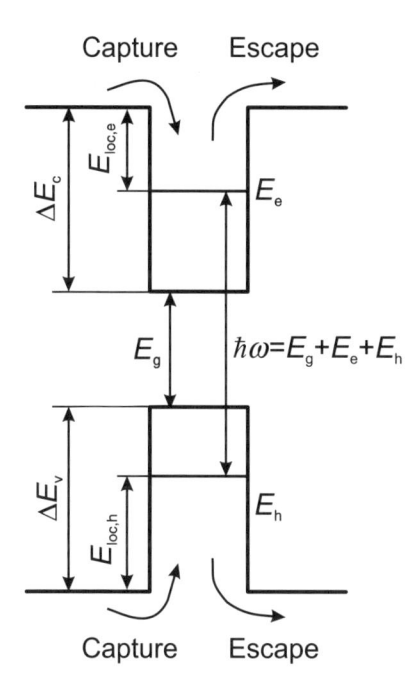

$$\frac{1}{\tau_D}\langle f_n f_p \rangle = v_n[n(1-\langle f_n\rangle)-n_1\langle f_n\rangle] = v_p[p(1-\langle f_p\rangle)-p_1\langle f_p\rangle] \quad (5.93)$$

Fig. 5.38 Schematic energy diagram for the quantum dot and barrier

where f_n and f_p are the occupation probabilities for electrons and holes. $\langle \ldots \rangle$ denotes the average over all dots, n and p are the carrier densities in the barrier, n_1 and p_1 determine the rate of thermal escapes of electrons and holes from the quantum dots into the barrier, v_n and v_p are the rates of carrier capture into the dots from the barrier, which can be written as $v = \sigma v_{\text{th}}$, σ being the carrier capture cross section and v_{th} the thermal velocity of carriers. Actually, the recombination current in the quantum dots cannot be properly expressed solely using electron and hole densities, which is discussed in detail in Section 5.5.2.3. The characteristic times for thermally induced escapes of an electron or a hole from the quantum dot into the barrier is given by

$$\tau_{e,n} = \frac{1}{v_n n_1} \quad \text{and} \quad \tau_{e,p} = \frac{1}{v_p p_1} \tag{5.94}$$

respectively (Asryan and Suris, 1996), analogous to the physics of Shockley–Read trapping centers.

The ratio of escape and recombination rates controls whether the dots have a thermal equilibrium population:

(a) Obviously, if the carrier recombination is much faster than the escape a strong nonequilibrium situation arises and the population does not depend on the electron or hole energy; i.e. dots of different sizes have the same population. In this case the dots are uncoupled and do not exchange carriers prior to recombination.

(b) In the reverse case, when a carrier is scattered back into the barrier (with a probability that depends on the localization energy) many times before it recombines, a thermal equilibrium between the barrier and the quantum dot ensemble is established. Larger quantum dots, providing greater localization, become more strongly populated than smaller ones, allowing for easier escape.

Typical situations for real quantum dots are in between these two extreme cases. In thermal equilibrium the thermal release time τ_e from a trap (for which no other carrier loss mechanism exists) is given by (Lampert and Mark, 1970)

$$\tau_e \approx 10 \text{ ps} \exp\left(\frac{E_{\text{loc}}}{k_B T}\right) \tag{5.95}$$

when a typical three-dimensional density of states is used. The relation (Eq. 5.95) is visualized in Fig. 5.39 for different temperatures and localization energies. At room temperature a typical localization energy of 120 meV gives rise to $\tau_e = 1$ ns, which is also a typical radiative recombination lifetime. Thus for typical quantum dots, τ_e and τ_D can easily be the same for elevated temperatures.

The two limiting cases are now treated. If the characteristic escape time is small compared to the radiative recombination in the quantum dot, $\tau_{e,n}, \tau_{e,p} \ll \tau_D$, the carriers can redistribute between different quantum dots and thus establish a thermal

MODELING OF QUANTUM DOTS

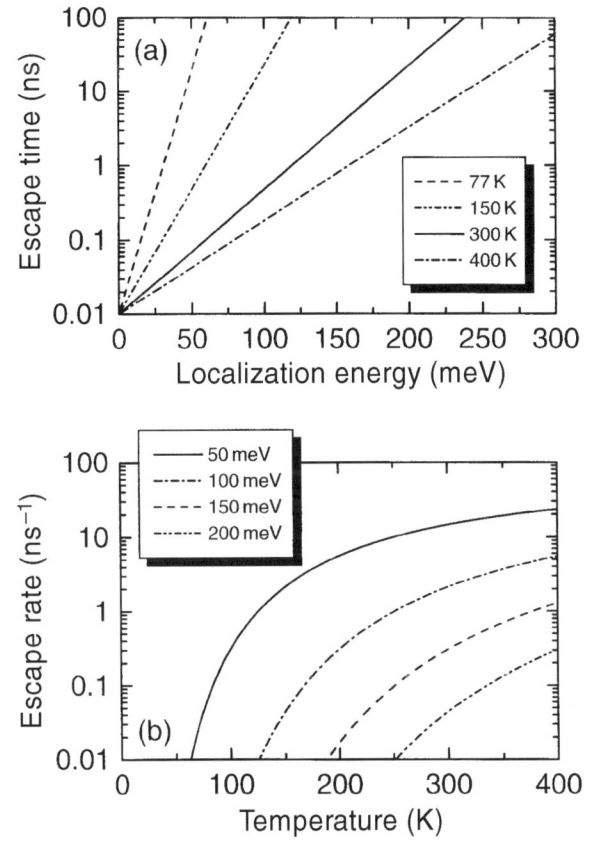

Fig. 5.39 Thermal equilibrium escape time from a trap (for which no other loss mechanism exists) as a function of localization energy and temperature, illustrating Eq. (5.95)

distribution. In this case the population is given by Fermi–Dirac distribution functions with quasi-Fermi levels E_{F_c} and E_{F_v}, respectively:

$$f_n(E_e, E_{F_c}) = \left[1 + \exp\left(\frac{E_e - E_{F_c}}{k_B T}\right)\right]^{-1}, \quad f_p(E_h, E_{F_v}) = \left[1 + \exp\left(\frac{E_h - E_{F_v}}{k_B T}\right)\right]^{-1} \tag{5.96}$$

Under these conditions the recombination rate in the quantum dots can be neglected in Eq. (5.93) and the free carrier densities are expressed as

$$\frac{n}{n_1} = \frac{\langle f_n \rangle}{1 - \langle f_n \rangle} \quad \text{and} \quad \frac{p}{p_1} = \frac{\langle f_p \rangle}{1 - \langle f_p \rangle} \tag{5.97}$$

The bulk electron density is given as

$$n = N_c \frac{2}{\sqrt{\pi}} F_{1/2}\left(\frac{E_{F_c} - \Delta E_c}{k_B T}\right) \quad \text{with} \quad N_c = 2\left(\frac{m_c k_B T}{2\pi \hbar^2}\right)^{3/2} \quad (5.98)$$

If the Boltzmann approximation holds (i.e. that the quasi-Fermi level E_{F_c} is several $k_B T$ below the barrier),

$$n = N_c \exp\left(\frac{E_{F_c} - \Delta E_c}{k_B T}\right) \quad (5.99)$$

Similar formulas hold for the holes. Therefore we get ($E_{\text{loc}} < 0$)

$$n_1 = N_c \exp\left(\frac{E_e - \Delta E_c}{k_B T}\right) = N_c \exp\left(\frac{E_{\text{loc},e}}{k_B T}\right)$$

and (5.100)

$$p_1 = N_v \exp\left(\frac{E_h - \Delta E_v}{k_B T}\right) = N_v \exp\left(\frac{E_{\text{loc},h}}{k_B T}\right)$$

The other extreme case is present when the radiative recombination is much faster than the thermal reemission of carriers from the quantum dot, i.e. $\tau_{e,n}$, $\tau_{e,p} \gg \tau_D$. In this case the carrier density in the three-dimensional barrier is given by

$$np = \frac{1}{\tau_D} \frac{1}{v_n v_p} \frac{f_n^2 f_p^2}{(1 - f_n)(1 - f_p)} \quad (5.101)$$

The distribution functions $f_{n,p}$ represent nonequilibrium populations and do not (or only weakly) depend on the dot energy and thus dot size.

The intermediate case, when both escape and recombination play a role, is much harder to access. It has also to be considered that electrons and holes have different localization energies. A further complication is introduced by excited states which can serve as intermediate levels for escape and capture.

5.5.2 POPULATION STATISTICS

Upon external excitation, e.g. by an injection current or by carriers created from photoabsorption, the quantum dot level(s) become populated. In the following section an adequate theoretical description of cw and transient-level population is discussed.

5.5.2.1 Conventional rate equation models

Conventional rate equation models are mean field theories in the sense that the population of the ith (nondegenerate) electronic level is described with the ensemble averaged population probabilities f_i. For an ensemble of N_D quantum dots γ_i^j denotes

MODELING OF QUANTUM DOTS 159

whether the ith electronic level of the jth dot ($j = 1, \ldots, N_D$) is filled ($\gamma_i^j = 1$) or not ($\gamma_i^j = 0$). Then f_i is given by

$$f_i = N_D^{-1} \sum_{j=1}^{N_D} \gamma_k^j$$

More or less detailed conventional rate equation systems have been used to model populations of and recombination from quantum dot levels (e.g. see Adler *et al.*, 1996; Mukai *et al.*, 1996b; Grosse *et al.*, 1997). By modeling the inter-level scattering with such rate equation models quantitative conclusions have been drawn for the scattering time determining the energy relaxation (phonon bottleneck; see Section 5.7.2). In the following we discuss the simple 'trickle-down' model which leads to the typical, incorrect, result of conventional rate equation models, namely the vanishing occupation of excited states for fast energy relaxation.

Two non-degenerate energy levels for electron–hole pairs (made up from an electron and a hole in single-particle states with the same quantum numbers) are assumed (Fig. 5.40). The population of the levels are f_1 and f_2, with $0 \leq f_i \leq 1$. Electron–hole (eh) pairs are captured with a generation rate (excitation) G into the upper level. The radiative lifetime τ_r of the eh pairs in the ground and the excited states is assumed to be identical. The relaxation from level 2 to 1 is governed by the intrinsic relaxation time τ_0. The rate equation model (Fig. 40a) then yields

$$\frac{df_2}{dt} = -\frac{f_2}{\tau_r} - \frac{f_2(1-f_1)}{\tau_0} + G$$
$$\frac{df_1}{dt} = -\frac{f_1}{\tau_r} + \frac{f_2(1-f_1)}{\tau_0}$$
(5.102)

This model is a simplified two-level version of the model presented by Mukai *et al.* (1996b) for five levels with degeneracies. The term $(1-f_1)$ avoids 'overfilling' of the state, also called 'Pauli blocking'. It serves only as a typical example; additional complexity does not alter its principal results. An analytical expression can be obtained for the solution of Eqs. (5.102) in the stationary case. In the limit $\tau_0 \to 0$, the solution for $G < 1/\tau_r$ is $f_1 = G\tau_r$, $f_2 = 0$. In the case $1/\tau_r < G < 2/\tau_r$, the

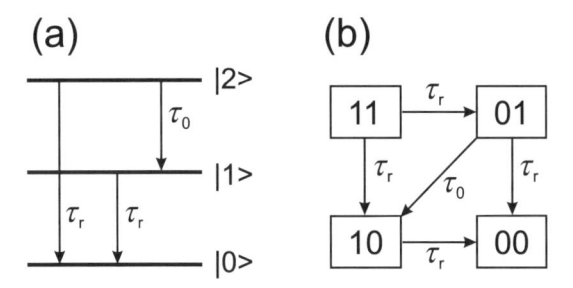

Fig. 5.40 Schematic representation of (a) the conventional rate equation model and (b) the Master equation for the microstates for a two-level system

solution is $f_1 = 1$, $f_2 = G\tau_r - 1$. We note that for an infinitely fast energy relaxation (or inter-level scattering) rate equation models generally yield the unreasonable result of vanishing population of excited levels as long as a lower level is not completely filled. Let us look at a simple case and assume that the external excitation is low, $G < 1/\tau_r$. The (large) total number of quantum dots is N_D. Thus $f_1 N_D$ quantum dots are filled with one electron–hole pair; the rest is empty. If now an additional eh pair is captured from the reservoir by the quantum dot ensemble, two things can happen. If it is captured by an empty dot, f_2 remains zero. However, there is a finite chance that it is captured into one of the $f_1 N_D$ dots which are already filled with one eh pair, making $f_2 > 0$. Thus conventional rate equation models cannot describe correctly the occupation of excited states or simulate finite inter-level scattering times.

5.5.2.2 Master equations for microstates

The alternative to conventional rate equation models are master equations formulated for the microstates of the quantum dot ensemble. The ensemble is described by the number of quantum dots N_k that have the same individual population k, defined by a particular set of γ_i (Grundmann and Bimberg, 1997b). The sum over all possible different populations must then be the total number of quantum dots, $\sum_k N_k = N_D$. When a randomly occurring capture or recombination process causes one specific dot to go from a configuration k_1 to another k_2, the change of ensemble state is exactly described by $N_{k_1} \to N_{k_1} - 1$ and $N_{k_2} \to N_{k_2} + 1$. If it is irrelevant what particular dots have a particular configuration (e.g. for a nonspatially resolved spectroscopic experiment), the only approximation in this random population (RP) model is that eventually differential equations are used for the time evolution of the N, which is valid for large N_D.

The correct description of the above two-level system is achieved using the microstates of the system (Fig. 5.40b). These are given by dots with the two levels filled with (n,m) electron–hole pairs (made up from an electron and a hole in single-particle states with the same quantum numbers), i.e. (0,0) for empty dots, (1,0) and (0,1) for partially filled dots, and (1,1) for completely filled dots. The probability of finding a dot with a specific microstate in the ensemble is given by w_{00}, w_{10}, w_{01}, and w_{11}, respectively, fulfilling $w_{00} + w_{10} + w_{01} + w_{11} = 1$. The Master equations (without external excitation) are

$$\frac{dw_{00}}{dt} = \frac{w_{10}}{\tau_r} + \frac{w_{01}}{\tau_r}$$
$$\frac{dw_{10}}{dt} = -\frac{w_{10}}{\tau_r} + \frac{w_{11}}{\tau_r} + \frac{w_{01}}{\tau_0}$$
$$\frac{dw_{01}}{dt} = -\frac{w_{01}}{\tau_r} + \frac{w_{11}}{\tau_r} - \frac{w_{01}}{\tau_0}$$
$$\frac{dw_{11}}{dt} = -2\frac{w_{11}}{\tau_r}$$

(5.103)

MODELING OF QUANTUM DOTS

We note that at no point is 'Pauli blocking' (Bockelmann et al., 1996; Grosse et al., 1997) explicitly introduced. Implicitly it is included by not considering 'overfilled' microstates, like (2,0) in the present example.

The level populations f_1 and f_2 used in the conventional rate equations can be expressed as $f_1 = w_{10} + w_{11}$ and $f_2 = w_{01} + w_{11}$. The above Master equations then read for the f_i:

$$\frac{df_1}{dt} = -\frac{f_1}{\tau_r} + \frac{w_{01}}{\tau_0}, \quad \frac{df_2}{dt} = -\frac{f_2}{\tau_r} - \frac{w_{01}}{\tau_0} \quad (5.104)$$

The inter-level scattering is modeled by w_{01}, while in the conventional rate equation system the inter-level relaxation terms translate into

$$\frac{f_2(1-f_1)}{\tau_0} \rightarrow \frac{(w_{01} + w_{00})(w_{01} + w_{11})}{\tau_0} \quad (5.105)$$

which obviously does not contain reasonable physics. The term is quadratic in w_{01} and includes, due to the ensemble average scheme, empty (w_{00}) and completely filled (w_{11}) dots, which do not contribute to the inter-level scattering rate.

For the limit of low temperatures, when excited levels are not thermally populated, the RP theory has been worked out by Grundmann and Bimberg (1997b). It is found that in the limit of a large number of quantum dot levels which can be occupied by carriers, the probability w_n of dots having n carriers is given by the Poisson statistic

$$w_n^M = \frac{\lambda^n}{n!}\exp(-\lambda) \quad \text{with } \lambda = \langle n \rangle \quad (5.106)$$

where the average number of carriers in the ensemble $\langle n \rangle$ is determined by the level of external excitation. If only radiative recombination from the quantum dot ensemble is considered, $\langle n \rangle = G\tau_r/N_D$.

Modeling of either capture of eh pairs or separate, independent capture of electrons and holes results in different population statistics (Grundmann and Bimberg (1997b). In the latter case charged dots become possible.

Finite inter-level scattering can be modeled within the RP model when

$$p_n^M = \binom{n}{M} = \frac{n!}{M!(n-M)!}$$

possible distributions of the n carriers over the M available levels are considered individually. An important conclusion from the RP model for the interpretation of luminescence spectra for increasing excitation density is that, even for infinitely fast energy relaxation, the first excited state becomes populated before the ground state saturates. However, the second and higher excited states should only show up in the luminescence spectrum when energy relaxation is delayed, e.g. by a 'phonon bottleneck' effect (Fig. 5.41).

For modeling the population of quantum dots at finite temperatures the reemission of carriers from confined levels into the wetting layer and barrier has also to be considered. Using detailed balance arguments the thermal population of excited states at low excitation densities is obtained.

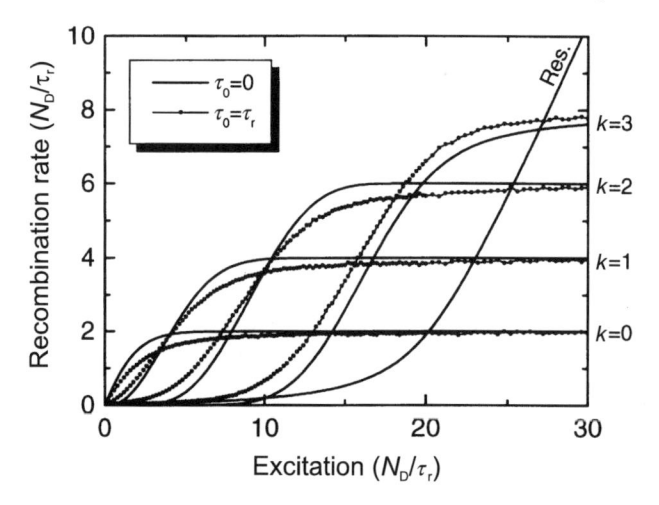

Fig. 5.41 Intensity of luminescence from QD states (two-dimensional harmonic oscillator model with four states, $k = 0, 1, 2, 3$ with degeneracies $g = 2k + 2$) as a function of excitation for different inter-level scattering times ($\tau_0 = 0$, $\tau_0 = \tau_r$). All quantum dot transitions and the barrier (Res.) are modeled to have the same radiative lifetime τ_r. The random population causes the excited level to gain intensity before the first transition is saturated, even for infinitely fast energy relaxation ($\tau_0 = 0$). Finite inter-level scattering time causes the ground state to saturate more slowly and excited states to gain intensity at lower excitation. The scatter in the points for $\tau_0 \neq 0$ is due to the Monte Carlo method used. (After Grundmann and Bimberg, 1997b)

5.5.2.3 Recombination current

The recombination current in a quantum dot ensemble cannot be described using only the average electron and hole densities. Throughout the literature the bulk recombination rate (van Roosbroeck and Shockley, 1954) is used for the recombination current which is essentially a *bimolecular* expression

$$j = \frac{2eN_\mathrm{D}}{\tau_\mathrm{r}} f_\mathrm{e} f_\mathrm{h}$$

having a maximum of $j = 2eN_\mathrm{D}/\tau_\mathrm{r}$ for the fully occupied ground state, τ_r being the exciton lifetime.

We start with a simple example to show that this scheme does not hold for quantum dots. Two quantum dot ensembles with $N_\mathrm{D} = 2$ dots which have the *same* average carrier density (in units/dot) $f_\mathrm{e} = f_\mathrm{h} = \frac{1}{4}$ are compared (Fig. 5.41). The associated recombination currents are extremely different, $j_{\mathrm{r},\mathrm{I}} = e/\tau_\mathrm{r}$ for ensemble *I* and $j_{\mathrm{r},\mathrm{II}} = 0$ for ensemble II (Grundmann and Bimberg, 1997c). Obviously, charged dots offer a lower recombination current per carrier. Therefore the precise distribu-

MODELING OF QUANTUM DOTS

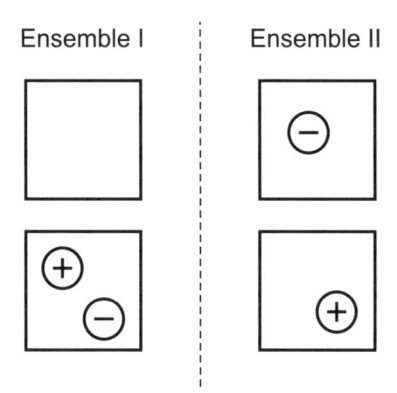

Fig. 5.42 Comparison of two quantum dot ensembles (containing two quantum dots each). Both ensembles exhibit the same carrier density, but different recombination current

tion of carriers over the dots has to be known in order to build the sum for the recombination current j_r properly:

$$j_r = \frac{N_X}{\tau_r} + \frac{N_{X^-}}{\tau_r} + \frac{N_{X^+}}{\tau_r} + \frac{2N_{XX}}{\tau_r} \quad (5.107)$$

where N_X denotes the number of dots filled with an exciton, N_{X^-} and N_{X^+} the number of dots with negatively or positively charged excitons, and N_{XX} those with biexcitons. The lifetimes of X^-, X^+, and XX have been taken as in the strong confinement limit, e.g. $\tau_{XX} = \tau_r/2$ (see Section 5.3.1.1). If excited states are populated with carriers additional terms enter Eq. (5.107).

5.5.2.4 Transients

Conventional rate equation models have been used for the interpretation and fit of cw and time-resolved experiments on quantum dots by many authors (e.g. see Mukai et al., 1996b; Grosse et al., 1997). For the simple two-level model, we compare in Fig. 5.43 the results of the conventional rate equation model with the correct result from the Master equations in order to explain the main flaw of the RE model. The initial distribution ensures that both levels are completely filled, i.e. $f_1 = 1$, $f_2 = 1$, and $w_{11} = 1$, respectively.

For $\tau_0 \to 0$ the rate equation model yields an essentially nonexponential transient for the excited state and a kink in the transient of the ground state (Grundmann et al., 1997). The rate equation model can generate transients with a somehow reasonable shape only for $\tau_0/\tau_r > \frac{1}{3}$. In the RE model the decay of the excited state is always too fast, requiring the artificial introduction of sufficiently large values of τ_0. The only feature modeled correctly by the RE model is the asymptotic decay constant of the ground state τ_r. Only for $\tau_0 \to \infty$ does the rate equation model give the same result

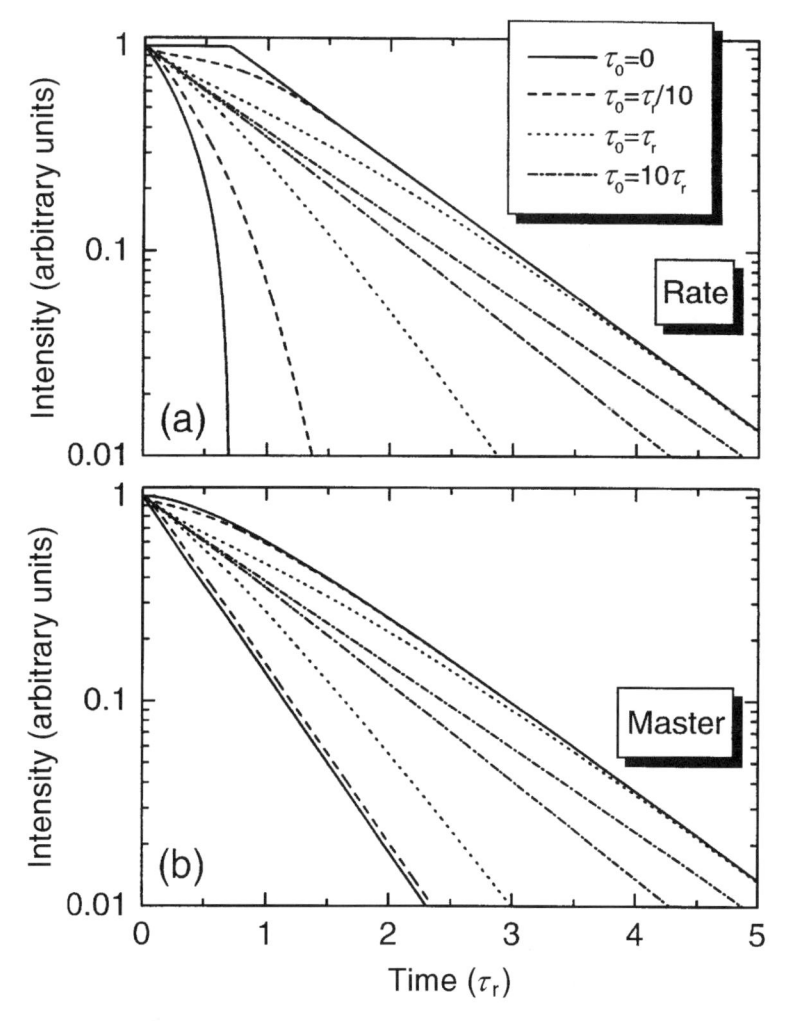

Fig. 5.43 Transients for the conventional rate equation model and Master equation for different values of $\tau_0/\tau_r = 0$, 0.1, 1, and 10 and initially a completely filled two-level system. Upper (lower) curves in each plot are intensity of state $|2\rangle$ ($|1\rangle$)

as the Master equation model, namely two states independently decaying with the time constant τ_r.

For $\tau_0 = 0$ no inter-level scattering processes have to be considered and the carriers in a dot are always in the energetically lowest possible states. The asymptotic transient of the first excited state after complete initial filling, for example, is then governed by dots in the $(2,1,\ldots)$ state. This microstate obviously decays with the time constant $1/\tau_{(2,1)} = 1/\tau_2 + 2/\tau_1$, where τ_2 (τ_1) denotes the exciton recombination time constant on the first excited (ground) state (Grundmann et al., 1997). If the exciton lifetimes on both levels are identical, the asymptotic behavior of the first excited state is given by $\tau_1/3$.

For finite values of τ_0, the asymptotic decay constant τ_∞^2 of the first excited state is given by the following equations in the MEM model, which is illustrated in Fig. 5.44a.

$$\frac{1}{\tau_\infty^2} = \begin{cases} \dfrac{1}{\tau_2} + \dfrac{2}{\tau_1}, & \tau_0 < \tau_1 \\ \dfrac{1}{\tau_2} + \dfrac{2}{\tau_0}, & \tau_0 \geq \tau_1 \end{cases} \quad (5.108)$$

We reach the important conclusion that a finite value of τ_0 has only an impact on the *asymptotic* transient of the excited state for $\tau_0 > \tau_1$, i.e. when the inter-level scattering time is larger than the ground state exciton recombination lifetime. As can be seen in Fig. 5.44b, for fast inter-level scattering ($0 \leq \tau_0 \leq \tau_1/3$) the (2,1,0,0,0) dots, having the ground state filled with the biexciton, govern the decay. In an intermediate regime, $\tau_1/3 \leq \tau_0 \leq 2\tau_1/3$, the (1,1,0,0,0) dots, having a single exciton in the ground state, are most probable in the asymptotic distribution of microstates for which the excited state is populated. For slow inter-level scattering, $\tau_0 \geq 2\tau_1/3$, the (0,1,0,0,0) microstate with an empty ground state is most probable at large times. However, for $\tau_0/\tau_1 \approx 1$ the true asymptotic values are only reached for large delay times. Within the typical dynamic range of 10^1–10^4 of time-resolved luminescence experiments no simple formula can be given to obtain τ_0 directly from the observable slopes, which are depicted in Fig. 5.44a together with the asymptotic values.

5.5.3 AUGER EFFECT

The carrier relaxation into quantum dots is largely controlled by phonon emission, which is discussed in more detail in Section 5.7. The Auger effect also contributes to the population dynamics of quantum dot levels. As pointed out by Efros *et al.* (1995), Auger-like processes contribute to the relaxation of excitons into the low quantum dot levels. A similar effect is provided by Coulomb scattering, discussed in Bockelmann and Egeler (1992). Since energy is transferred to a carrier which leaves the quantum dot the Auger effect is rather a 'depopulation' effect.

5.6 STATIC EXTERNAL FIELDS

The single-particle energies and the Coulomb interaction are modified in the presence of external fields. An electric field leads to the quantum confined Stark effect (QCSE), with the consequence of red shifts of the absorption peaks. A magnetic field leads to shifts in the energy levels and lifts degeneracies. Apart from zero-field magnetic quantum number degeneracies of excited states, the spin degeneracy, also of the ground state, is lifted (spin splitting).

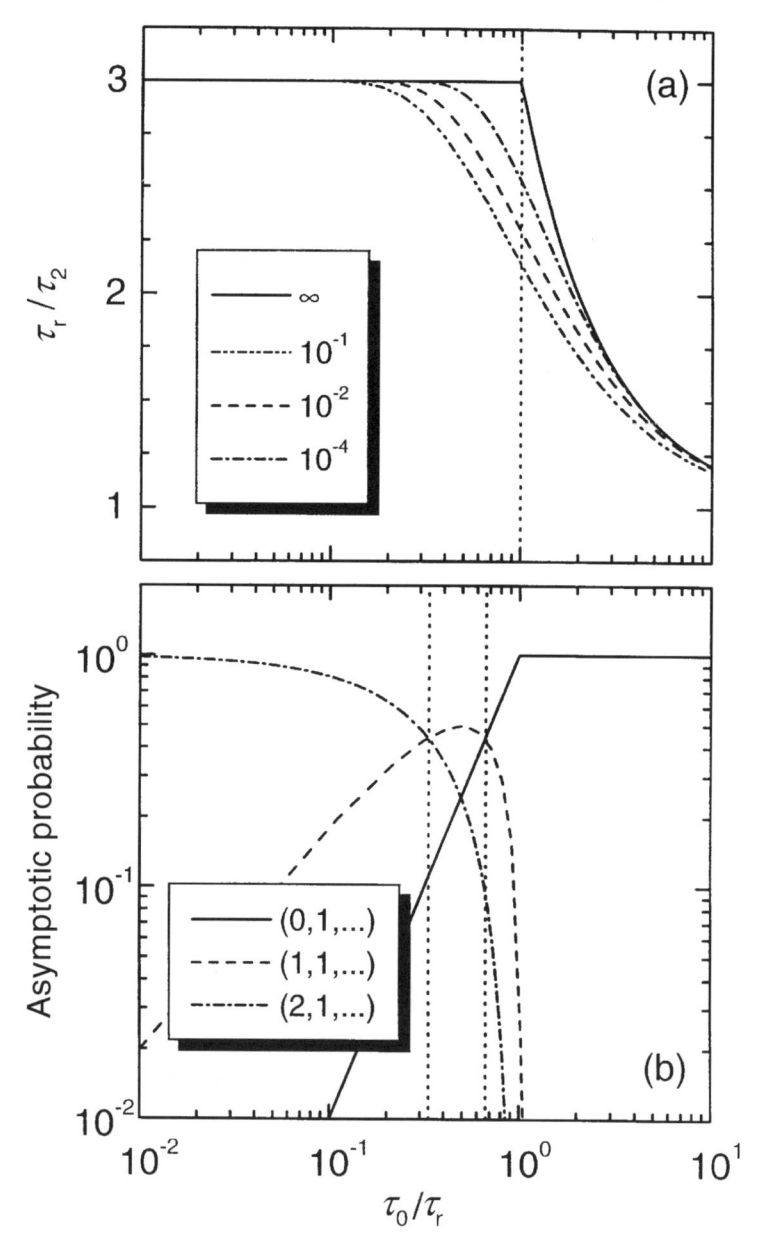

Fig. 5.44 (Local) exponential slope τ_2 of the transient of the first excited state as a function of the inter-level relaxation constant. Solid line: asymptotic value at large delay times; other lines: value at the time when the signal from the excited state has decayed from 2 (plateau region) to 10^{-N} in an experiment with finite dynamic range

5.6.1 ELECTRIC FIELDS

5.6.1.1 Exciton in single quantum dot

Similar to the problem of the atomic Stark effect (Coulomb potential in an electric field) the addition of a potential energy V_F due to the electric field **F** to the confinement potential

$$V_F(\mathbf{r}) = -q\mathbf{F} \cdot \mathbf{r} \tag{5.109}$$

which takes arbitrarily large negative values in sufficient distance from the dot, leads to nonstationary solutions. For low field strength the carrier loss from the inside of the dot occurs very slowly. Technically, the density of states of the continuous spectrum around the zero-field confined state has to be determined. A red shift and broadening of the resonance are expected. (We note that the calculated red shift is *larger* when the (attractive) Coulomb interaction is neglected (single-particle energies). This may seem a paradox at first and is caused by the contribution of excited states to the exciton.) Due to the reduced electron–hole overlap the oscillator strength is expected to decrease with increasing electric field. The reduced symmetry of the problem causes the angular momentum to be no longer a good quantum number. However, the (z) component along the field direction is still conserved.

The problem has been treated with various approximations and methods for different geometries. Chiba and Ohnishi (1988) neglected the center of mass motion of the exciton, assuming that the center of mass is at all times identical to the center of the sphere. The effective one-particle Hamiltonian is numerically solved with further approximations. Figure 5.45 depicts the result for 6.4 nm radius spherical GaAs/Al$_{0.3}$Ga$_{0.7}$As dots in different electric fields (Chiba and Ohnishi, 1989).

The problem has also been solved by variational methods for infinite confinement potentials, where the solutions remain stationary, for spheres (Nomura and Kobayashi, 1990a, 1990b) and a parabolic potential (Jaziri, 1994). The matrix diagonalization approach has also been used (Esch *et al.*, 1990; Wen *et al.*, 1995). It was found that the results obtained for the ground state exciton with the matrix diagonalization approach agree very well with the results of the variational calculation of Nomura and Kobayashi (1990a). The Stark red shift decreases with decreasing sphere radius. The excitonic absorption coefficient of a linear array of quantum dots with finite square-well potential under the influence of a uniform electric field has been calculated by Pacheco and Barticevic (1997).

The quantum confined Stark effect has been calculated by Susa (1996) for quantum disks in an electric field along the cylinder axis with finite barriers and variable size using a variational procedure based on that used by Le Goff and Stébé (1992). The calculated spectral Stark red shift is shown in Fig. 5.45 for various disk size parameters and field strengths. The red shift of a disk becomes smaller with decreasing quantum disk diameter and is smaller than that of a quantum well of the same thickness. However, the oscillator strength per area of a disk is much larger than that of a quantum well. Thus for a large area or volume filling factor with disks favorable operation of electro-optical devices is expected (Sahara *et al.*, 1996; Susa,

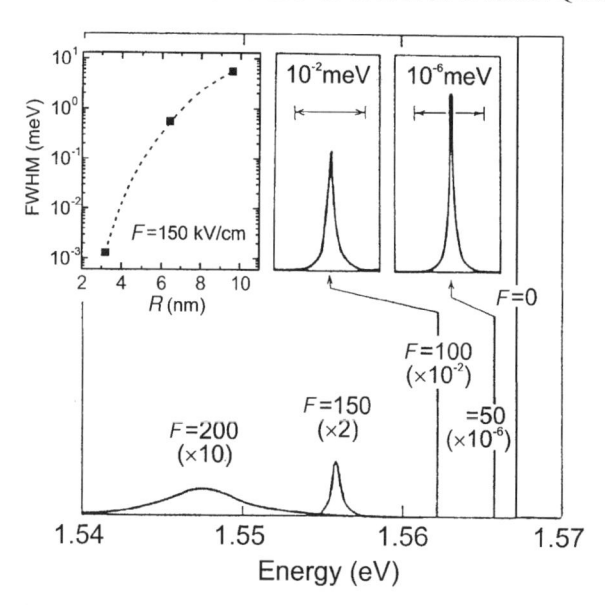

Fig. 5.45 Calculated energy spectrum of the exciton in a spherical GaAs/Al$_{0.3}$Ga$_{0.7}$As quantum dot (radius 6.4 nm) for electric field strengths $F = 0$, 50, 100, 150, and 200 kV/cm. Left inset depicts dot size dependence of broadening for $F = 150$ kV/cm, dashed line is guide to the eye. (After Chiba and Ohnishi, 1989)

1996). Results for pyramidal InAs/GaAs QDs can be found in Jiang and Singh (1998) for field directions perpendicular and parallel to the growth direction. In the latter case, the Stark effect is relatively weak; depending on the field polarity blue and red shifts are predicted to be possible.

5.6.1.2 Exciton in coupled quantum dots

A coupled (symmetric) quantum dot pair with a bias (electric field along the pair (z) direction) has been discussed in Bryant (1993a). The quantum dots are modeled as thin square plates in the xy plane with one state due to quantization in the z direction. The exciton oscillator strength as a function of applied bias and coupling strength between the dots (thickness of barrier) is depicted in Fig. 5.47. Dots with 100 nm lateral size show almost no lateral confinement effects and represent almost the case of two coupled quantum wells.

When the Coulomb effects are dominant (for wide tunnel barriers), the exciton ground state changes its character twice. For zero bias the ground state is the coherent superposition of the direct excitons in both dots, with nearly twice the oscillator strength of a direct exciton f_d. When inter-dot tunneling becomes important, the zero-bias oscillator strength becomes $f < 2f_d$ and is larger for the strong lateral confinement. For a large range of sizes, coupling strength and bias

MODELING OF QUANTUM DOTS

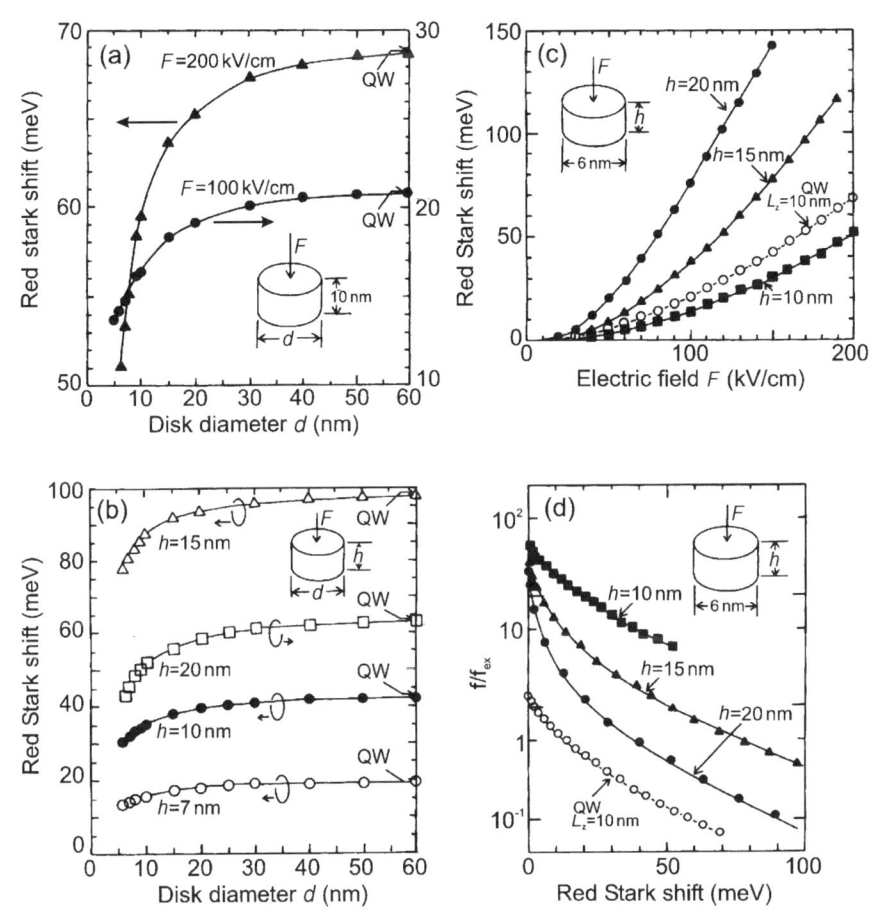

Fig. 5.46 Stark shift of quantum disks. (a) Dependence on disk diameter for fixed field strength $F = 100$ and $200\,\text{kV/cm}$, (b) dependence on disk diameter for $F = 150\,\text{kV/cm}$ and various disk heights $h = 7, 10, 15$, and $20\,\text{nm}$, (c) dependence on electric field strength for disks with 6 nm radius and various heights $h = 10, 15$, and $20\,\text{nm}$ and a $10\,\text{nm}$ thick two-dimensional quantum well. (d) Oscillator strength (in units of the quantum well exciton oscillator strength at zero field) as a function of the red Stark shift for the $10\,\text{nm}$ reference quantum well and disks with 6 nm diameter and various heights $h = 10, 15$, and $20\,\text{nm}$. (After Susa, 1996)

coherence ($f > f_d$) persists. At large bias the ground state is spatially indirect (electrons and holes are localized in the two different dots) and the oscillator strength becomes small. For intermediate bias, the ground state is the lowest-energy direct exciton. For stronger lateral confinement the direct exciton persists up to higher bias. Excitons in asymmetric coupled dots under bias are discussed in Bryant (1992).

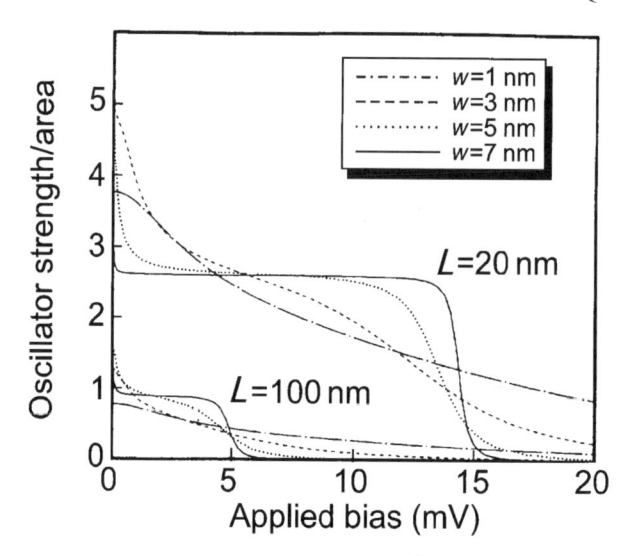

Fig. 5.47 Oscillator strength normalized per dot area L^2 of the exciton ground state in a pair of symmetric laterally confined quantum wells (quantum disks) with base side $L = 20$ and 100 nm. Dependence on applied bias for different tunnel barrier widths of 1, 3, 5, and 7 nm. (After Bryant, 1993a)

5.6.2 MAGNETIC FIELDS

The Hamiltonian of N (charged) particles in the presence of a magnetic field, an external confinement potential, and Coulomb interaction is given as

$$H = \sum_{i=1}^{N} \left\{ \frac{\hbar^2}{2m_i} [i\hbar \nabla_i - e\mathbf{A}(\mathbf{r}_i)]^2 + V_i(\mathbf{r}_i) \right\} + \frac{1}{2} \sum_{\substack{i,j=1 \\ (i \neq j)}}^{N} \frac{q_i q_j}{4\pi\varepsilon_0 \varepsilon_r} \frac{1}{|\mathbf{r}_i - \mathbf{r}_j|} \quad (5.110)$$

where parabolic bands are assumed. V_i is the external confinement potential for the ith particle and $\mathbf{A}$ is the vector potential $\mathbf{A}(\mathbf{r}) = \frac{1}{2}\mathbf{B} \times \mathbf{r}$.

5.6.2.1 One particle

The problem of one electron in an in-plane (x,y) parabolic confinement potential $V(r) = \frac{1}{2} m \omega_0^2 r^2$ with a perpendicular magnetic field $\mathbf{B}_{\|z}$ is analytically solvable (Fock, 1928; Darwin, 1930; Dingle, 1952). The rotational symmetry of the Hamiltonian with respect to the z axis allows the wavefunction to be separated into a radial and an angular part; thus, for $M = 0, \pm 1, \pm 2, \ldots$, $\psi(\mathbf{r}) = u_M(r) \exp(iM\phi)$. For the radial function $u_M(r)$ the Schrödinger equation

$$\left[-\frac{\hbar^2}{2m}\left(\frac{\partial^2}{\partial r^2} + \frac{1}{r}\frac{\partial}{\partial r}\right) + \frac{1}{2}m\left(\frac{1}{4}\omega_c^2 + \omega_0^2\right)r^2 + \frac{\hbar^2}{2m}\frac{M^2}{r^2} - \frac{1}{2}M\hbar\omega_c \right] u_M = E u_M \quad (5.111)$$

has to be solved; $\omega_c = eB/(mc)$ denotes the cyclotron frequency. The solutions involve Laguerre polynoms. The eigenenergies are, for $n = 0, 1, 2, \ldots$,

$$E_{nM} = (2n + 1 + |M|)\hbar\left(\tfrac{1}{4}\omega_c^2 + \omega_0^2\right)^{1/2} - \tfrac{1}{2}M\hbar\omega_c \qquad (5.112)$$

For $\omega_0 \to 0$, i.e. a vanishing confinement potential, or $\omega_c \gg \omega_0$, i.e. a large magnetic field, the Landau solution is obtained. In this limit the energy is $E_{nM} \to (n' + \tfrac{1}{2})\hbar\omega_c$ with $n' = n + (|M| - m)/2$. The essential impact of the confinement potential is that the energy of states with positive M actually depends on M, while in the Landau case it is independent of M.

The energies of single electron states in the presence of a perpendicular magnetic field of lens-shaped self-assembled quantum dots have been calculated by Wojs and Hawrylak (1996). A numerical solution for the actual shape of the dots is compared to the states of the infinite parabolic potential discussed above with the same ω_0 in Fig. 5.17. The parabolic potential seems to be a good approximation for the given geometry.

The one- and two-hole states in parabolic (unstrained) QDs in the presence of a perpendicular magnetic field have been calculated by Pedersen and Chang (1997) on the basis of a four-band Luttinger Hamiltonian. The electron and hole states in parabolic QDs with a finite square potential in vertical direction in the presence of an axial magnetic field have been calculated by Li and Xia (1998). Conduction band non-parabolicity is taken into account and the holes are treated within a four band Luttinger approximation; apparently biaxial strain is assumed. Large strain and strong magnetic field are found to decrease the effect of mixing between heavy hole and light hole.

5.6.2.2 One exciton

The problem of a two-dimensional exciton in a magnetic field perpendicular to the confinement plane is discussed in Akimoto and Hasegawa (1967). For $M = 0$, analytical solutions are obtained for low and high magnetic fields.

For strong magnetic fields the Landau regime is reached (E_X^b is the Rydberg energy given in Eq. 5.55), $\gamma \approx 3$:

$$\frac{E_n}{E_X^b} = (2n+1)\frac{\hbar\omega_c}{E_X^b} - \gamma\left[\frac{\hbar\omega_c}{(2n+1)E_X^b}\right]^{1/2} \qquad (5.113)$$

At low magnetic fields the energy in terms of E_X^b is

$$\frac{E_n}{E_X^b} = -\frac{1}{(n+1/2)^2} + \frac{5}{8}\left(n+\frac{1}{2}\right)^4\left(\frac{\hbar\omega_c}{E_X^b}\right)^2 + \cdots \qquad (5.114)$$

The energy increases proportionally to B^2, the diamagnetic shift. By treating the part H_D of the Hamiltonian

$$H_D = \frac{1}{8}m\omega_c^2 r^2 = \frac{1}{8}\frac{e^2 B^2}{mc^2}r^2 \tag{5.115}$$

as a perturbation, the widely used formula for the diamagnetic shift ΔE_D is obtained:

$$\Delta E_D = \frac{e^2}{8c^2}\left(\frac{\langle r^2 \rangle_e}{m_e} + \frac{\langle r^2 \rangle_h}{m_h}\right)B^2 = \frac{e^2}{8\mu c^2}\langle r^2 \rangle B^2 \tag{5.116}$$

where μ is the reduced mass of the electron–hole pair. $\langle r^2 \rangle$ describes the extension of the wavefunction perpendicular to the magnetic field direction.

5.6.2.3 Spin splitting

The two states $|j, \pm m_j\rangle$ due to the orbital magnetic momentum of the Bloch functions split for several reasons already at the zero magnetic field (Pfeffer and Zawadzki, 1995). The bulk inversion asymmetry, present in III–V and II–VI semiconductors, leads to a lifting of the degeneracy for finite **k**. Additionally, structural inversion asymmetry contributes to spin splitting (Bychkov and Lashba, 1984; Gammon et al., 1996a; Stier et al., 1996). Additional splitting is caused in the presence of a magnetic field. The magnitude ΔE of splitting for an exciton is proportional to the electron and hole g factors. Spin splitting of the conduction band states is negligible (0.03 meV/T) for lens-shaped self-assembled InGaAs/GaAs dots (Wojs and Hawrylak, 1996). The main effect is carried by the holes due to spin–orbit interaction (Bayer et al., 1995) (see Section 6.7.1.4).

5.6.2.4 Few electron systems

Considerable effort has been devoted to calculate the electronic states for zero-dimensionally confined few electron systems in magnetic fields. Solution of the Hamiltonian (Eq. 5.110) for many particles is only possible using numerical methods. A few electron system in a parabolic potential has been calculated, for example, by Maksym and Chakraborty (1990, 1992), Ruan et al. (1995) and Maksym (1996). States of different symmetry represent the ground state at different magnetic fields; magic angular momenta and spin combinations occur. For parabolic confinement the optical excitations of a many-body system are predicted to be exactly the same as those of a single-particle system, which has been experimentally confirmed (Sikorski and Merkt, 1989). This theorem is known as the generalized Kohn theorem (Kohn, 1961).

The states of a four-electron system in a parabolic confinement are discussed in detail in Chengguang et al. (1996). The particular differences introduced by a rectangular (finite) potential well are discussed in Ugajin (1995, 1996).

Theoretical results for few electron systems in QDs have been reviewed by Johnson (1995) and Beenakker (1997). An exact solution for the Thomas-Fermi

MODELING OF QUANTUM DOTS

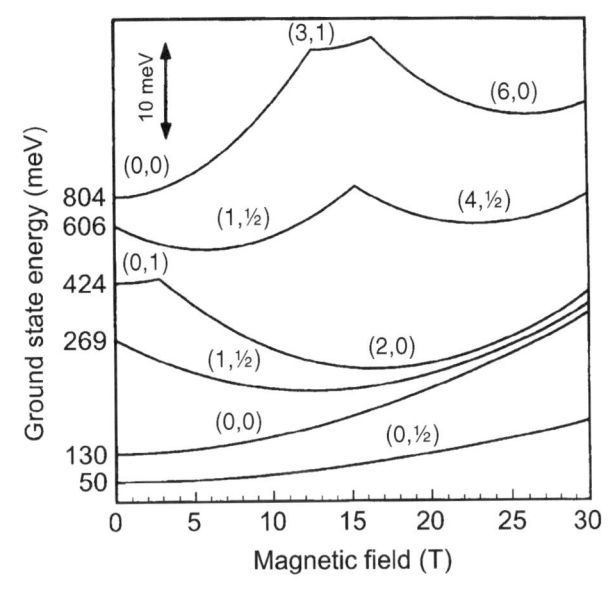

Fig. 5.48 Many-electron system in a hemispherical quantum dot in the magnetic field. Ground state energies for different number of electrons N. Curves are vertically shifted and offset at zero field is given on the left axis. Configurations of the states (total angular momentum, total spin) are indicated. (After Wojs and Hawrylak, 1996)

approximation (for dots with many electrons) in a parabolic QD has been given by Pino (1998).

For the many electron states in a lens-shape self-assembled dot it was found (Wojs and Hawrylak, 1996) that for up to three electrons the ground states, also at magnetic fields up to 30 T, are given by the lowest kinetic energy noninteracting states. A transition induced by the magnetic field is predicted for $N=4$ electrons (filled s shell and half-filled p shell). At a zero or low ($B < 2.8$ T) magnetic field the ground state is given by the state $S=1$ with maximum spin according to Hund's rule in order to lower the total energy. For increasing magnetic field the degeneracy of the p states is more and more removed, making it eventually energetically more favorable to pay a price in exchange energy for the population of the lowest kinetic energy with $S=0$ (Fig. 5.48).

5.6.2.5 Coulomb coupling of dots

The coupling of two parabolic quantum dots via Coulomb interaction is treated in Chakraborty *et al.* (1991). Many-body effects due to the breaking of the circular symmetry, namely anti-crossing behavior of magneto-optical transitions, are calculated in agreement with experimental observations (Demel *et al.*, 1990).

A model for an arbitrary number of colinear, two-dimensional parabolic dots filled with two electrons each is reported in Benjamin and Johnson (1995). Intra-dot coupling leads to the occurrence of entangled ground states at finite magnetic fields. The coupling of dots could be used for ultra-small logic gates and quantum cellular automata (Lent and Tougaw, 1993; Lent et al., 1993a, 1993b; Paz and Mahler, 1993; Chen and Porod, 1995).

5.7 PHONONS

5.7.1 PHONON SPECTRUM

The phonon spectrum of nanostructures is modified as compared to that of the bulk material due to size quantization, the presence of interfaces and inhomogeneous strain for the case of non-lattice-matched materials. In short-period superlattices folded acoustical phonons were observed by Colvard et al. (1980). Confined optical (Jusserand et al., 1984) and interface optical (Sood et al., 1985) modes were also discovered. The phonon quantization in ultra-small structures leads generally to a decrease in the longitudinal-optical (LO) phonon energy. In a compressively strained quantum well the phonon energy is shifted toward higher energies (see Section 5.1.4). In thin ($L_z \approx 1$ ML) InAs/GaAs quantum wells both effects cancel (Wang et al., 1996b).

In quantum dots with a typical size of several nanometers the quantization plays no role for the phonon energy. Thus the energy shift due to strain is prevalent. For an InAs/GaAs pyramid the shift of phonon energies has been calculated by Grundmann et al. (1995c) in the approximation of slowly varying strain; the LO phonon is predicted to shift by $+2.2$ meV from the bulk value of 29.9 meV (Aoki et al., 1984), mainly due to the hydrostatic strain.

The presence of interfaces leads to new phonon modes, which have been calculated in Knipp and Reinecke (1992) in the dielectric continuum approximation for different dot shapes. Calculations for GaAs spherical dots in $Al_xGa_{1-x}As$ were reported in de la Cruz et al. (1995).

5.7.2 ELECTRON–PHONON INTERACTION

The carrier–phonon interaction and charge carrier relaxation are drastically modified in quantum dots with respect to bulk material or even quantum wells:

(1) The inter-sublevel (inter-subband) relaxation process dominant in three-, two-, and one-dimensional structures, LO phonon emission via Fröhlich interaction, is forbidden unless the sublevel separation matches $\hbar\omega_{LO}$ (at the zone center).
(b) Deformation potential scattering via longitudinal-acoustic (LA) phonons in bulk material is weak compared to the Fröhlich interaction and becomes even weaker with decreasing dot size (Bockelmann and Bastard, 1990).

The consequences of such reduction of energy loss mechanisms (the 'phonon bottleneck' effect) were thought to be frustrated or at least delayed population of quantum dot ground states if capture occurs only via localized excited states. Subsequently, localized excited states have a longer occupation time, and thus can be subject to nonradiative recombination or easier evaporation to barrier states at higher temperatures. High-frequency device operation might be hindered by slow capture.

However, this consideration does not include the increasing role of interface phonons. Moreover, with decreasing structural size additional scattering mechanisms arise. Knipp and Reinecke (1996) discussed the importance of the 'ripple' mechanism, scattering due to the modification of confinement potential and spatially varying mass by the displacement field of acoustic phonons. The scattering rate increases by orders of magnitudes for decreasing quantum dot size and finally becomes stronger than the contribution from deformation potential scattering. However, the absolute scattering rates remain small.

Following the notation of Inoshita and Sakaki (1992), the phonon scattering probability is evaluated using Fermi's golden rule and the Hamiltonians for the Fröhlich interaction or deformation potential scattering. In first order, i.e. considering LO or LA emission, the phonon scattering time τ_{ph} is

$$\tau_{\text{ph}}^{-1} = \frac{2\pi}{\hbar} \sum_q |M_q^{\text{if}}|^2 (N_q + 1)\delta(E_0 - \hbar\omega_q) \quad (5.117)$$

where i and f denote the initial and final states, and N_q is the Bose distribution function:

$$N_q = \left[\exp\left(\frac{\hbar\omega_q}{k_B T}\right) - 1\right]^{-1} \quad (5.118)$$

The matrix elements can be written as

$$M_q^{\text{if}} = \langle i|e^{i q \cdot r}|f\rangle a_q, \quad a_q = \begin{cases} D\sqrt{\hbar q/2\rho c \Omega} & \text{(LA mode)} \\ M/q\sqrt{\Omega} & \text{(LO mode)} \end{cases} \quad (5.119)$$

where Ω is the system volume, D is the deformation potential, ρ is the density, and c the sound velocity. The Fröhlich coupling constant can be obtained from

$$M = \left[2\pi e^2 \hbar \omega_{\text{LO}} \left(\frac{1}{\varepsilon_\infty} - \frac{1}{\varepsilon_0}\right)\right]^{1/2} \quad (5.120)$$

The q dependence of the matrix elements is such that for $q \to 0$ the behavior is dominated by a_q, while for large q the form factor governs, which decays quickly to zero if q becomes larger than $Q = \pi/R_0$. For a typical dot the size is still much larger than the lattice constant, and therefore only phonons with small q contribute to the transition probability. Together with the δ function rapid energy relaxation is expected only in a limited range around E_0.

The significance of multiphonon processes was pointed out by Inoshita and Sakaki (1992). Processes of the type LO ± LA have a scattering probability given by

$$\tau_{ph}^{-1} = \frac{2\pi}{\hbar} \sum_k \sum_q \left| \sum_s \left(\frac{M_q^{is} M_k^{sf}}{E_i - E_s - \hbar\omega_q} + \frac{M_k^{is} M_q^{sf}}{E_i - E_s \mp \hbar\omega_k} \right) \right|^2$$
$$\times (N_q + 1)\left(N_k + \frac{1}{2} \pm \frac{1}{2}\right) \delta(E_0 - \hbar\omega_q \mp \hbar\omega_k) \qquad (5.121)$$

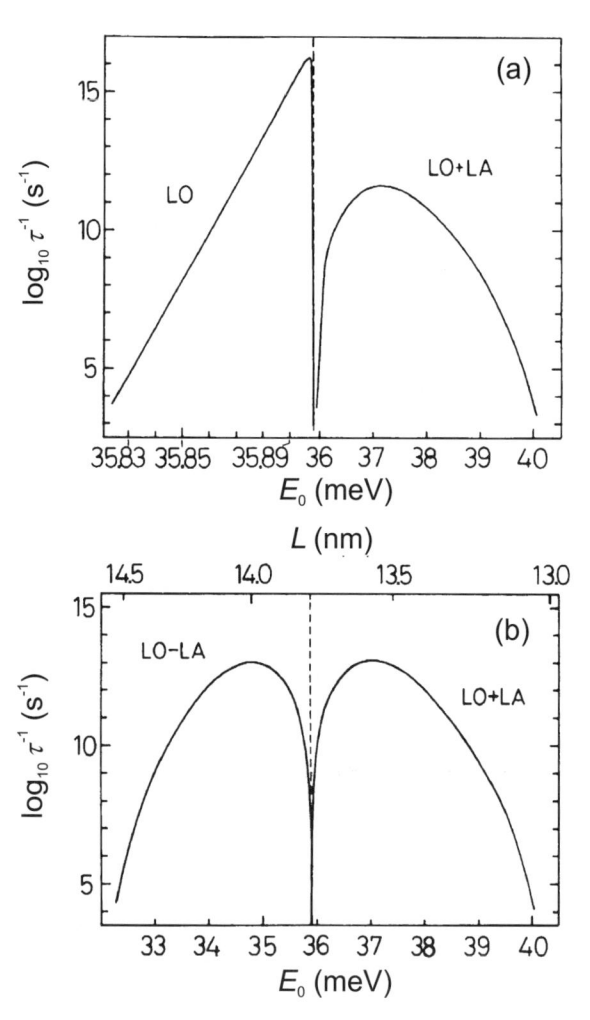

Fig. 5.49 Electron relaxation rate $1/\tau_{ph}$ for E_0 in the vicinity of $\hbar\omega_{LO} = 35.9 \text{ eV}$ for (a) $T = 0$ K and (b) $T = 300$ K. Note the special abscissa in (a) for $E_0 < \hbar\omega_{LO}$. (After Inoshita and Sakaki, 1992)

where the upper (lower) sign corresponds to LA emission (absorption). The quantities q and k refer to LO and LA modes, respectively. The summation over s excludes the states i and f.

The numerical results of Inoshita and Sakaki (1992) are shown in Fig. 5.49 for $T = 0$ and 300 K. At low temperatures (Fig. 5.49a) the LO emission is the leading relaxation mechanism, but only in a narrow energy window of about 38 μeV for $\tau_{ph} < 100$ ps. The inclusion of the LO + LA mechanism yields a rather broad peak at the high-energy side, with a maximum emission rate exceeding $\tau_{ph}^{-1} = 10$ ps. In a window of about 2.1 meV (1.3 meV) the emission time is smaller than 100 ps (10 ps). At room temperature (Fig. 5.49b) the LO peak is nearly the same as for 0 K and thus is not shown in the plot. Now the LO−LA process gives rise to a peak on the low-energy side of the LO phonon, which is almost perfectly mirror symmetric to the LO + LA process because $N_q \approx 10^2 \gg 1$. The energy interval for which the phonon emission time is smaller than 100 ps (10 ps) is increased to 5.1 meV (2.5 meV).

Nakayama and Arakawa (1996) point out that Fermi's golden rule does not strictly apply any more because the coupling between the electrons and the LO phonons is strong. The direct solution of the time-dependent Schrödinger equation using coupled mode equations yields larger capture rates. For a phonon lifetime τ_{ph} of even 2.5 ps a relaxation rate of 10^{10} s^{-1} is realized in an energy range of about 50 meV. For a phonon lifetime of 10 fs the rate exceeds 10^{11} s^{-1} for almost the whole energy range. ps multi-phonon scattering processes are also predicted by Král and Khás (1998a, 1998b).

5.8 QUANTUM DOT LASER

A quantum dot laser built from *ideal* quantum dots, i.e. an ensemble having identical electronic properties, where the carriers are strongly confined has very intriguing properties, as has been predicted theoretically in the early 1980s (Arakawa and Sakaki, 1982; Asada *et al.*, 1986). Particular advantages are ultra-low threshold current density, independent of temperature, and high material and differential gain.

The traits of a *real* structure, like finite inhomogeneous broadening of the ensemble properties, a particular volume coverage with dots, inhibited carrier capture, and relaxation and re-evaporation due to finite localization, etc., partly balance those advantages. In this section we will focus on the requirements for realistic quantum dot arrays serving as the basis for lasers, which are superior to quantum well lasers.

Since quantum dot lasers are fabricated very similarly to quantum well lasers, with the main difference being that the optically active medium consists of quantum dots, we start with an introductory section covering some standard properties of quantum well lasers.

5.8.1 BASIC PROPERTIES OF QUANTUM WELL LASERS

In this section some standard characteristics and elements of quantum well lasers, such as optical confinement, threshold conditions, modal and material gain, are introduced. The basic structure of a quantum well edge emitting laser is depicted in Fig. 5.50. The optical wave propagates parallel to the plane of the p–n junction. The active layer (quantum well of thickness L_z) is embedded in an optical waveguide (material of smaller refractive index) for the wavelength λ/n_r in the semiconductor material. Such a structure is called a separate confinement heterostructure (SCH) because the confinement of the charge carriers and of the optical wave are engineered individually. Optimized waveguides often include a graded index layer (GRIN-SCH). Carriers are injected vertically through the structure, e.g. by contacting the top and bottom surfaces. The ratio of total light intensity and light intensity in the active layer characterizing the overlap between the quantum well and the optical mode is called the optical confinement factor Γ:

$$\Gamma = \frac{\int_{L_z/2}^{-L_z/2} |E(z)|^2 \, dz}{\int_{-\infty}^{\infty} |E(z)|^2 \, dz} \qquad (5.122)$$

In an optimized single quantum well structure Γ is of the order of $2L_z n_r/\lambda$. The cavity (Fabry–Perot resonator) in the edge emitter geometry is defined by the front and back facets, which have reflectivities R_1 and R_2 of about 30% for GaAs- and InP-based devices when fabricated by cleavage. With high-reflection coatings values of 99% or higher can be achieved.

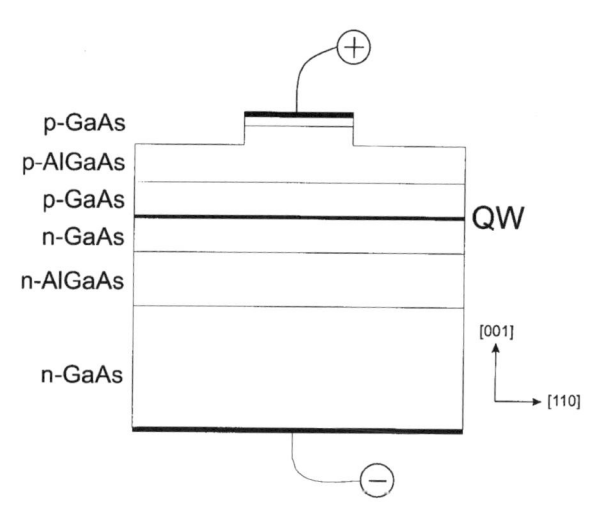

Fig. 5.50 Schematic view on the output facet of the AlGaAs/GaAs quantum well laser

One decisive property of a laser is the current density j_t necessary to obtain *transparency*; i.e. total material gain and absorption and scattering losses in the waveguide compensate. Obviously the threshold current, for which additionally the losses at the mirrors have to be overcome, has to be at least that large. The gain at threshold g_{thr} is written in terms of the internal loss coefficient α_i of the active layer and the loss coefficient α_c in the cladding layers:

$$\Gamma g_{thr} = \Gamma\alpha_i + (1-\Gamma)\alpha_c + \frac{1}{2L}\ln\left(\frac{1}{R_1 R_2}\right) = \alpha_{tot} \qquad (5.123)$$

where $2L$ is the round trip length in the cavity and α_{tot} is called total loss. Γg is called the *modal* gain. The modal gain has to overcome the waveguide and mirror losses. The latter can be varied and decreased by facet coating and by using longer cavities. Systematic variation of the cavity length is one of the important experimental means to characterize lasers. Transparency current and gain are obtained by extrapolating to $L \to \infty$.

In vertical cavity surface emitting lasers (VCSELs) the optical wave is emitted perpendicular to the surface. The structure consists of the active quantum well between two high-reflectivity mirrors, realized by Bragg reflectors. Due to the short cavity length of the device (< 1 μm) the optical losses at the facets are dominant; thus the reflectivity has to be above 99%, requiring about 20 layers for the mirror stacks in the AlAs/GaAs system.

The gain spectrum $g(\hbar\omega)$ is obtained in a similar way as the recombination lifetimes in Section 5.3.1.5. With the same notation we can note for one pair of two-dimensional sub-levels (Yan *et al.*, 1990; Chuang *et al.*, 1995)

$$g(\hbar\omega) = \frac{\pi e^2}{m_0^2 \varepsilon_0 c n_r \omega} \int M^2 \rho_{red}(\varepsilon)[f_c(\varepsilon, E_{F_c}) - f_v(\varepsilon, E_{F_v})] \frac{\Gamma_{in}/\pi}{(\hbar\omega - \varepsilon)^2 + \Gamma_{in}^2} d\varepsilon \qquad (5.124)$$

The matrix element is $M^2 = M_{QW}^2 F^2 M_b^2$ and in the notation of Section 5.3.1.5 $\Gamma_{in} = \hbar/\tau_{in}$ is the homogeneous broadening due to the intra-band relaxation time. (As pointed out in Yan *et al.* (1990) some authors use improperly normalized Lorentzians, which are wrong by $1/\pi$.) For $\tau_{in} \to \infty$, the last factor in Eq. (5.124) can be replaced by $\delta(\hbar\omega - \varepsilon)$. Typical intra-band relaxation times are 0.1 ps (Knox *et al.*, 1986). The populations of electrons (holes) in the conduction and valence band are $f_c = f_n$ and $f_v = 1 - f_p$ (Eq. 5.96), respectively, usually described by Fermi–Dirac distributions. E_{F_c} and E_{F_v} denote the quasi Fermi levels and ρ_{red} is the reduced density of states. (Formula Eq. (5.124) assumes that a factor of 2 is included in the density of states for spin degeneracy.)

From this formula the gain spectrum can be calculated. Handy analytical formulas for gain and other properties of quantum well lasers can be found in Maksym (1996). Equation (5.124) also holds for systems of other dimensionality when the appropriate densities of states are utilized. The net gain condition $f_c(\hbar\omega, E_{F_c}) -$

$f_v(\hbar\omega, E_{F_v}) > 1$ at the optical transition energy is the (population) inversion condition and can also be written as

$$E_{F_c} - E_{F_v} > \hbar\omega \qquad (5.125)$$

At this point we recall that the index of refraction $n(\omega)$ and the extinction coefficient $\kappa(\omega)$ (the absorption/gain is $\alpha = 4\pi\kappa/\lambda$) are parts of the complex dielectric function

$$\sqrt{\varepsilon(\omega)} = n(\omega) + i\kappa(\omega) \qquad (5.126)$$

The real and imaginary part of $\varepsilon = \varepsilon' + i\varepsilon''$ are then given by $\varepsilon' = n^2 - \kappa^2$ and $\varepsilon'' = 2n\kappa$. The real and imaginary parts are connected via the Kramers–Kronig relations, which can be derived on very general grounds, mainly causality between electrical field and dielectric polarization:

$$\varepsilon'(\omega) = \varepsilon'(\infty) + \frac{2}{\pi} P \int_0^\infty \frac{s\varepsilon''(s)}{s^2 - \omega^2} ds, \qquad \varepsilon''(\omega) = -\frac{2\omega}{\pi} P \int_0^\infty \frac{\varepsilon''(s)}{s^2 - \omega^2} ds \qquad (5.127)$$

where P denotes the principal value of the integral. A detailed discussion of the dielectric function in solids can be found in Stern (1963).

Due to increasing populations of higher states in the band and evaporation of carriers from the quantum well into the barrier layers with increasing temperature, the carrier density at a given energy decreases and the threshold current increases. The temperature dependence of j_{thr} is described by an exponential close to a base temperature T_1, introducing the characteristic temperature T_0:

$$j_{\text{th}}(T) = j_{\text{th}}(T_1) \exp\left(\frac{T - T_1}{T_0}\right) \qquad (5.128)$$

Usually the design objective is to have as little variation of the threshold current density with temperature as possible, i.e. to maximize T_0.

Instead of a single quantum well one can employ multiple quantum wells (N wells). The optical confinement factor Γ, and thus the modal gain, increases approximately N times. In case the losses are too large to allow lasing for a single quantum well, a multiple quantum well structure may have sufficient modal gain. If the single quantum well already exhibited lasing, the use of N wells will increase the transparency current density by a factor of N because more states have to be filled to obtain inversion.

Once the laser is above threshold, the light output will increase linearly with the current. The slope is called external differential quantum efficiency and is close to almost 100% for record quantum well lasers (Zhang, 1994a). The internal quantum efficiency is obtained by extrapolating the external efficiency for $L \to \infty$. At very large drive current the light output may start to grow slower and finally saturate when the carrier refill time is identical with the stimulated recombination time (gain saturation).

5.8.2 THE IDEAL QUANTUM DOT LASER

Arakawa and Sakaki (1982) and Asada *et al.* (1986) calculated and compared the gain and threshold current density for bulk (double heterostructure), quantum well, wire and box lasers for the lattice-matched material systems GaAs/AlGaAs and In$_{0.53}$Ga$_{0.47}$As/InP. Distinct effects of dimensionality on the lasing properties have been predicted and motivated extensive experimental and further theoretical research on lasers with low-dimensional gain media.

The gain spectra of bulk, quantum well, wire and dot lasers based on In$_{0.53}$Ga$_{0.74}$As/InP are compared in Fig. 5.51 (Asada *et al.*, 1986). For this calculation a homogeneous (relaxation) broadening with $\tau_{in} = 0.1$ ps ($\Gamma = 6.6$ meV) has been used, and no inhomogeneous broadening of the two-, one-, and zero-dimensional systems is considered. The material gain spectrum for quantum dots is much narrower than for higher-dimensional systems and the maximum gain is larger. The maximum material gain for In$_{0.53}$Ga$_{0.74}$As/InP structures is shown in Fig. 5.52a (Asada *et al.*, 1986) as a function of carrier density. For a given carrier density of 3×10^{18} cm^{-3} the gain of quantum boxes is about 20 times higher than for bulk crystals and 10 times higher than for quantum wells. In this calculation the effects of excitonic interaction and variations of the intra-band relaxation time from the bulk value have been neglected. It was pointed out in Section 5.3.1.5 that

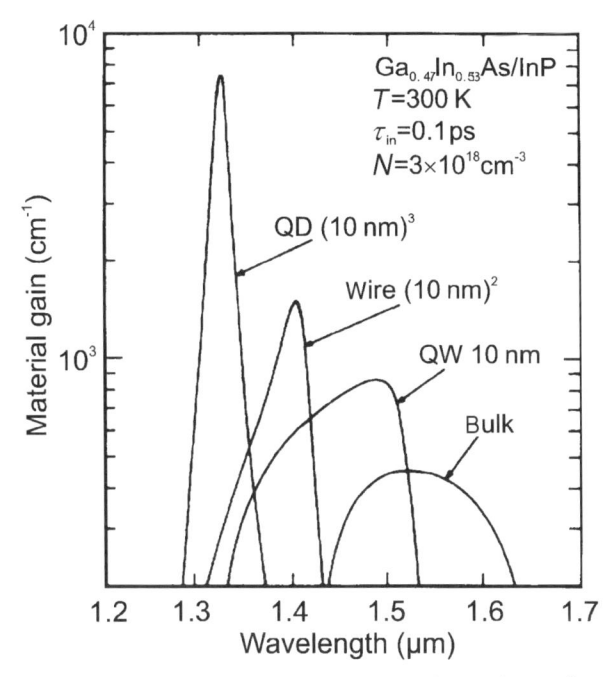

Fig. 5.51 Gain spectra for In$_{0.53}$Ga$_{0.47}$As/InP quantum box, wire, well, and bulk (double heterostructure) at $T = 300$ K, electron density 3×10^{18} cm^{-3} and $\tau_{in} = 0.1$ ps. Confined lateral dimensions are 10 nm each. (After Asada *et al.*, 1986)

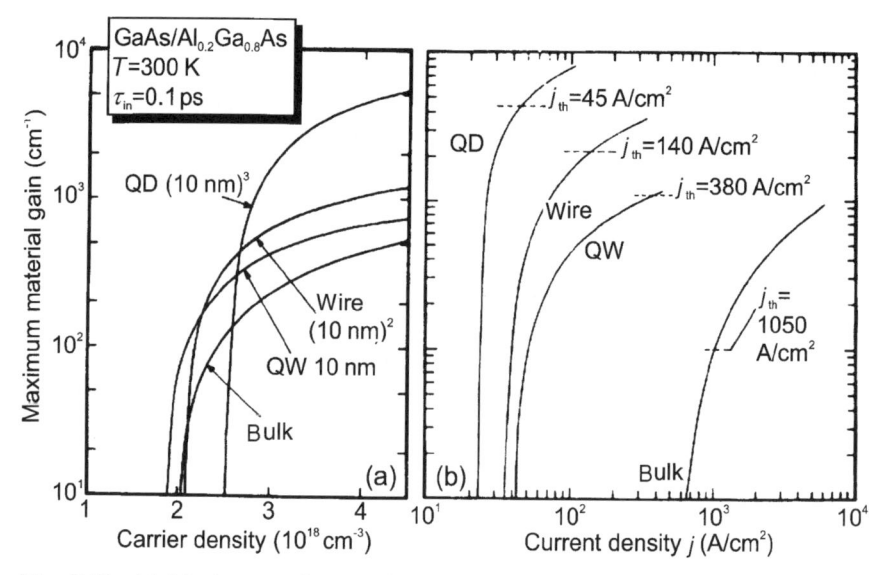

Fig. 5.52 (a) Maximum gain as a function of electron density for $In_{0.53}Ga_{0.47}As/InP$ quantum box, wire, well, and bulk at 300 K for $\tau_{in} = 0.1$ ps. Confined lateral dimensions are 10 nm each. (b) Maximum gain as a function of injection current density for GaAs/$Al_{0.2}Ga_{0.8}As$ quantum box, wire, well, and bulk (conventional double heterostructure) at 300 K for $\tau_{in} = 0.1$ ps. The dashed lines for each curve denote the threshold. (After Asada *et al.*, 1986)

Coulomb correlation will increase the quantum dot oscillator strength compared to the strong confinement (uncorrelated) limit. Since carriers in quantum dots face less scattering processes than carriers in a quantum well, the homogeneous broadening will be smaller, increasing the peak gain linearly with τ_{in}. For a quantum well laser the increase from $\tau_{in} = 0.1$ ps to $\tau_{in} = \infty$ has a rather weak effect and increases the peak gain only by about 50% (Asada *et al.*, 1984).

In Fig. 5.52b threshold current densities for the GaAs/AlGaAs system are compared, showing that a minimum value is obtained for quantum box structures. The parameters used for that calculation were a loss of $10\,cm^{-1}$ and a surface coverage with dots of 0.25.

For the ideal quantum dot laser, assuming infinite barriers for the carrier confinement, the threshold becomes independent of the temperature, $T_0 = \infty$. In Fig. 5.53 the temperature dependence of the threshold for ideal bulk material, quantum well, wire and dot lasers is compared (Arakawa and Sakaki, 1982).

5.8.3 THE REAL QUANTUM DOT LASER

In a real quantum dot laser several issues have to be taken into account. Due to fabrication imperfections or intrinsic limitations, the spectra of the quantum dot

MODELING OF QUANTUM DOTS
183

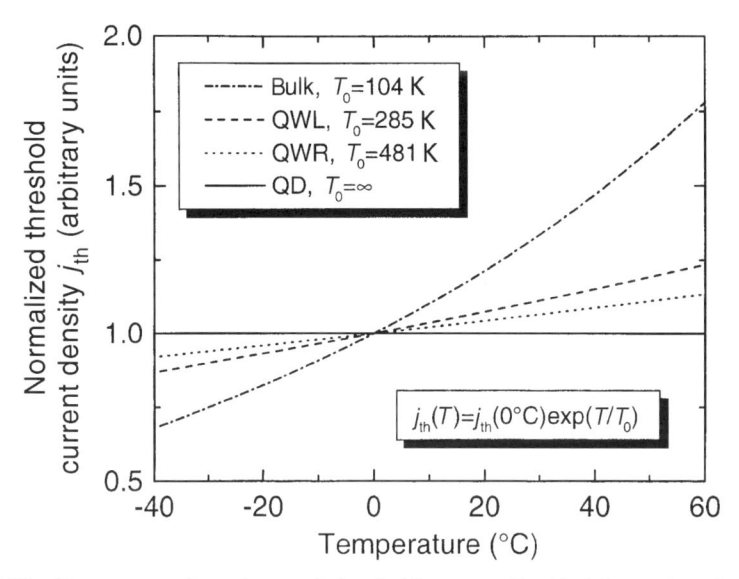

Fig. 5.53 Temperature dependence of threshold current for ideal lasers based on bulk material, quantum well, wire, and dot. (After Arakawa and Sakaki, 1982)

ensemble are inhomogeneously broadened (see Section 5.4.2). Size fluctuations especially will lead to a recombination spectrum with a width of the order of several tens of millielectronvolts, orders of magnitude broader than the expected homogeneous linewidth. For typical laser dimensions of several micrometers in stripe width and several hundred micrometers in cavity length and quantum dot densities of several $10^{10}\,\mathrm{cm}^{-2}$, there are about 10^6 quantum dots in the active zone. Depending on the broadening, the peak material gain can be very high. However, the optical confinement factor for quantum dot arrays is very small. The resulting modal gain might be small. In real lasers all confinement potentials are finite, making the escape of carriers into the barrier possible. Also, the energy separation of a quantized state becomes finite, enabling a thermal population of excited states.

5.8.3.1 Optical confinement factor

For a quantum dot array the optical confinement factor is of the order of total dot volume to total waveguide volume. It can be split into the in-plane and vertical confinement factors Γ_{xy} and Γ_z, respectively:

$$\Gamma = \Gamma_{xy}\Gamma_z \qquad (5.129)$$

The in-plane confinement factor is given by the number of quantum dots N_D of average in-plane size A_D for an area A:

$$\Gamma_{xy} = \frac{N_D A_D}{A} = \zeta \qquad (5.130)$$

where ζ is also called the area coverage with dots. If in x and y directions different coverage is present and Γ_{xy} can be split into $\Gamma_x\Gamma_y$. The vertical confinement factor is given by the vertical overlap of quantum dots and optical mode, averaged over the plane of area A:

$$\Gamma_z = \frac{1}{A}\int\int_A\int_{\text{dot}} |E(z)|^2 \, dz / \int_{-\infty}^{\infty} |E(z)|^2 \, dz \qquad (5.131)$$

For a typical quantum dot array of 4×10^{10} quantum dots per cm^2, with an approximate size of $7 \times 7 \times 2$ nm^3, the area coverage is $\zeta = 0.02$ and the vertical confinement factor is about $\Gamma_z = 7 \times 10^{-3}$ for a waveguide thickness of 150 nm. Thus the total optical confinement factor is $\Gamma = 1.4 \times 10^{-4}$.

In the case of stacked quantum dot arrays with several sheets of uncoupled dots, the vertical confinement increases approximately linearly with the number of dot layers. In vertically stacked, electronically coupled quantum dots, additionally the oscillator strength increases.

5.8.3.2 Gain and Threshold

With the reduced density of states for the ith single quantum dot with ground state quantization energy $E_{0,i}$,

$$\rho_{\text{red}}^{\text{QD},i}(\varepsilon) = \frac{2}{V_0}\delta(\varepsilon - E_g - E_{0,i}) \qquad (5.132)$$

the material gain of the quantum dot ensemble is (the summation extends over all dots)

$$g(\hbar\omega) = \frac{\pi e^2}{m_0^2 \varepsilon_0 c n_r \omega}$$
$$\times \int \sum_i M^2 \frac{2}{V_0} \delta(\varepsilon - E_g - E_{0,i})[f_c(\varepsilon, E_{F_c}) - f_v(\varepsilon, E_{F_v})] \frac{\Gamma_{\text{in}}/\pi}{(\hbar\omega - \varepsilon)^2 + \Gamma_{\text{in}}^2} \, d\varepsilon \qquad (5.133)$$

The matrix element M_b was discussed in Section 5.3.1.5 and is $M_b^2 = p_{cv}^2/3$. (In the strong confinement limit assuming envelope wavefunction overlap, $F^2 = 1$. A factor of 2 for spin degeneracy has been included in the density of states.) This sum over about 10^6 quantum dots is obviously difficult to handle. In many cases it makes sense to assume a Gaussian dot volume size distribution with a particular relative standard deviation ξ and average volume V_0, corresponding to an average dot ground state quantization energy E_0 (see Section 5.4.2). Then the energetic broadening function is given by

$$P(\varepsilon, \sigma_E) = \frac{1}{\sigma_E\sqrt{2\pi}} \exp\left[-\frac{(\varepsilon - E_g - E_0)^2}{2\sigma_E^2}\right] \qquad (5.134)$$

with a spectral width (see Eq. 5.83) of $\sigma_E \approx 2\xi \bar{E}_0$. The gain can then be written as

$$g(\hbar\omega) = \frac{\pi e^2}{m_0^2 \varepsilon_0 c n_r \omega} \\ \times \int M_b^2 \frac{2}{V_0} P(\varepsilon, \sigma_E)[f_c(\varepsilon, E_{F_c}) - f_v(\varepsilon, E_{F_v})] \frac{\Gamma_{in}/\pi}{(\hbar\omega - \varepsilon)^2 + \Gamma_{in}^2} d\varepsilon \quad (5.135)$$

Comparing this equation with Eq. (5.124) shows that the inhomogeneously broadened quantum dot ensemble exhibits a decreased (ground state) peak density of states:

$$\rho_{red}^{Ensemble}(\varepsilon) = \frac{2}{V_0} P(\varepsilon, \sigma_E) \quad (5.136)$$

In the case $\Gamma_{in} \ll \sigma_E$, the Lorentzian can be approximated by a δ function again and the material gain for the quantum dot ground state transition in an inhomogeneous ensemble is

$$g(\hbar\omega) = C_g P(\hbar\omega, \sigma_E)[f_c(\hbar\omega, E_{F_c}) - f_v(\hbar\omega, E_{F_v})], \quad C_g = \frac{\pi e^2}{m_0^2 \varepsilon_0 c n_r \omega} M_b^2 \frac{2}{V_0} \quad (5.137)$$

The peak material gain is inversely proportional to the energetic broadening due to ensemble fluctuations. If, for simplicity, the homogeneous broadening is assumed as Gaussian $P(\varepsilon, \gamma)$, σ_E in Eq. (5.136) can be replaced by $\sqrt{\sigma_E^2 + \gamma^2}$.

In a simple model it is assumed that all injected carriers are captured by the quantum dots and that overall charge neutrality exists, i.e. $f_v = 1 - f_c$ and $f_c - f_v = 2f_c - 1$. Further, it is assumed that all dot properties such as lifetime, carrier capture, etc., are independent of the ground state energy. In this case the total number N of electrons (and holes)

$$N = 2f_c N_D \quad (5.138)$$

is equally distributed amongst all dots, i.e. a nonequilibrium population (NTC, nonthermal coupling) exists. The shape of the material gain spectrum of the quantum dot ensemble is then independent of the pumping level† and temperature (see Fig. 5.54a):

$$g(\hbar\omega) = \frac{N - N_D}{N_D} C_g P(\hbar\omega, \sigma_E) \quad (5.139)$$

The gain increases *linearly* (Fig. 5.55, solid line) from the value $-C_g P$ with increasing carrier density and saturates at $+C_g P$ when all quantum dot ground states are filled with two electrons and two holes ($N = 2N_D$) (Vahala, 1988).

---

† Small energy shifts of the ground state transition due to biexciton binding energy are neglected since they are small compared to the typical inhomogeneous broadening.

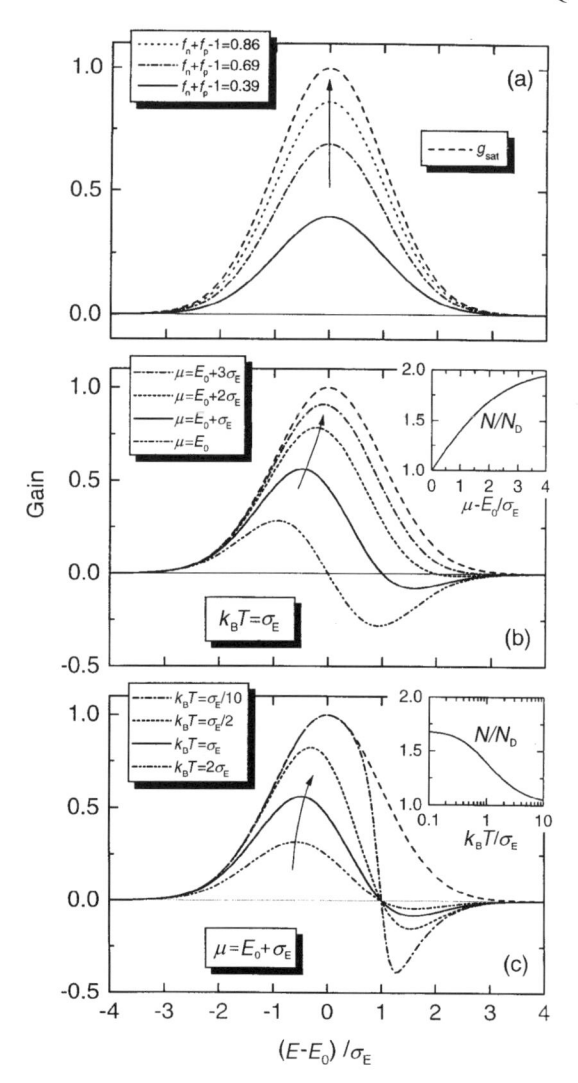

Fig. 5.54 Gain spectra (in units of maximum saturated modal gain, $\zeta\Gamma_z C_g/(\sqrt{2\pi}\sigma_E)$), for a quantum dot array, (a) for vanishing thermal escape from the dots (NTC, nonequilibrium population) and different inversion levels, (b) and (c) for the equilibrium dot population (TC, Fermi statistic). (b) Gain as a function of chemical potential ($\mu = \mu_e + \mu_h$, $\mu_e = \mu_h$) for $k_B T = \sigma_E$, where the inset depicts carrier number as a function of chemical potential. (c) Gain as a function of temperature for $\mu = \sigma_E$, where the inset depicts carrier number as a function of temperature for $\mu = \sigma_E$. Dashed line in all graphs is the spectrum of saturated gain (After Grundmann and Bimberg, 1997c)

MODELING OF QUANTUM DOTS 187

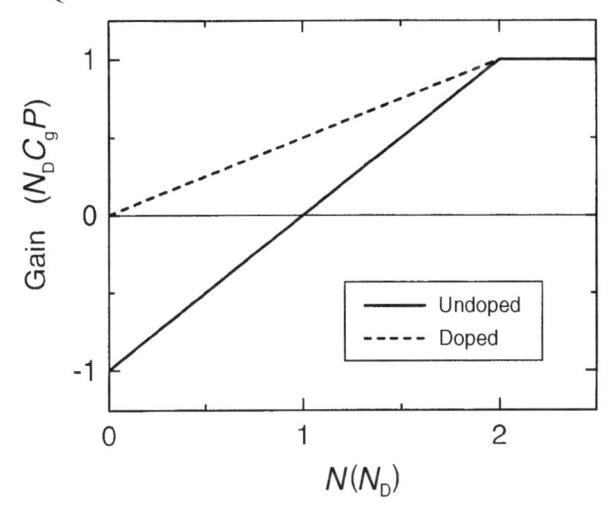

Fig. 5.55 Gain of quantum dots (in units of $C_g P$, see text) as a function of carrier density N in units of quantum number N_D. Strictly, for undoped dots N is the averaged electron and hole density $(n_e + n_h)/2$, for the doped case N represents the minority carrier density. (After Vahala, 1988)

As pointed out in Grundmann and Bimberg (1997b), it is nontrivial to obtain the correct relation between the gain and the injection current. The same gain value can be obtained for different populations that have the same carrier density but different (recombination) current (as detailed in Section 5.5.2.3). For example, dots filled with two electrons (20) and dots filled with an exciton (11) both provide zero gain. The (20) dots, however, show no recombination current, whereas the (11) dots have a recombination rate of $1/\tau_D$, i.e. one electron–hole pair per exciton lifetime τ_D.

In the mean field spirit of conventional rate equation models the transparency current I_{tr} and transparency current density j_{tr} are ($N_{tr} = N_D$)

$$I_{tr} = \frac{2eN_D}{\tau_D} f_c(1 - f_v) = \frac{1}{2} \frac{eN_D}{\tau_D} \left(\frac{N_{tr}}{N_D}\right)^2 = \frac{1}{2} \frac{eN_D}{\tau_D} \quad (5.140)$$

$$j_{tr} = \frac{I_{tr}}{A} = \frac{1}{2} \frac{e}{A_D \tau_D} \zeta \left(\frac{N_{tr}}{N_D}\right)^2 = \frac{1}{2} \frac{e}{A_D \tau_D} \zeta \quad (5.141)$$

A more detailed theory, taking into account the random character of carrier capture and recombination (Grundmann and Bimberg, 1997c), yields a factor of 5/8 instead of 1/2 in Eq. (5.140), when *independent* electron and hole capture is assumed (see Fig. 5.56a). For capture of electron–hole pairs (only neutral dots exist) the pre-factor is 1. In the following we will use the pre-factor 5/8, being aware of the fact that, depending on the carrier capture processes, the threshold current may be (by less than a factor of two) larger. The case of dots that have an initial occupation with carriers without injection, e.g. due to doping, is discussed further below.

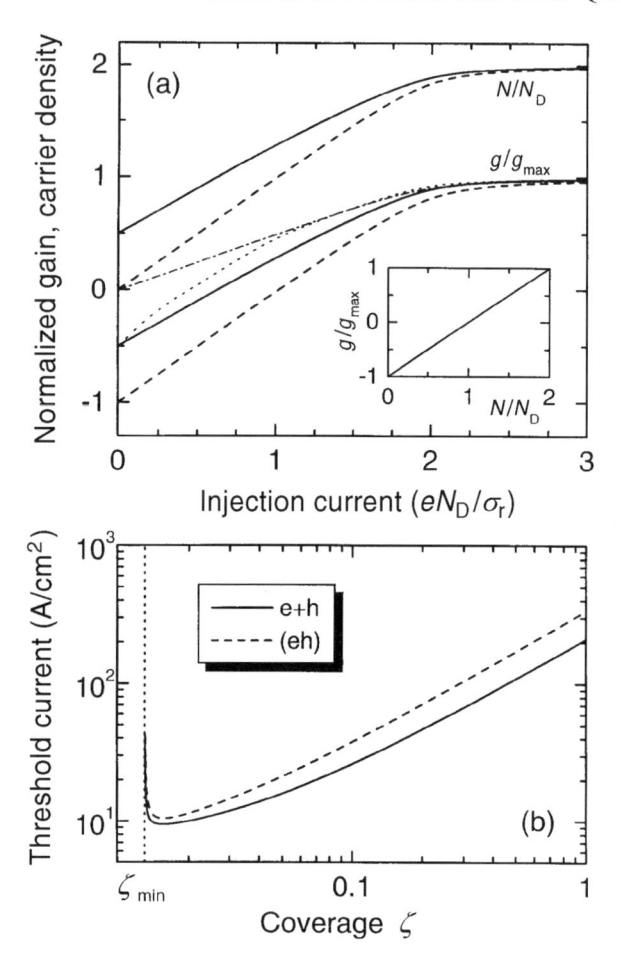

Fig. 5.56 (a) Gain and carrier density in a quantum dot ensemble as a function of injection current for separate (solid lines) and simultaneous (dashed lines) capture of electron and holes. The gain is given in units of the maximum gain $C_g P$ (see text). A barrier (wetting layer) which provides an additional recombination channel is included. We have assumed identical recombination time constants τ_b and τ_D in the barrier and dot, respectively, and a capture time into the dot of $\tau_c = \tau_D/100$. The inset shows the relation between gain and carrier density in the dot ensemble, being identical for both capture models. The dash–dotted (dotted) line shows the gain for doped dots with an average population of 2 (1) electron or hole without external injection current. The dashed line for the carrier density also gives the recombination rate (in units of N_D/τ_D) for simultaneous capture. The recombination rates for the other models are very close to that curve. (b) Threshold current for both capture models as a function of coverage for a typical dot ensemble of 4×10^{10} dots/cm^2 discussed in the text; $\tau_D = 1$ ns and a total loss of $\alpha_{tot} = 10$ cm^{-1}. (After Grundmann and Bimberg, 1997c)

With $\sqrt{A_D} = 7$ nm, $\zeta = 0.02$ (corresponding to 4×10^{10} dots/cm^2), and $\tau_D = 1$ ns the transparency current amounts to $j_{tr} = 4$ A/cm^2. For 2 nm high dots with $\sigma_E = 20$ meV, the saturated maximum material gain per dot (at $\hbar\omega = E_g + E_0$) is $\hat{g}_{mat}^{sat} = C_g/(\sqrt{2\pi}\sigma_E)$, which amounts to about 1×10^5 cm^{-1}. This value is huge, but the maximum saturated modal gain $\hat{g}_{mod}^{sat} = \zeta\Gamma_z\hat{g}_{mat}^{sat}$ is rather small; with $\Gamma_z \approx 0.7\%$, $g_{mod} = 14$ cm^{-1}, which is only somewhat larger than losses found in typical lasers (2–10 cm^{-1}).

In order to overcome a given total loss α_{tot}, the carrier density at threshold Nth has to be at least (lasing starts at $\hbar\omega = E_g + E_0$)

$$N_{th} = N_D\left(1 + \frac{1}{\zeta}\frac{\sqrt{2\pi}\sigma_E\alpha_{tot}}{\Gamma_z C_g}\right) \tag{5.142}$$

Since Nth cannot be larger than $2N_D$ (when only gain of the ground state is considered), the minimum area dot coverage is

$$\zeta_{min} = \frac{\sqrt{2\pi}\sigma_E\alpha_{tot}}{\Gamma_z C_g} \tag{5.143}$$

The threshold current is approximately given by (see Fig. 5.55a)

$$j_{th} \approx \frac{1}{2}\frac{e}{A_D\tau_D}\zeta\left(1 + \frac{\zeta_{min}}{\zeta}\right)^2 \tag{5.144}$$

For the parameters of the above quantum dot array and a total loss $\alpha_{tot} = 10$ cm^{-1}, $\zeta_{min} = 0.013$, the threshold current density (Fig. 5.56b) is about $j_{th} = 10$ A/cm^2. For $\zeta < \zeta_{min}$, the maximum saturated gain cannot overcome the total losses and lasing does not occur†, i.e. $j_{th} = \infty$. For $\zeta \gg \zeta_{min}$ the threshold current density approaches the transparency current density (Eq. 5.141) $j_{th} \to j_{tr}$ for infinite cavity length.

When spontaneous recombination in the waveguide of thickness b is included in the model with a bimolecular recombination coefficient B and carrier densities n and p (see Section 5.3), an additional current

$$j_b = ebBnp \tag{5.145}$$

is introduced, where B denotes the bimolecular recombination coefficient. Miyamoto et al. (1989) used a monomolecular process which gives qualitatively similar results. With the results from Section 5.3 an additional contribution to the threshold current is introduced (using Eq. 5.101):

$$j_{th} \approx \frac{1}{2}\frac{e}{A_D\tau_D}\zeta\left(1 + \frac{\zeta_{min}}{\zeta}\right)^2 + \frac{1}{4}\frac{ebB}{v_n v_p \tau_D}\frac{(1 + \zeta_{min}/\zeta)^2}{(1 - \zeta_{min}/\zeta)^2} \tag{5.146}$$

---

† The model does not consider the contribution of excited quantum dot states to gain.

A similar dependence of the threshold current on the filling factor has been obtained by Nambu and Asakawa (1995) for $B = 0$ but taking into account the filling of higher states, which is discussed in the next section.

For the case of strong thermal carrier reemission from the quantum dots (and recapture), an equilibrium Fermi (Eq. 5.96) distribution (TC, thermal coupling) is established for the quantum dot population; i.e. larger dots have a higher filling factor and achieve inversion more quickly. The shape of the gain spectrum of the whole ensemble then changes with pumping level and temperature (see Fig. 5.53b and c):

$$g(\hbar\omega) = C_g P(\hbar\omega, \sigma_E)[f_c(\hbar\omega, E_{F_c}) - f_v(\hbar\omega, E_{F_v})] \quad (5.147)$$

The maximum gain value depends on the carrier distribution in the quantum dot ensemble (Fig. 5.57). As can be seen already from Fig. 5.54a, in the case of no thermal coupling the gain maximum is always at $E_{max} = E_0$ and increases proportionally to $f_n + f_p - 1$. For a Fermi distribution of the carriers, the energy position of the gain maximum is $E_{max} \leq E_0$ and depends on the temperature and the Fermi energy.

If only gain from the ground state is considered, the gain maximum is always larger for a thermal (TC) than for a nonthermal coupling (NTC) distribution, for a given carrier density. If the population of excited states E_n is considered, this is not generally the case. Using the absorption spectrum (i.e. working with parabolic bands and neglecting the biexciton binding energy) of Fig. 5.36, the threshold current density and the energy E_{max} of the maximum gain can be calculated for spherical quantum dots (Fig. 5.58). The disorder parameter $\xi = \sigma_R/R_0$ determines the ratio of

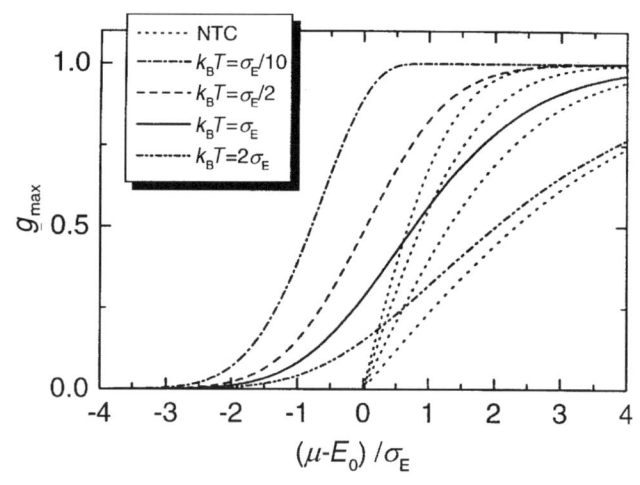

Fig. 5.57 Maximum gain (in units of maximum saturated modal gain, $\zeta\Gamma_z C_g/(\sqrt{2\pi}\sigma_E)$, for a quantum dot array. Fermi distributions for various temperatures are compared to nonequilibrium (NTC) population of dots (dashed lines) for the same carrier number for a given value of μ

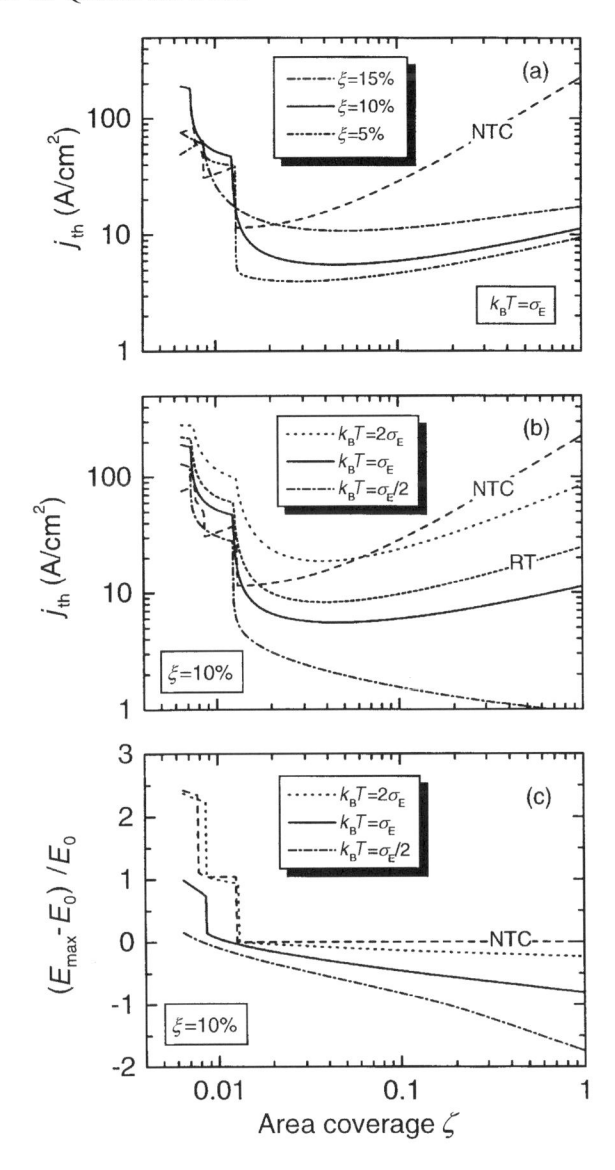

Fig. 5.58 Quantum dot laser properties with inclusion of gain from excited states, assuming spherical dots and parabolic bands. (a) Threshold current density for different disorder parameters $\xi = \sigma_E/(2E_0)$ for $k_B T = \sigma_E = 20$ meV. Dashed line is the NTC limit when only the ground state is considered. (b) Threshold current density for $\xi = 10\%$ for different temperatures. Dashed line is the NTC limit when excited states are taken into account. (c) Energy of gain maximum for $\xi = 10\%$ and different temperatures in units of $E_0 = \sigma_E/(2\xi) = 100$ meV. For small area coverage lasing occurs on excited states (After Grundmann and Bimberg, 1997c)

energetic broadening and sublevel separation via Eq. (5.83), $2\xi = \sigma_E/E_0$. From Fig. 5.58a it is obvious that a greater disorder increases the threshold current for a given area coverage by dots. For a small disorder, $\xi < 10\%$, the thermal carrier distribution leads to smaller threshold currents, especially for large area coverage ζ, compared to a nonthermal population. For $\zeta < \zeta_{min}$, sufficient gain on excited states enables lasing also for low coverage (or higher total losses).

In Fig. 5.58b and c the temperature of the Fermi distribution is varied. Particularly for large area coverage, low temperatures lead to small threshold current. From the dependence of the energy of the gain maximum E_{max} on the area coverage it is evident that for $\zeta < \zeta_{min}$ lasing occurs on excited states. Depending on the localization energy, in addition to excited states the carrier density in the barrier and the optical waveguide has to be considered.

For k stacked layers of identical, uncoupled quantum dots, the previous formulas remain valid if the variables are properly scaled. The minimum coverage is decreased by a factor k. The maximum gain is increased by a factor of k, thus reducing the problem of gain saturation. Since for large coverage the thermal distribution resulted in significantly lower threshold currents than the NTC population, this is even more true for stacked dots. For vertically coupled but in-plane decoupled quantum dots an intermediate situation of NTC and TC arises (Schmidt et al., 1996b; Grundmann and Bimberg, 1997c).

When the quantum dots are populated by two electrons or two holes per quantum dot, then f_c or f_v are always 1 or 0, respectively, and the ensemble gain is always larger than for the undoped case (Fig. 5.55, dashed line):

$$g(\hbar\omega) = \frac{N}{2N_D} C_g P(\hbar\omega, \sigma_E) \tag{5.148}$$

where N/N_D describes here the *minority* carrier density in the dot array. The gain increases linearly from 0 to $+N_D C_g P$ and saturates (at the same level as for the undoped array) for $N > 2N_D$.

Even without injection the quantum dots are already at the transparency limit (Vahala, 1988), i.e. $j_{tr} = 0$. The threshold current (for strong capture) is plotted in Fig. 5.56a according to

$$j_{th} = 2\frac{e}{A_D \tau_D}\frac{\zeta_{min}^2}{\zeta} + \frac{1}{4}\frac{ebB}{v_n v_p \tau_D}\frac{(1 + \zeta_{min}/\zeta)^4}{(1 - \zeta_{min}/\zeta)^2} \tag{5.149}$$

Because there is no 'penalty' to achieve inversion, the threshold for doped arrays with higher coverage is significantly smaller.

More detailed modeling of quantum laser properties has to include the size dependence of lifetimes and capture rate, thermal evaporation, as well as details of the microscopic gain mechanism (see Section 5.8.3.3).

5.8.3.3 Microscopic gain mechanism

Detailed gain spectra obtained from numerical solutions for exciton and biexciton wavefunctions, additionally taking into account valence band mixing and the surface polarization effect, have been calculated by Hu et al. (1996) for CdSe quantum dots. Gain dynamics are computed by solving the Fourier transformed multilevel Bloch equations.

In a model with parabolic bands without Coulomb interaction and valence band mixing, only transitions between electron and hole states with equal quantum numbers are dipole-allowed. Valence band mixing additionally allows the transition between the ground state and $|1s,1d_{3/2}\rangle$ (Xia, 1989). The Coulomb interaction allows a series of additional transitions between biexcitons and excitons (Hu et al.,

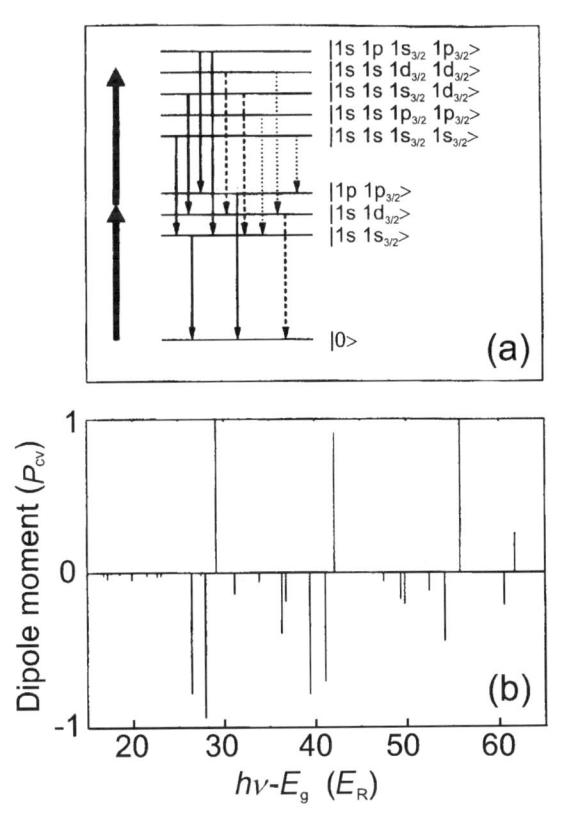

Fig. 5.59 (a) Exciton and biexciton levels of a quantum dot. $|0\rangle$ is the ground state, $|1s,1s_{3/2}\rangle$ is the exciton ground state, and $|1s,1s,1s_{3/2},1s_{3/2}\rangle$ is the lowest biexciton ground state. The solid downward arrows denote the dipole allowed spontaneous emission channels. Dashed (dotted) arrows denote channels becoming dipole, allowed due to the inclusion of Coulomb interaction (valence band mixing) into the calculation. (b) Dipole moments between exciton states and ground state (upper half) and biexciton states and exciton states (lower half). Energy is in units of $E_R = 13.6$ meV. Material parameters are $R_0 = 0.5\, a_B$ ($a_B = 5.3$ nm), $m_e^* = 0.2$, $\gamma_1 = 5.0$, $\mu = 0.75$, $\varepsilon_2 = 10$, $\varepsilon_1 = 1$. (After Hu et al., 1996)

1990). In Fig. 5.59 a simplified scheme of the allowed transitions and their dipole matrix elements is shown. The allowed transitions can be obtained from symmetry arguments; obtaining their relative strength requires a numerical calculation.

When initially biexciton states are populated, optical gain can be realized by stimulated transitions from biexciton to exciton states. Due to the positive biexciton binding energy (see Section 5.3.1.8), the stimulated emission from the ground state biexciton $|1s,1s,1s_{3/2}\rangle$ to the ground state exciton $|1s,1s_{3/2}\rangle$ has a lower energy than the linear absorption of the ground state. Other comparably weaker transitions from biexciton to excited exciton states have even lower energy, e.g. the $|1s,1s,1s_{3/2}\rangle$ to $|1p,1p_{3/2}\rangle$ transition (Hu et al., 1996). The resulting linear absorption and gain spectra are depicted in Fig. 5.60. The first three absorption resonances are due to transitions between the ground state and $|1s,1s_{3/2}\rangle$, $|1s,1d_{3/2}\rangle$, and $|1p,1p_{3/2}\rangle$. The gain spectra, calculated for different initial biexciton populations, are dominated by the biexciton to exciton transitions.

In the strong confinement regime the gain (in units of the maximum gain) and recombination current (in τ_X^{-1}) for the different populations of the ground state are given by the values in Table 5.3 (Grundmann and Bimberg, 1997c).

In the previous chapter it was assumed that the biexciton binding energy is much smaller than the inhomogeneous broadening. In the case where the biexciton binding energy is larger than the inhomogeneous broadening, exciton and biexciton transi-

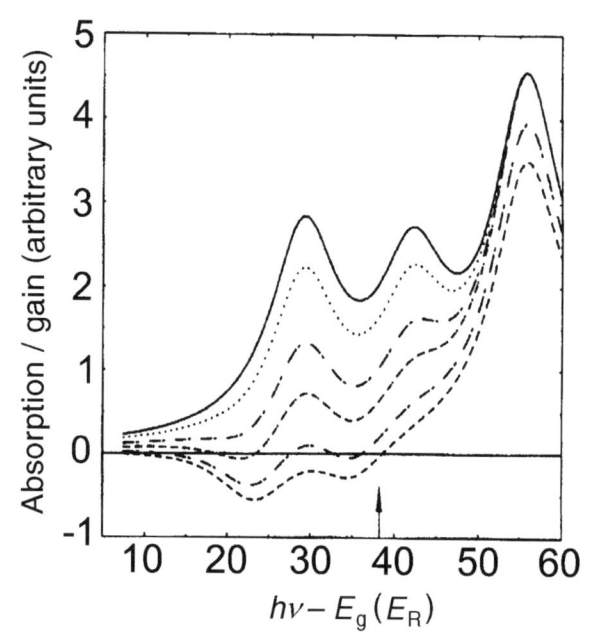

Fig. 5.60 Linear absorption (solid) and bleached absorption/gain (dashed) with various biexciton or exciton populations. Material parameters are the same as in Fig. 5.59. The homogeneous dephasing time is chosen as 10 fs. From top to bottom, dashed curves correspond to increasing biexciton or exciton population. The arrow shows one-half of the maximum biexciton energy that was populated in the calculation. (After Hu et al., 1996)

MODELING OF QUANTUM DOTS

Table 5.3 Dependence of the carrier density $n_c = (n+m)/2$ (so that a completely filled dot has $n_c = 2$), absorption coefficient α (scaled so that an empty dot has $\alpha = 1$), gain g (in units of the maximum gain, thus varying between -1 and 1), and recombination rate R (in units of $1/\tau_r$, i.e. one electron-hole pair per lifetime) and the energy E_{ph} of the emitted photon, for a quantum dot ground state having different populations with n electrons and m holes. For all populations $g = n_c - 1$ is valid. X denotes the exciton and XX the biexciton. (After Grundmann and Bimberg, 1997c)

n	m	n_c	α	g	R	E_{ph}
0	0	0	1	-1	0	—
1	0	0.5	-0.5	0	0	—
0	1	0.5	0.5	-0.5	0	—
1	1	1	0.5	0	1	X
2	0	1	0	0	0	—
0	2	1	0	0	0	—
2	1	1.5	0	0.5	1	X$^-$
1	2	1.5	0	0.5	1	X$^+$
2	2	2	0	1	2	XX

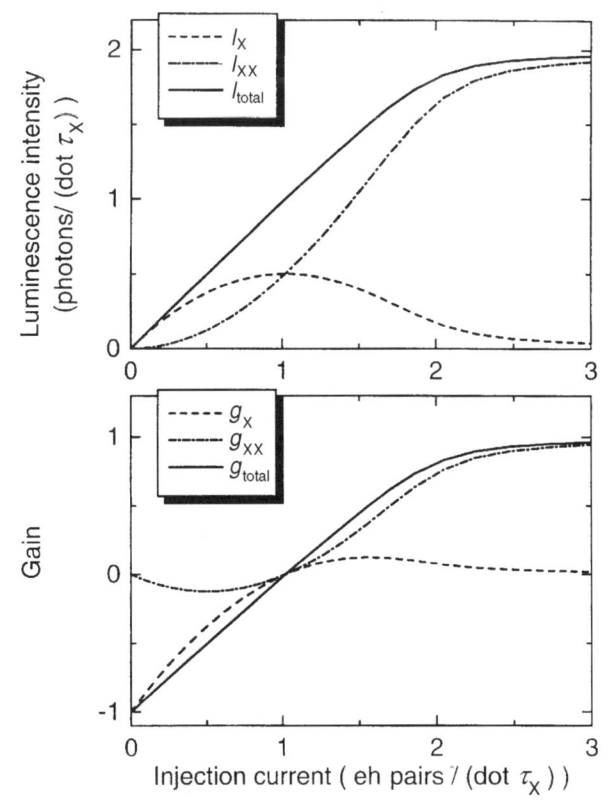

Fig. 5.61 Total, excitonic, and biexitonic luminescence intensities and gains of a quantum dot ground state versus external injection current. A capture time from the barrier $\tau_c = \tau_X/100$ and a barrier recombination channel with $\tau_b = \tau_X$ have been assumed

tions are observed separately. This can be the case for II–VI compounds (Strassburg et al., 1998). Following the stochastic population theory outlined in Section 5.5.2.2, the luminescence intensity and gain of the exciton and biexciton transition are shown in Fig. 5.61.

5.8.3.4 Dielectric effects

The dielectric function of the host material is changed by the presence of quantum dots around the optical transition energies (see Eq. 5.127). The modification depends, of course, on the energetic positions of the QD levels, their oscillator strength, the homogeneous and inhomogeneous broadening, and the carrier population (Grundmann and Bimberg, 1997c), as well as on the volume filling factor and the lateral arrangement (Kayanuma, 1993; Kalosha et al., 1998).

Strong changes in the real part of the index of refraction due to the QD oscillator strength have been observed for both InAs (Aigouy et al., 1997) and CdZnSe QDs (Lendentsov et al., 1996c), leading to waveguiding on the low energy side of the exciton transition (Alferov et al., 1994). The refractive index in the case of empty dots and carrier injection is compared in Fig. 5.62.

Similarly, chirp (Section 8.2.3.2) is due to the time-dependent modulation of the index of refraction during laser pulse emission. In microcavities the (empty) resonator mode interacts with the dispersion of the quantum dots, creating coupled modes when the QD transition energy is close (but not necessarily identical) to the empty cavity mode (Section 8.2.4).

The incoherent sum over all quantum dots for the total gain as written in Eq. (5.133) is strictly valid only for dilute arrays of irregularly placed dots. In dense and

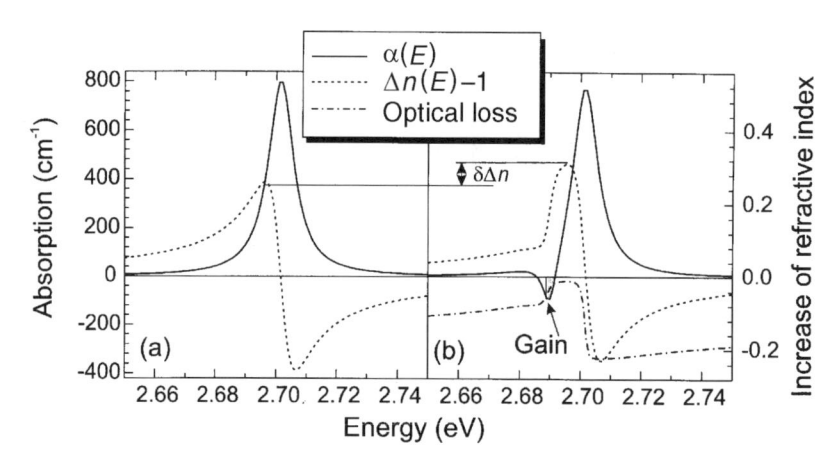

Fig. 5.62 Principal of excitonic waveguiding. Quantum dot absorption, change of index of refraction, and optical loss in the case of (a) no carriers and (b) carrier injection where a gain region forms on the low energy side. Calculations are made for the CdSe/ZnSe system. (After Krestnikov et al., 1998)

regular arrays electrodynamic coupling between dots arises, even when the size of and distance between dots are smaller than the wavelength of the resonance energy (Kayanuma, 1993; Kalosha *et al.*, 1998). For dense and periodic arrays the polariton becomes an eigenstate of the photon/material system (Hopfield, 1958), leading, for example, to a forbidden gap around the resonance energy in which the reflection coefficient is unity. Such effects are weakened and eventually destroyed by inhomogeneous broadening and geometric disorder.

6 Electronic and Optical Properties

Optical experiments are particularly well suited to reveal the unique electronic properties of zero-dimensional systems. The δ-function density of states manifests itself in the occurrence of very narrow spectral lines emitted by single dots. The size dependence of shift and splitting of the energy states are directly visualized by the optical spectra; cw-luminescence, absorption, reflection and excitation spectroscopy probe ground and excited zero-dimensional electronic states, state filling, and energy relaxation amongst other properties. Carrier dynamics, namely capture, relaxation and recombination, are experimentally accessed using time-resolved spectroscopy. Local luminescence probe techniques prove to be particularly useful for the investigation of single quantum dots (Gustafsson et al., 1998).

Sections 6.1 to 6.6 briefly review the results of such experiments on various zero-dimensional systems fabricated with a number of different technologies, discussed in Section 6.2. Most attention is, however, devoted to self-organized quantum dots. For this class of quantum dots larger quantization effects, higher localization energies and sublevel separations are observed than for any other class of dots. In addition, their area density of up to several $10^{11}\,\mathrm{cm}^{-2}$ at which they can be reproducibly produced with comparatively little effort makes them extremely attractive for many different future photonic and electronic devices taking advantage of the unique electronic properties. Optical properties of type-I self-organized quantum dots are discussed in Section 6.7. Results on single layers of dots, and stacked, doped, and annealed dots are presented. The properties of type-II quantum dots are discussed in Section 6.8.

6.1 ETCHED STRUCTURES

The initial attempts towards fabrication of small nanostructures (<100 nm) using lithography and dry etching techniques were haunted by small quantum efficiencies. Etching introduced surface damage which led to optically dead layers. The observed decrease of luminescence efficiency from *pseudo*-wires with decreasing length by Benisty et al. (1991) was originally attributed to an intrinsic slow-down of carrier capture and energy relaxation, the so-called phonon bottleneck (see Section 5.7.2), leading to vanishing quantum efficiency. In the meantime it has become clear that nonradiative recombination related to process-induced defects is at its origin.

Optical spectra of quantum 'disks' dry etched from quantum well superlattices were presented in Kash et al. (1986) and Reed et al. (1986). A slight modification of the spectra but no clear evidence for zero-dimensional effects was reported.

ELECTRONIC AND OPTICAL PROPERTIES

By employing lithography and wet chemical etching it is now possible to fabricate quantum dots with much improved optical properties (Steffen *et al.*, 1994, 1996c). The energy of the ground and first excited state transition increases with decreasing diameter of the free-standing InGaAs/GaAs dots (Fig. 6.1) (Steffen *et al.*, 1996b; Michel *et al.*, 1997). With increasing excitation intensity the excitonic transition saturates and a biexciton line at lower energies appears. At still higher excitation intensity the exciton line disappears and the biexciton luminescence saturates. At the high-energy side of the exciton line transitions due to the excited states appear (Forchel *et al.*, 1996a; Steffen *et al.*, 1996a) (Fig. 6.2). The excitation-dependent integrated intensities of the various recombination lines in a single 60 nm diameter dot agree very well with the predictions made by the random population (RP) theory (Fig. 6.3) (see Section 5.5.2.2). We note that the maximum of the intensity of the exciton is about one-fifth of the saturation value of the biexciton as predicted by RP theory. The maximum of the exciton intensity occurs at the predicted value of excitation of one electron–hole pair per exciton lifetime per dot. Multiexciton complexes were investigated systematically by Bayer *et al.* (1998).

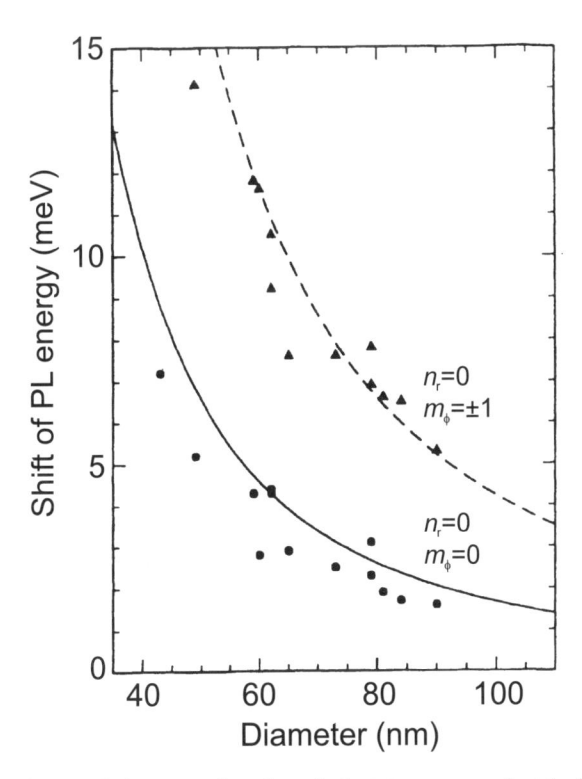

Fig. 6.1 Dependence of the ground and excited state energy of etched InGaAs/GaAs quantum dots as a function of dot diameter. Symbols are experimental data, lines are model calculation. (After Forchel *et al.*, 1996a)

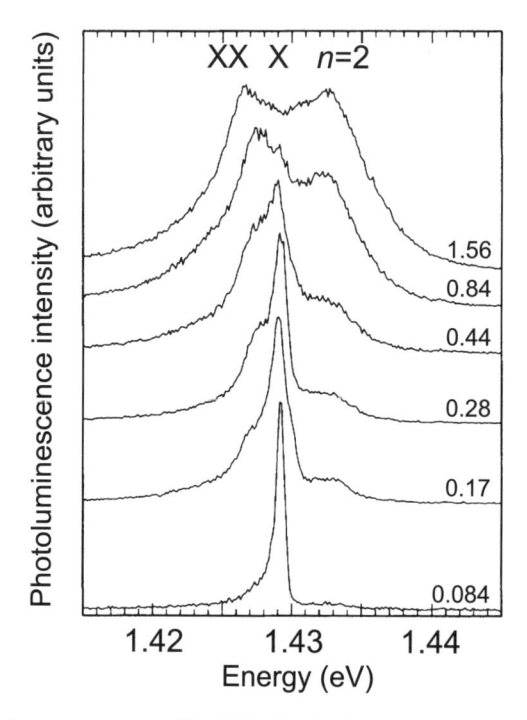

Fig. 6.2 Photoluminescence spectra ($T = 2$ K) of a single-etched InGaAs/GaAs quantum dot with 60 nm diameter for different excitation densities. (After Steffen et al., 1996a)

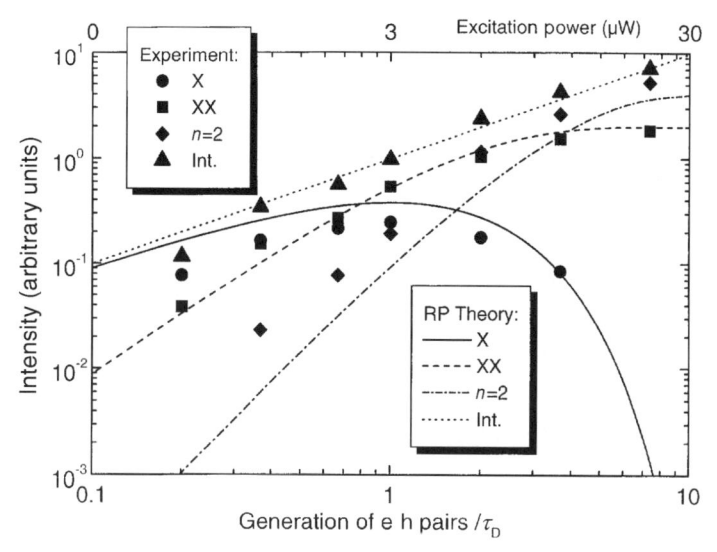

Fig. 6.3 Dependence of integrated exciton (X), biexciton (XX), excited state ($n = 2$) and total (Int.) luminescence intensity of a single-etched InGaAs/GaAs quantum dot (after Forchel et al., 1996a) with theoretical prediction from random population theory in terms of carrier generation rate per exciton lifetime (Grundmann and Bimberg, 1997b)

ELECTRONIC AND OPTICAL PROPERTIES 201

6.2 LOCAL INTERMIXING OF QUANTUM WELLS

'Quantum well boxes' have been made by local implantation-enhanced interdiffusion of a 5 nm GaAs/Al$_{0.35}$Ga$_{0.65}$As quantum well, among others by Cibert *et al.* (1986). After creating circular mesas in a metal mask, Ga$^+$ ions were implanted followed by subsequent rapid thermal annealing, resulting in enhanced interdiffusion of Ga and Al atoms across the interface. The resulting lateral potential is governed by the atom diffusion profile and described by a sum of error functions. In Fig. 6.4 a cathodoluminescence image of an array of such boxes for 280 nm diameter metal dots is shown together with a spectrum taken at the position of one of these boxes. The lowest energy peak is blue-shifted from the quantum well position slightly below 1.610 eV by 15 meV to 1.625 eV and was attributed to the ground state transition of a zero-dimensional system. At least part of that shift is, however, due to aluminum diffusion underneath the masked area. The occurrence of peaks at higher energies was attributed to excited states. However, the linewidth of the peaks from individual boxes remained as broad as the original quantum well (≈ 9 meV). For this

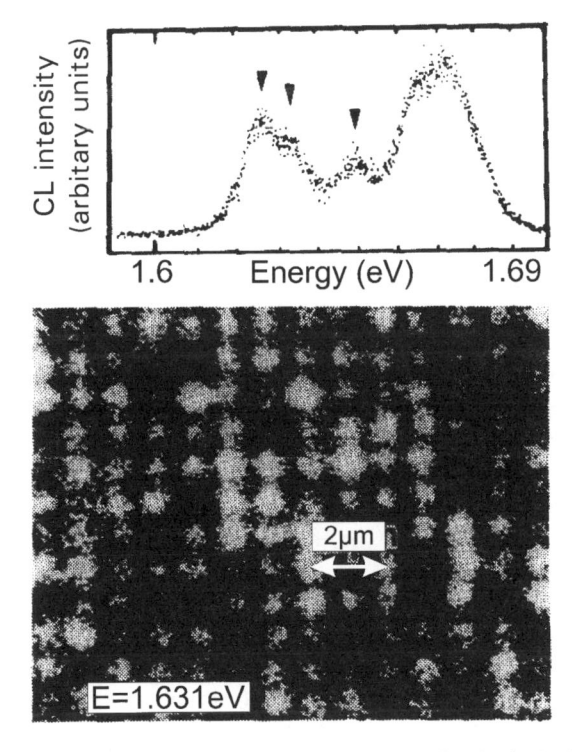

Fig. 6.4 Low-temperature cathodoluminescence spectrum of a single GaAs/AlGaAs quantum well box with mask size 280 nm and the monochromatic cathodoluminescence image at $E = 1.631$ eV. Arrows denote transitions in quantum boxes. (Reproduced by permission of American Institute of Physics from Cibert *et al.*, 1986)

structure the sum of electron and hole localization energy was claimed to be $E_{loc} \approx 45$ meV with a sublevel separation of $\Delta E \approx 9$ meV.

A similar method was employed by Brunner *et al.* (1992). Isolated single quantum dots were fabricated by laser-induced local interdiffusion of a 3 nm GaAs/ Al$_{0.35}$Ga$_{0.65}$As quantum well. Square-shaped Si$_3$N$_4$ mesas of different base length w were fabricated to locally protect the quantum well from interdiffusion produced by local heating and subsequent thermal interdiffusion (Schlesinger and Kuech, 1986) caused by an Ar$^+$ laser beam. The length w was varied between 1 μm and 250 nm. Using microscopic photoluminescence the spectra of single dots were detected. Figure 6.5 visualizes mesa size-dependent peak shifts and the occurrence of excited states. The linewidths found at low excitation power are below 0.5 meV. Figure 6.6 shows the Si$_3$N$_4$ mask size dependence of ground state transition energy and the separation ΔE between the two peaks at lowest energies. A maximum separation of $\Delta E = 10$ meV was observed for a localization energy of about $E_{loc} \approx 55$ meV. The population of excited states, even at very low excitation density, was interpreted as an indication of slowed-down energy relaxation in accordance with theoretical predictions by Bockelmann (1993). Coulomb scattering has been identified as the dominant relaxation mechanism, at least at moderate carrier densities (several excitons per dot) (Bockelmann *et al.*, 1996). The quantum dot luminescence exhibits a short rise time (<10 ps) independent of the excitation excess energy, being below or above the LO phonon energy. At highest carrier densities the spectra exhibit at short delay times after excitation-significant spectral broadening (Fig. 6.7).

6.3 STRESSORS

Masks on a sample surface induce an attractive potential via the strain† extending into the layers underneath (Kash *et al.*, 1988, 1991) and are called 'stressors'. Similar effects can also be achieved using buried stressors (Xu and Petroff, 1991). For quantum wire structures induced by amorphous carbon stressors subband separations of $\Delta E = 2.4$ meV were obtained (Kash *et al.*, 1991) for a localization of about $E_{loc} \approx 20$ meV. Using a laterally patterned In$_{0.35}$Ga$_{0.65}$As quantum well as the stressor on top of a GaAs/AlGaAs multiquantum well structure, localization energies of up to 12.3 meV were obtained (Tan *et al.*, 1993). As expected, the strain-induced localization decays rapidly with increasing distance from the stressor. A typical stressor-induced potential is depicted in Fig. 6.8. The valence band has a rather complex structure and provides only shallow confining potentials. The conduction band is parabolic in the center and exhibits an energy *barrier* of a few millielectronvolts around the edge. This barrier hinders carrier capture at very low temperatures but can be overcome at higher temperatures (Zhang *et al.*, 1995).

---

† In order to create an *attractive* potential in the conduction band the strain has to be tensile.

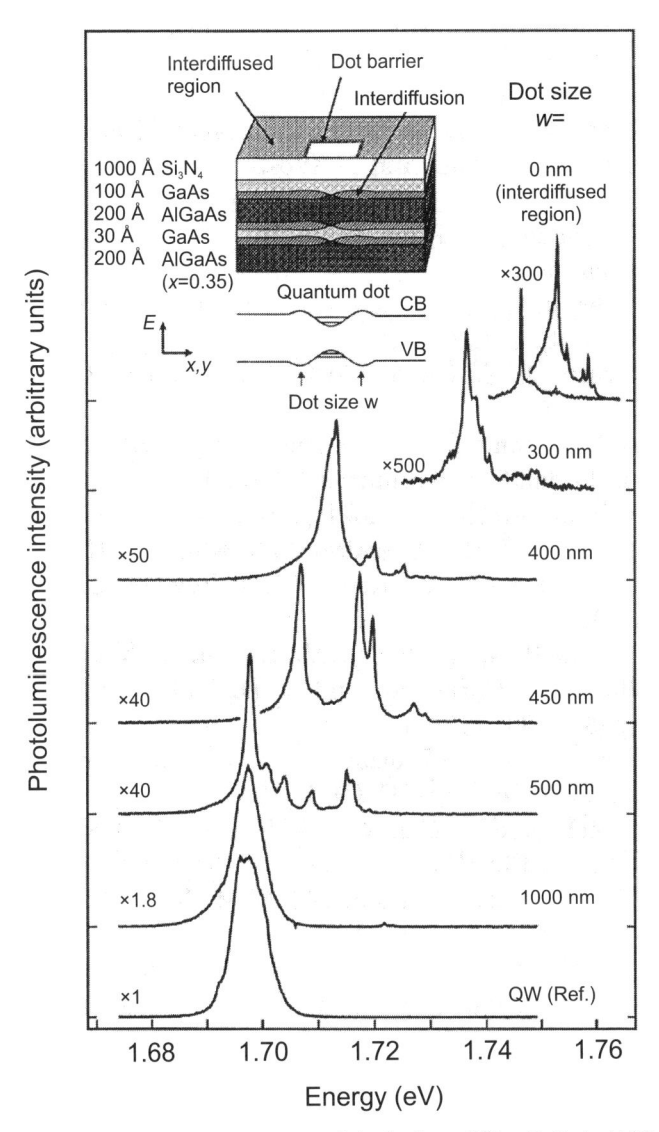

Fig. 6.5 Microphotoluminescence spectra of single (interdiffused) GaAs/AlGaAs quantum dots. Laser excitation was focused to a spot of 1.5 μm (1.96 eV, 1 μW). The inset shows a schematic diagram of the structure. (After Brunner *et al.*, 1992)

Larger localization energies and stronger confinement effects have been obtained using epitaxially grown dots as stressors for underlying quantum well structures (Lipsanen *et al.*, 1995; Sopanen *et al.*, 1995). A schematic view is provided in Fig. 6.9 from Tulkki and Heinämäki (1995) where strain and band structure calculations have been reported. Typical stressor size is 100 nm wide and 20 nm high; the

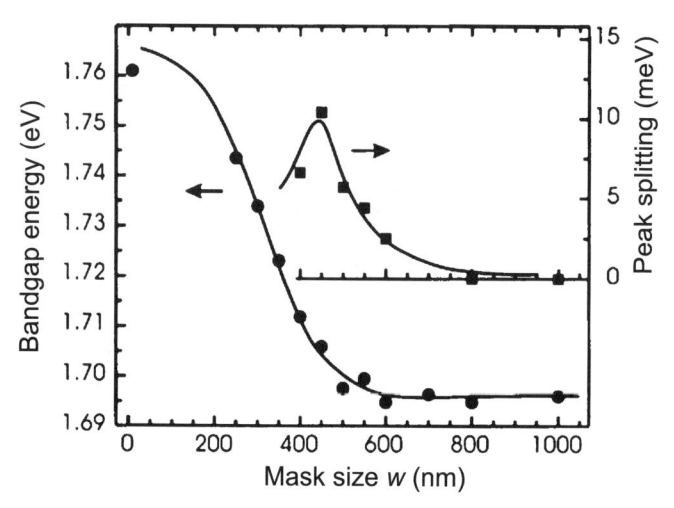

Fig. 6.6 Minimum photoluminescence peak energy and energy separation of the two lowest dominant peaks as a function of geometrical dot size (GaAs/AlGaAs). Solid lines are a guide to the eye. (After Brunner et al., 1992)

area stressor density is $2\times10^9\,\mathrm{cm}^{-2}$. A systematic variation of the confinement potential is obtained by growing structures with different distances d of the quantum well from the surface with the stressors. Theoretical results for different distances d and dot radius R are shown in Fig. 5.18.

Photoluminescence spectra at low excitation intensities of such strain-induced quantum dots are shown in Fig. 6.10. The largest localization energies (up to

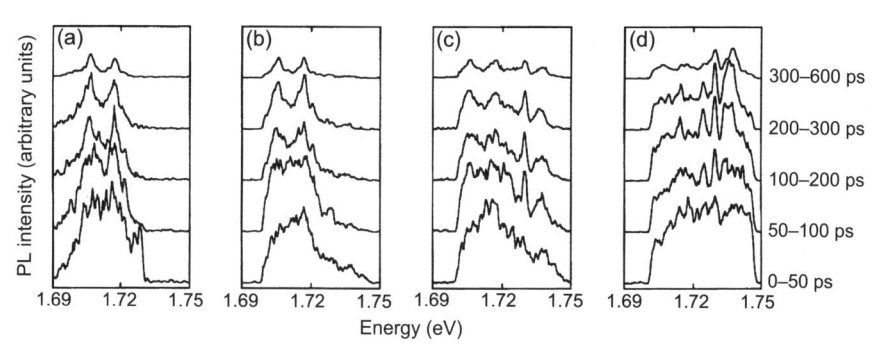

Fig. 6.7 Time-resolved photoluminescence spectra of single interdiffused quantum dot ($w=450$ nm) for various excitation powers $P=800$ nW (a,b), $2\,\mu$W (c), and $7\,\mu$W (d) focused on a 1.5 μm spot. Excitation energy is (a) 30 or (b–d) 59 meV above the lowest PL peak. (After Bockelmann et al., 1996)

ELECTRONIC AND OPTICAL PROPERTIES

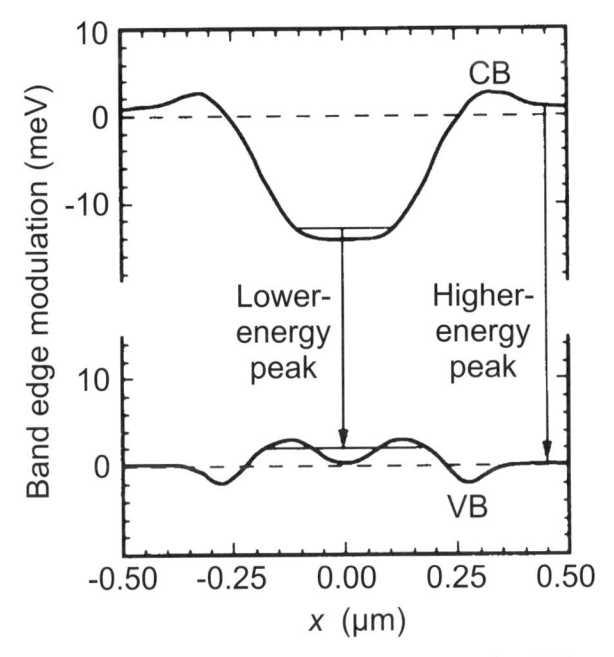

Fig. 6.8 Band edge modulation of the conduction and valence band 30 nm underneath a 120×120 nm² In$_{0.35}$Ga$_{0.65}$As stressor on AlGaAs/GaAs multiple quantum well structure. The dashed lines indicate the band positions before the stressor-induced bandgap modification. (Reproduced by permission of American Institute of Physics from Tan et al., 1993)

$E_{loc} \approx 100$ meV) result from the wells closest to the surface, since they experience the largest strain values (see also Fig. 6.11).

With increasing excitation density higher transitions become visible with nearly equal separation, consistent with a parabolic-like potential (Figs. 6.9 and 6.12). Sublevel separation can be up to $\Delta E = 20$ meV (Fig. 6.11) and decreases with increasing stressor size. The spectral line shape of the photoluminescence can be very well fitted by a multiple Gaussian. The calculations (Tulkki and Heinämäki, 1995) predict that transitions between electron and hole levels of equal quantum numbers dominate the spectrum. The spectral positions of optical transitions remain constant even at high excitation density when several sublevels are occupied. The width of the individual lines increases only little with the quantum number n of the transition. The broadening is thus mainly caused by fluctuations of the vertical (quantum well) confinement. Variations of the lateral width of the harmonic oscillator potential would cause a broadening proportional to n and seem to be of minor importance. This conclusion is corroborated by the fact that the peak widths are very similar for different values of d.

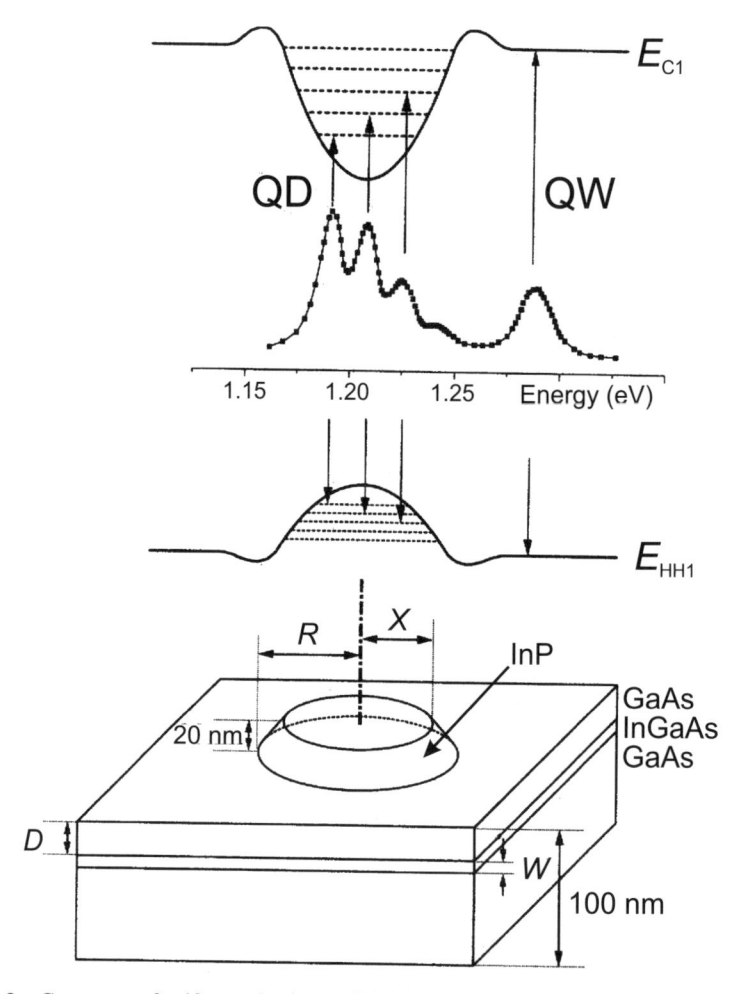

Fig. 6.9 Geometry of self-organized InP disk-like quantum dot on a GaAs/InGaAs/GaAs heterostructure, and schematic strain-induced band structure together with an experimental photoluminescence spectrum from Sopanen *et al.* (1995) and assignment of optical transitions. (After Tulkki and Heinämäki, 1995)

Small (<13 ps) carrier capture times upon nonresonant excitation in the barrier have been observed for such dots (Grosse *et al.*, 1996, 1997) and are attributed to Coulomb scattering for short times when a high carrier concentration is present. The experimental luminescence transients from an ensemble of such dots agree well with fits using master equations for the microstates (Grundmann *et al.*, 1997) (Fig. 6.13). The inter-sublevel scattering time ι_0 is determined to be similar to the radiative electron–hole pair recombination time constant of $\tau_r = 0.75$ ns. The slow inter-sublevel relaxation is manifested in a slowed-down transient of the first excited state.

ELECTRONIC AND OPTICAL PROPERTIES 207

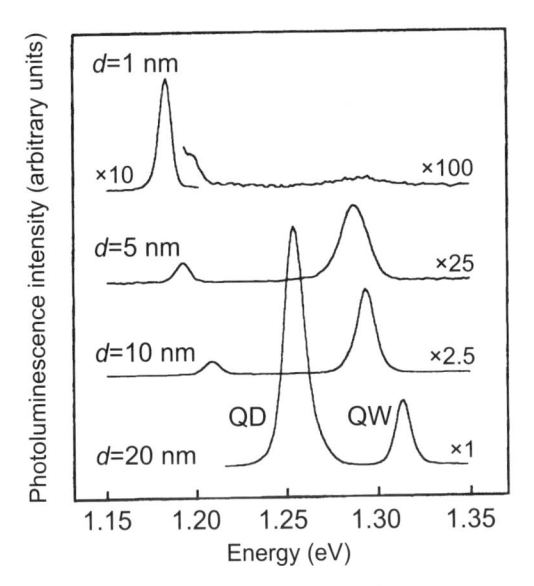

Fig. 6.10 Photoluminescence ($T = 12\,\text{K}$, $D = 15\,\text{W/cm}^2$) of strain-induced quantum dots with various separations d from the InP stressors on the surface. (After Lipsanen and Sopanen, 1995)

Since the level separation is about 11 meV and thus much smaller than the LO phonon energies, a phonon bottleneck for emission of acoustic phonons as calculated by Bockelmann (1993) seems to exist.

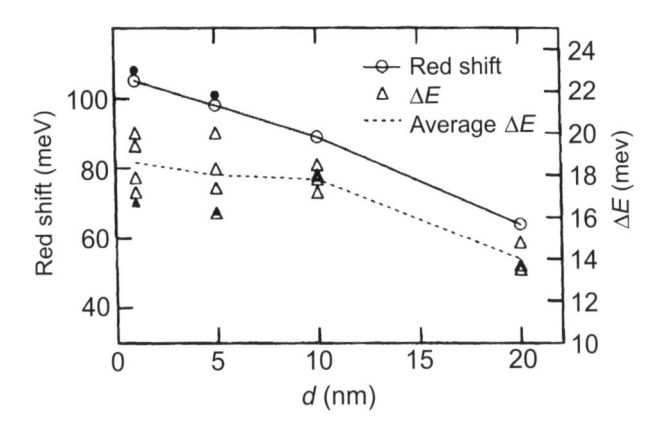

Fig. 6.11 Experimental red shift (open circles) and peak separation (open triangles) of adjacent peaks as a function of d. The lines are a guide to the eye. Calculated values for the red shift and the peak separation are given by full symbols. (After Lipsanen and Sopanen, 1995)

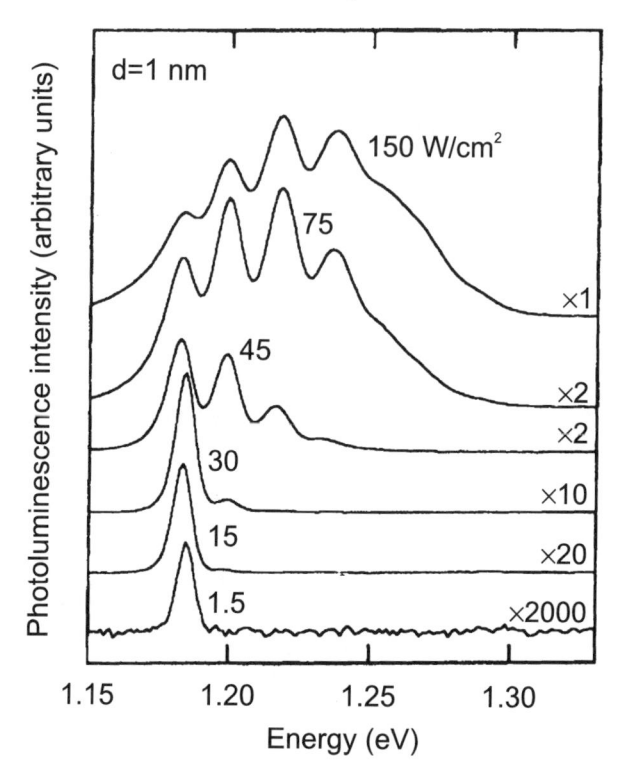

Fig. 6.12 Photoluminescence ($T = 12$ K) spectra of strain-induced quantum dots with $d = 1$ nm for various excitation densities. With increasing excitation, density transitions between excited states appear. (After Lipsanen and Sopanen, 1995)

6.4 SELECTIVE GROWTH

Using selective epitaxy and moreover 'substrate encoded size-reducing epitaxy' (see Section 2.1.6), a term coined by Madhukar et al. (1993), a number of attempts have been made to fabricate low-dimensional structures. For a review, Madhukar (1993) can be consulted.

Individual dots grown by selective MOCVD on GaAs (001) substrates covered by Si_3N_4 patterned with circular holes have been characterized by cathodoluminescence (Lebens et al., 1990). In Fig. 6.14a low-temperature spectra of single dots for feature sizes of 80–140 nm and an unpatterned control area are shown. A confinement-induced blue-shift of the emission is observable. Cathodoluminescence images of dot arrays (Fig. 6.14b) confirm the origin of the spectral lines from the dots.

Tetrahedral AlGaAs/GaAs/AlGaAs dots have been fabricated with MOCVD on (111)B GaAs substrate masked with SiO_2 (Fukui et al., 1991). Again a room temperature cathodoluminescence image of an array of such dots (Fig. 6.15) visualizes the local origin of luminescence. The recording wavelength is, however,

ELECTRONIC AND OPTICAL PROPERTIES 209

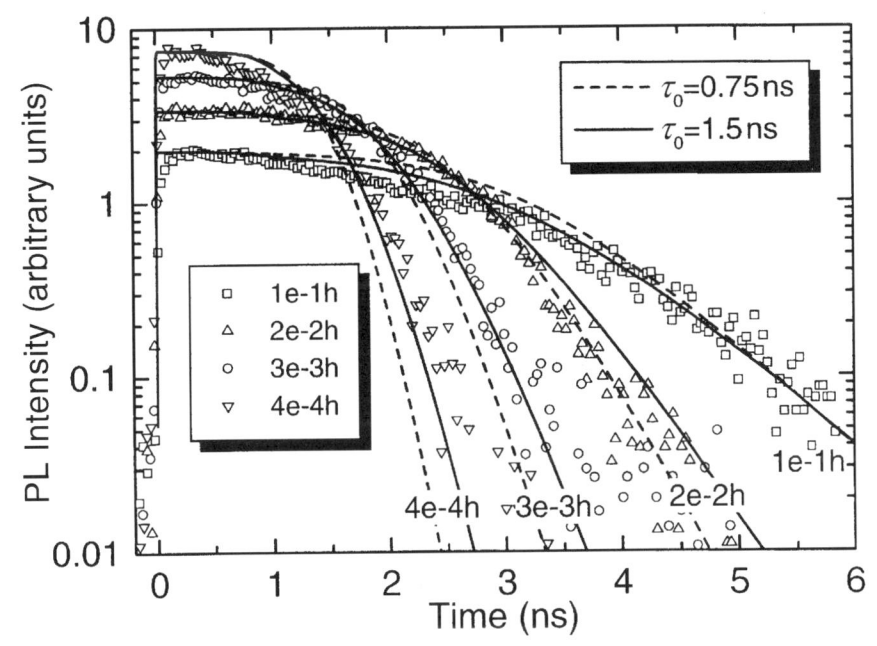

Fig. 6.13 Experimental transients of strain-induced quantum dots and fit with the master equation model for two values of the inter-sublevel scattering time constant $\tau_0 = 0.75$ and 1.5 ns. The radiative recombination time for the ground state is 0.75 ns. (After Grundmann et al., 1997)

882 nm, i.e. the GaAs band edge emission, indicating that no carrier confinement is achieved.

Pyramidally shaped dots of dimensions $25 \times 25 \times 15$ nm^3 were fabricated by selective MOCVD growth on SiO$_2$-masked (001) GaAs (Nagamune et al., 1994). The large area photoluminescence spectrum (Fig. 6.16a) exhibits a peak attributed to dots. This peak is blue-shifted by 19 meV from the GaAs bulk band edge, less than the expected confinement effect of 35 meV. The expected shift for a conventional 15 nm thick quantum well is 16 meV. The difference is attributed to the increased exciton binding energy which has been determined with magnetoluminescence (Fig. 6.16b and c) to be about 10 meV. While the luminescence peaks from carbon impurities, bulk material (substrate) and the thin side-wall quantum well show a linear shift with the magnetic field above 4 T (Landau regime) and the quantum dot related peak exhibits a diamagnetic shift up to 12 T due to the increased confinement.

The luminescence from single dots of size $190 \times 160 \times 12$ nm^3 fabricated with selective MOCVD has been analyzed in Nagamune et al. (1995) as a function of excitation power. The lateral exciton confinement energy, relative to the level of a 12 nm quantum well, is estimated from a cylindrical model as 8 meV, with

210

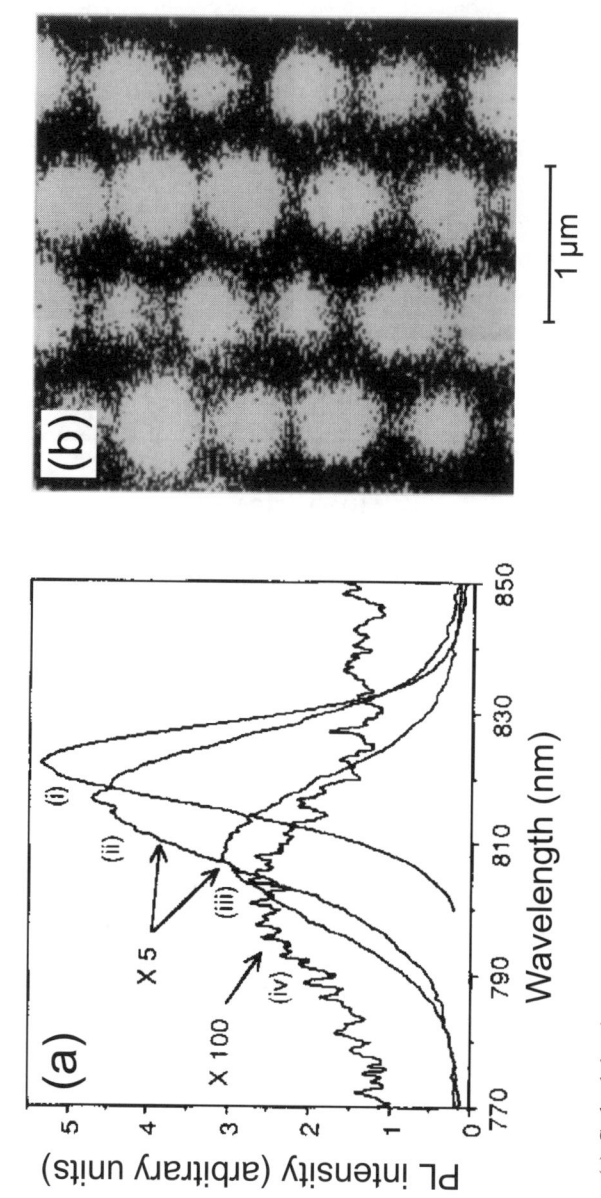

Fig. 6.14 (a) Cathodoluminescence spectra of single GaAs/AlGaAs quantum dots of three different sizes, fabricated with selective epitaxy. (i) The control region without quantum dots. The estimated dot diameters are (ii) 140 nm, (iii) 110 nm, and (iv) 80 nm. (b) Cathodoluminescence image of an array of 110 nm dots, recorded at the wavelength 809 nm, corresponding to the maximum of curve (iii) in (a) of the figure. (Reproduced by permission of American Institute of Physics from Lebens et al., 1990)

ELECTRONIC AND OPTICAL PROPERTIES

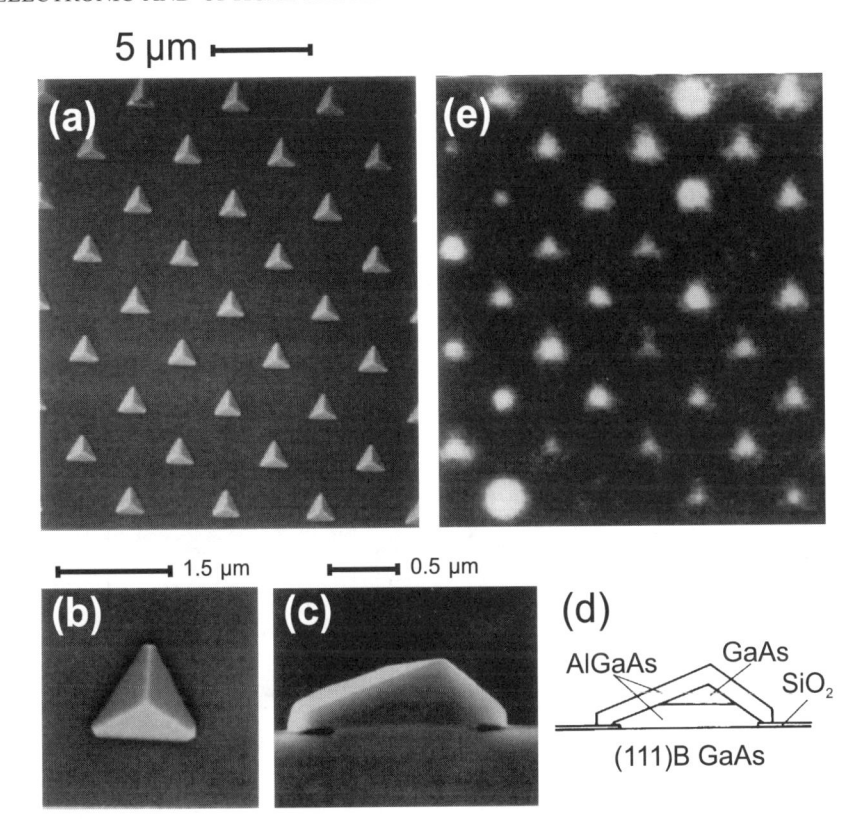

Fig. 6.15 (a) Plan-view SEM image and (e) cathodoluminescence image at room temperature of an array of tetrahedral GaAs/AlGaAs quantum dots. Recording wavelength is 822 nm, i.e. the GaAs bulk band edge. (b) and (c) Higher magnification plan-view and cross-sectional SEM images, respectively. (d) Schematic drawing of cross-sectional geometry. (Reproduced by permission of American Institute of Physics from Fukui *et al.*, 1991)

corresponding small sublevel separation. Due to state filling effects, the spectral width of the recombination spectrum largely depends on the excitation density (Fig. 6.17). Only at the lowest excitation power of 15 nW, for which the estimated population of a quantum dot is one electron–hole pair on time average, does the peak become narrow (0.9 meV). With increasing excitation power the peak broadens continuously due to the contribution of excited states which do not appear as separate lines.

The deposition of nominally 3.5 ML InAs on size-reduced GaAs (001) square mesas results in 75 nm square base InAs dots. The apparent InAs thickness determined from cross-sectional transmission electron microscopy is 8–9 nm. Cathodoluminescence imaging of the top of one such mesa structure (Fig. 6.18) revealed a peak due to confined InAs (Konkar *et al.*, 1995). The peak is about 60 meV broad, which can be attributed to extreme state filling.

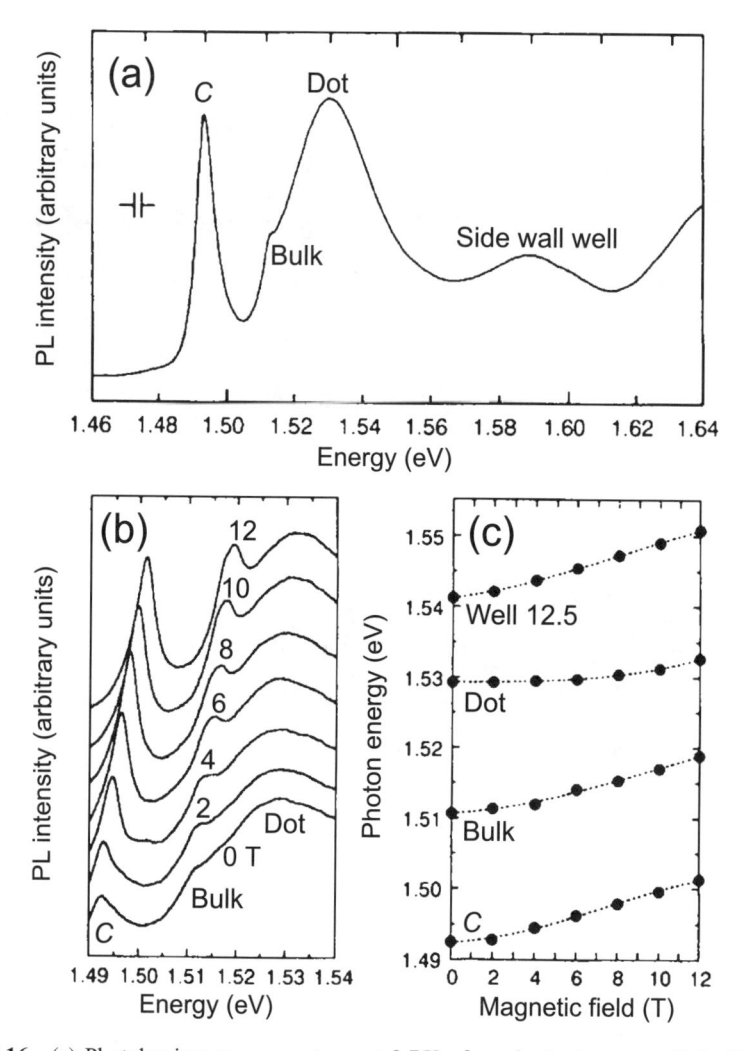

Fig. 6.16 (a) Photoluminescence spectrum at 8.7 K of a selectively grown GaAs/AlGaAs quantum dot. (b) Spectra at different magnetic fields perpendicular to the substrate. (c) Peak positions of the spectra as a function of the magnetic field. (Reproduced by permission of American Institute of Physics from Nagamune *et al.*, 1994)

6.5 EXCITONS LOCALIZED IN QUANTUM WELL THICKNESS FLUCTUATIONS

If the lateral dimension of a localizing feature, such as a quantum well thickness fluctuation, is much larger than the exciton diameter this center can be regarded as a region of thicker quantum well. If the size, however, is comparable to the exciton diameter, the localizing feature represents a shallow quantum dot (Fig. 6.19)

ELECTRONIC AND OPTICAL PROPERTIES 213

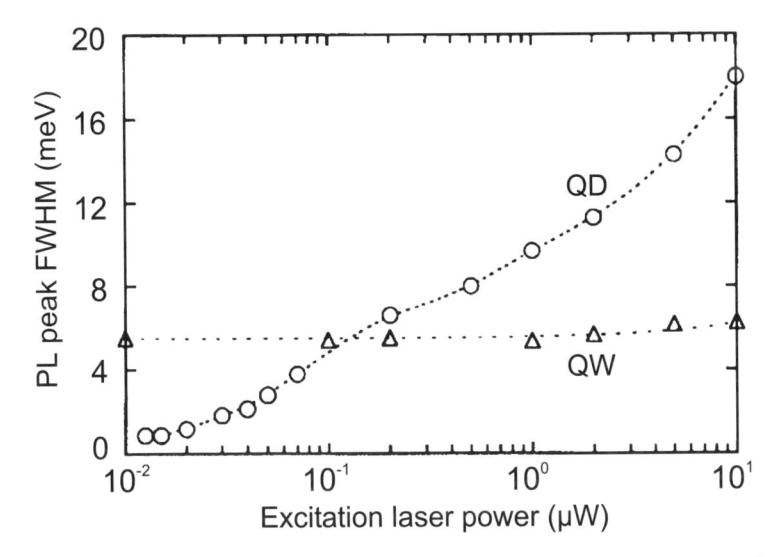

Fig. 6.17 Spectral width of the photoluminescence peak from a $190 \times 160 \times 12\,\text{nm}^3$ single quantum dot as a function of the excitation laser ($\lambda = 514.5\,\text{nm}$) power at 15 K. (Reproduced by permission of American Institute of Physics from Nagamune *et al.*, 1995)

(Christen and Bimberg, 1990; Christen *et al.*, 1990). The properties of the luminescence from localized excitons is very different from that of QW continuum states:

(a) slow spectral and spatial diffusion below the mobility edge (Hegarty and Sturge, 1985),
(b) strong violation of the k-selection rule (Christen and Bimberg, 1990; Christen *et al.*, 1990),
(c) strong increase of the radiative lifetime with respect to free excitons in quantum wells (Deveaud *et al.*, 1991; Sugiyama *et al.*, 1995) (see also Section 5.3.1.5).

All these results taken together demonstrate that in optical studies of quantum wells at low temperature the experimentalists deal mostly *not* with two-dimensional quantum well properties but with properties of arrays of shallow quantum dots. The corresponding density of states responsible for these transitions can be observed as a long-wavelength tail below the free quantum well exciton absorption edge. A detailed theoretical model of excitons and their cw and dynamic optical properties in the presence of quantum well interface roughness has been worked out by Zimmermann and Runge (1994, 1997).

μ-PL investigations revealed extremely sharp recombination lines at the low-energy side of GaAs/AlGaAs quantum well emissions (Brunner *et al.*, 1994a; Zrenner *et al.*, 1994). Similar results were obtained with scanning near-field microscopy (Hess *et al.*, 1994).

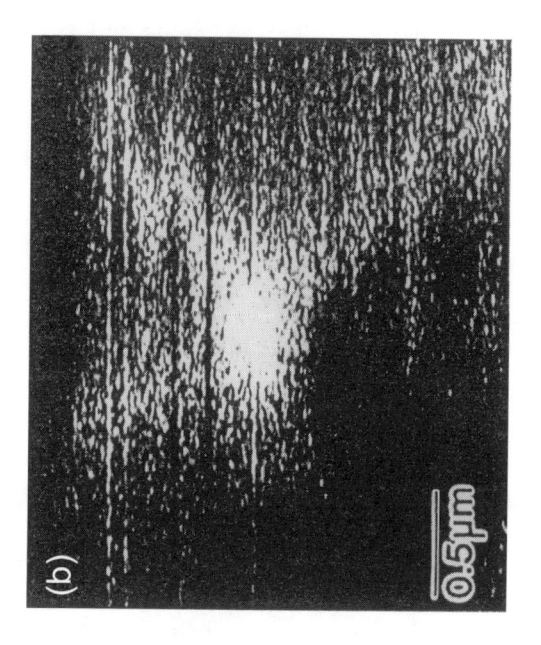

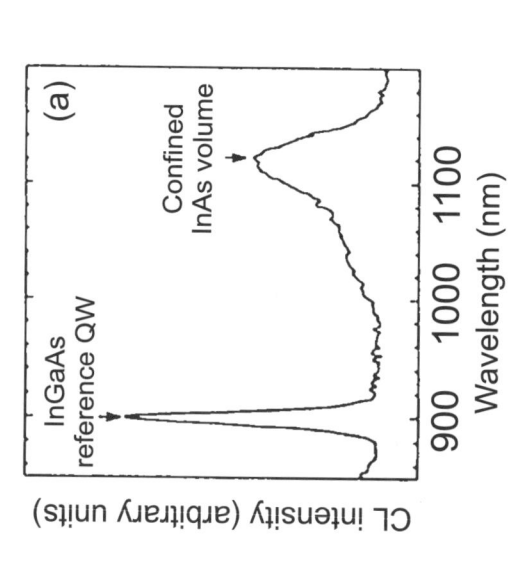

Fig. 6.18 (a) Cathodoluminescence spectrum averaged over the mesa top region. (b) Cathodoluminescence image of the mesa top recorded at 1120 nm. (After Konkar et al., 1995)

ELECTRONIC AND OPTICAL PROPERTIES 215

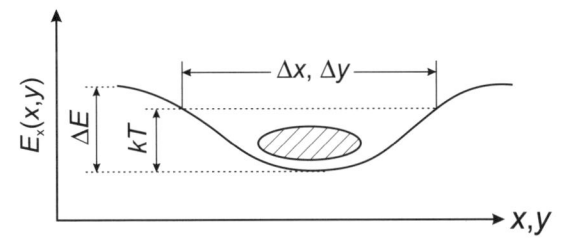

Fig. 6.19 Localization of the quantum well exciton in lateral energy fluctuations in the (xy) interface plane for $kT<\delta E$. (After Christen and Bimberg, 1990)

In Fig. 6.20 a microscopic photoluminescence spectrum with about 1.5 μm lateral resolution is shown (Brunner *et al.*, 1994a). Individual linewidths of single peaks are as small as 0.07 meV. The localization energy is of the order of $E_{\text{loc}} \approx 10$ meV. A typical area density of such features is a few per square micrometer. Excited states

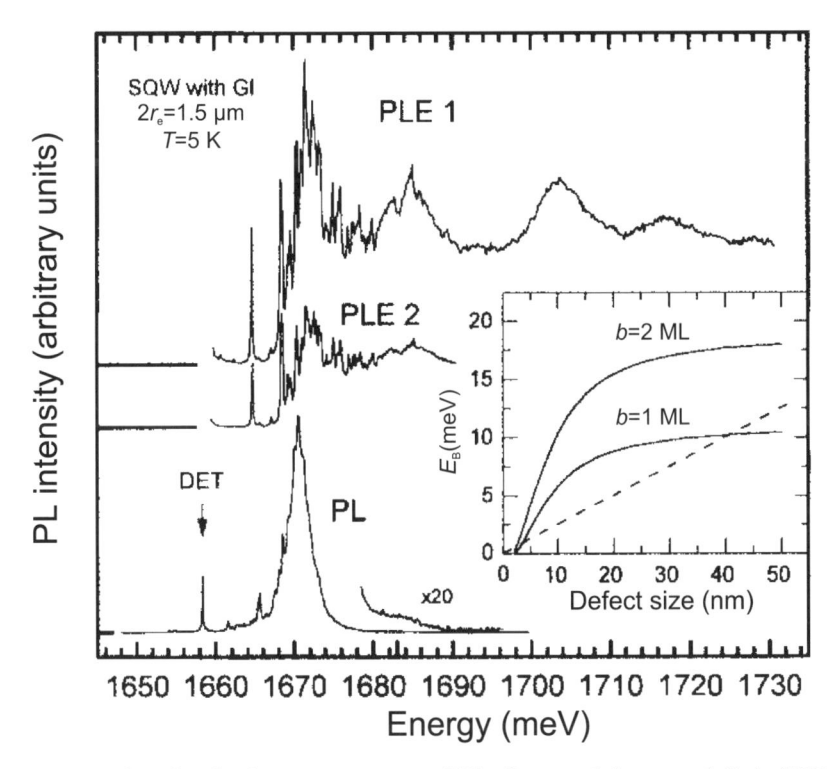

Fig. 6.20 Microphotoluminescence spectrum (PL) of a growth-interrupted GaAs/AlGaAs quantum well and photoluminescence excitation spectrum (PLE1) recorded with a spatial resolution of 1.5 μm at 5 K. PLE2 was measured several weeks later than PLE1 at half the excitation power. Inset: calculated binding energy of a quantum well exciton bound to a Gaussian shape protrusion of lateral radius and depth *b*. (Reproduced by permission of American Institute of Physics from Brunner *et al.*, 1994a)

are revealed in photoluminescence excitation spectra (Fig. 6.20); typical sublevel separation is $\Delta E \approx 6$ meV. Model calculations are consistent with one monolayer high interface islands of a lateral extension of several tens of nanometers.

Coincidence of luminescence from the first excited state and the lowest energy excitation spectroscopy peak was reported by Gammon *et al.* (1995) (Fig. 6.21), proving that objects with zero-dimensional density of states are indeed probed. In linearly polarized excitation spectra the degeneracy of the ground and excited states is observed to be lifted (Gammon *et al.*, 1996a) (Fig. 6.22), which is attributed to the impact of the asymmetries of the confining potential. In the same figure the homogeneous broadening of different higher transitions is shown as a function of energy separation from the ground state transition. The ground state transition has the smallest width, probably since it is subject to the least amount of scattering processes (for the temperature dependence see also Section 6.7.1.1).

Increasing the excitation density from 0.1 to 5 µW, a new sharp line appears about 5 meV below the exciton line (X) (Fig. 6.23a). It is attributed to the biexciton transition (XX) (Brunner *et al.*, 1994b). The exciton intensity increases originally

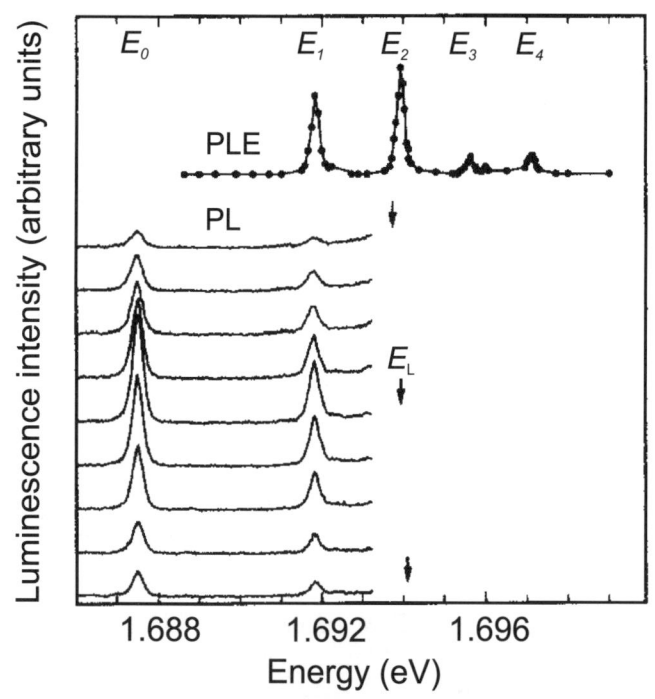

Fig. 6.21 Top: low-temperature photoluminescence excitation spectrum recorded with the detection energy on the ground state of the GaAs/AlAs quantum dot (E_0). The bottom series of photoluminescence spectra shows the luminescence (of ground state E_0 and first excited state E_1) upon scanning the excitation energy E_L in small equidistant steps through the second excited state E_2. (Reproduced by permission of American Institute of Physics from Gammon *et al.*, 1995)

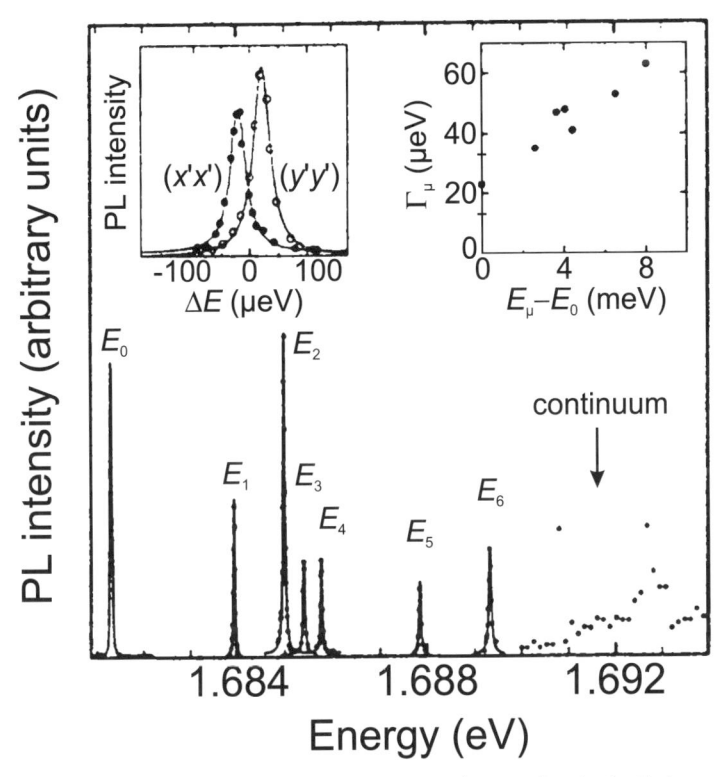

Fig. 6.22 Photoluminescence line (E_0) of the ground state of a single GaAs quantum dot formed in a QW thickness fluctuation and the associated photoluminescence excitation spectrum (measured with the ($y'y'$) polarization configuration) showing excited states E_μ. The arrow denotes the approximate onset of the 2D continuum. Each line actually consists of a doublet, as shown for the first excited state in the left inset. The homogeneous linewidth increases with energy separation $E_\mu - E_0$ from the ground state, as shown in the right inset. (After Gammon et al., 1996b)

linearly with excitation power I (Fig. 6.23b) and eventually saturates. The biexciton line increases proportionally to I^2 and remains sharp. For still higher excitation densities, a decrease of the exciton line and further increase and saturation of the biexciton line is expected (Grundmann and Bimberg, 1997b).

An interesting experiment is two-photon absorption, as illustrated in Fig. 6.24. Luminescence from the X and XX states is obtained when the sample is excited with laser light of energy $(E_X + E_{XX})/2$ of sufficient intensity (200 μW). Via an intermediate state a correlated biexciton is formed, which decays with a photon of energy E_{XX} and subsequently with a photon of energy E_X. The biexciton state can only be excited by linearly polarized light in order to allow the creation of electron–hole pairs of opposite spin orientation. The intensity dependence of X and XX recombination for the two-photon resonance is shown in Fig. 6.23b by open symbols. A linear and quadratic increase with excitation density is again observed.

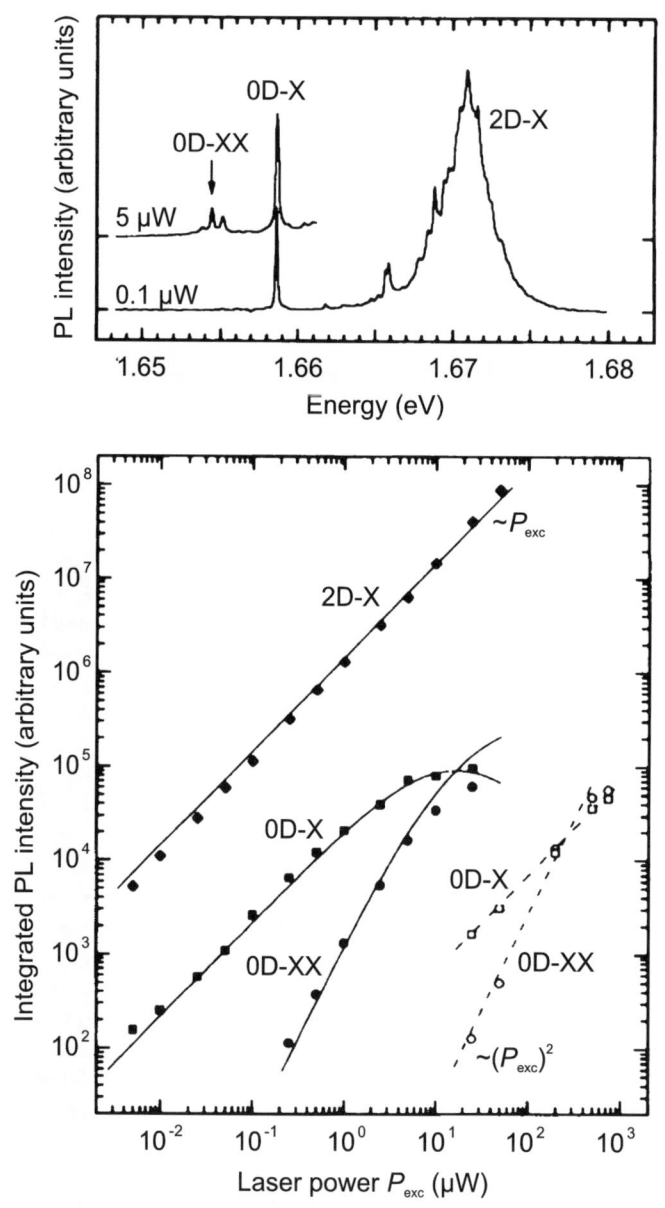

Fig. 6.23 (a) Microscopic photoluminescence spectrum of the same area of a growth-interrupted GaAs/AlGaAs quantum well at two different excitation powers 0.1 and 5 μW as indicated. (b) Excitation power dependence of the integrated photoluminescence intensity for the broad quantum well line (2D-X), the sharp line of the localized exciton (0D-X) and the biexciton line (0D-XX) from (a). Open symbols denote luminescence intensities on the X and XX lines for resonant two-photon absorption of the biexciton state. $T = 5$ K, probe size is 1.5 μm. (After Brunner *et al.*, 1994b)

ELECTRONIC AND OPTICAL PROPERTIES

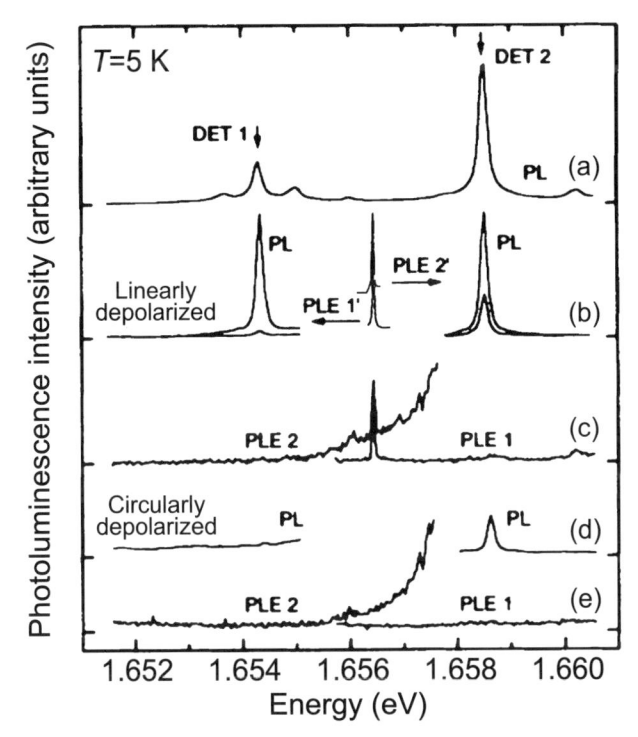

Fig. 6.24 (a) Photoluminescence of the biexciton (DET1) and exciton (DET2), localized in a GaAs/AlGaAs quantum well, upon resonant excitation at 1884.8 meV into the first excited state with excitation power $P_{exc} = =10\,\mu W$. (b) PL spectra excited by a laser beam of power $P_{exc} = 200\,\mu W$ at $E = 1656.5$ meV, which is linearly depolarized (top curve). The lower spectrum is at a detuning of the laser by 0.2 meV. The corresponding excitation spectra PLE1' and PLE2' have been recorded with a step size of 10 μeV. Equivalent excitation spectra PLE1 and PLE2 are depicted in (c) across a larger energy range. Spectra in (d) and (e) are equivalent to those in (b) and (c), respectively, but the laser beam and PL detection are circularly depolarized. (After Brunner et al., 1994b)

While most of the original work on such localized excitons had been devoted to the GaAs/AlGaAs system, recently similar results have been found for $Zn_{1-x}Cd_xSe$/ZnSe quantum wells for a Cd content $x \geqslant 0.3$ (Lowisch et al., 1996). The lattice mismatch present in this system leads to a corrugation of the otherwise flat quantum well. Cathodoluminescence spectra and images from quantum wells with a Cd content of less than 0.25 are unstructured. For $x \geqslant 0.3$, cathodoluminescence spectra detected at selected locations consist of a broad band superimposed by extremely sharp spikes with a resolution limited half-width of 0.5 meV. Monochromatic images detected on such sharp lines visualize the localization centers (Fig. 6.25).

Photoluminescence excitation on the high-energy side of the peak creates mobile excitons, as verified by the occurrence of emission energetically below the cut-off defined by the pump photons (Fig. 6.26a). When the excitation is directly within the

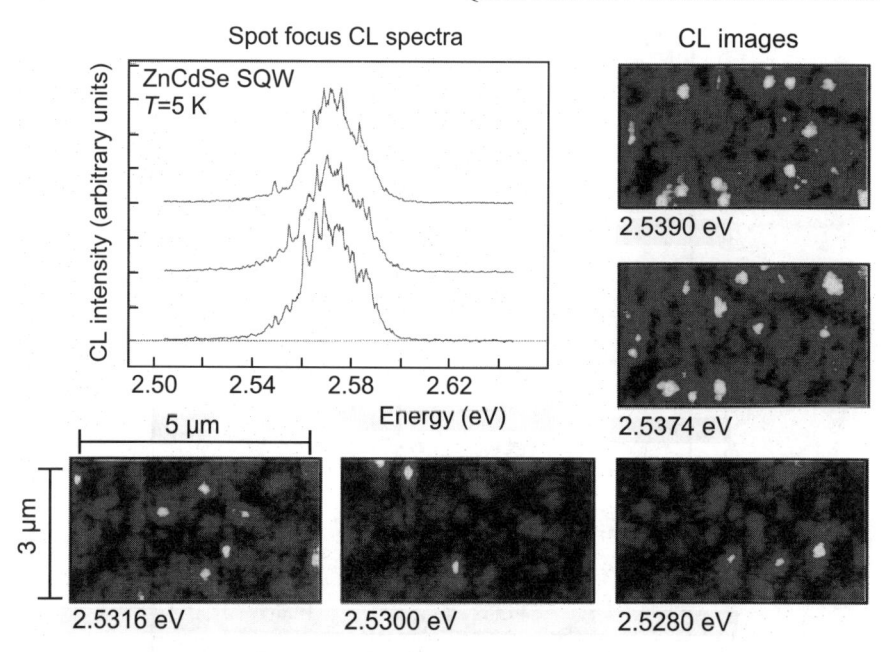

Fig. 6.25 Cathodoluminescence spectra of $Zn_{0.68}Cd_{0.32}Se/ZnSe$ quantum well ($L_z = 2.5$ nm) at 5 K for different excitation spots. Images: monochromatic plan-view CL images for various detection energies. (After Lowisch et al., 1996)

zero-dimensional density of states, spectral diffusion is entirely absent; a very sharp line appears, which exhibits no Stokes shift with respect to the excitation. Under cw conditions the photoluminescence signal cannot be separated from the stray light of the pump laser. Time-resolved experiments uncover a clear contribution of resonant emission after the decay of the pump laser (Fig. 6.26b). Recombination lifetimes between 200 and 250 ps are found. Charged excitons and other complexes have been observed in micro-PL experiments reported by Shiraki et al., (1998). Transitions between multiexciton levels of single self-organized InAs Qds have been reported by Dekel et al., (1998) for populations with up to eight excitons.

Excitons trapped in monolayer width fluctuations under the influence of a lateral electric field are investigated in Heller et al., (1998). Under the lateral electric field a red shift is found. This Stark shift is reduced at higher laser excitation power when the dot becomes populated with several excitons that screen the electric field.

6.6 TWOFOLD CLEAVED EDGE OVERGROWTH

Twofold cleaved edge overgrowth (2CEO) has been predicted to create electronic quantum dots at the juncture of three orthogonal quantum wells (Grundmann and

ELECTRONIC AND OPTICAL PROPERTIES 221

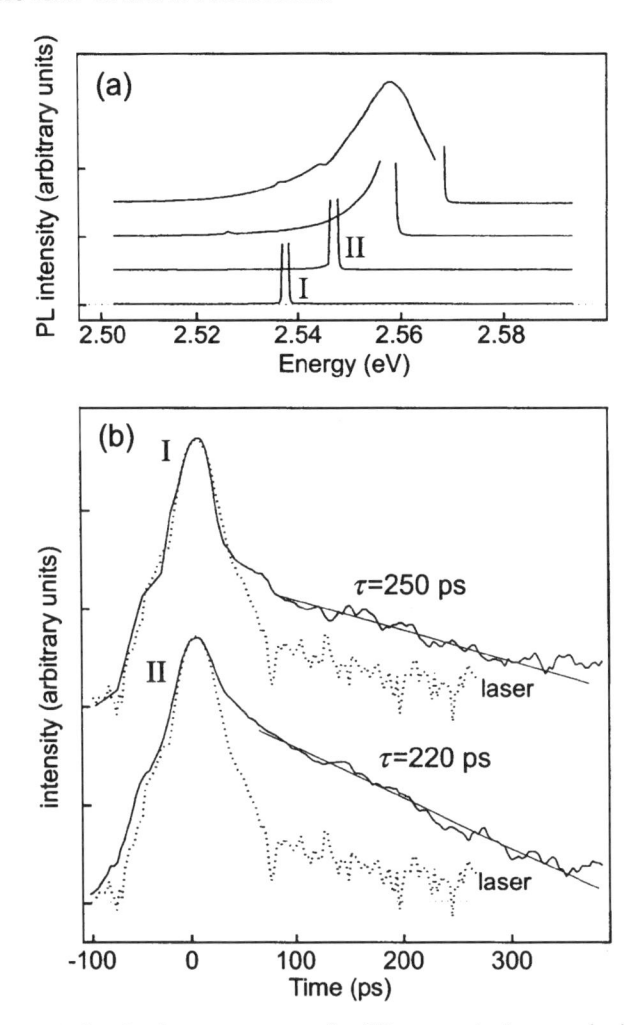

Fig. 6.26 (a) CW-photoluminescence spectra for different excitation energies in the quantum well density of states ($Zn_{0.68}Cd_{0.32}Se/ZnSe$). (b) Time decay of spectrally sharp emission under resonant excitation of zero-dimensional states (logarithmic intensity scale). Dotted curves represent the excitation laser pulse measured simultaneously. All data measured in cross-polarized geometry

Bimberg, 1997a) (see Section 5.2.7), with a typical localization energy of several millielectronvolts. Such quantum dots have been realized by Wegscheider *et al.* (1997) in the AlGaAs/GaAs system using MBE. In μ-PL spectra (Fig. 6.27) evidence has been found for zero-dimensionally confined excitons in addition to luminescence originating from the connected quantum wires and wells. The coupling of pairs of such dots is investigated in Schedelbeck *et al.* (1998) as a function of inter-dot distance. The splitting of the ground state transition is found to

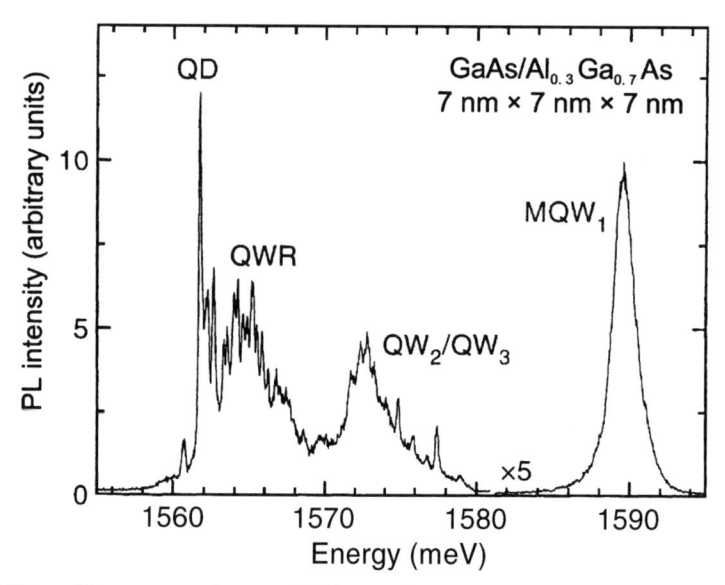

Fig. 6.27 μ-PL spectrum from a 2CEO sample consisting of 7 nm GaAs/Al$_{0.3}$Ga$_{0.7}$As quantum wells. The contribution due to the (001) multiple quantum well (MQW$_1$), the (110) quantum wells (QW$_2$, QW$_3$), the quantum wires (QWR), and a quantum dot (QD) is visualized. (After Wegscheider *et al.*, 1997)

increase with decreasing dot separation. Both levels are found to be radioactive (see Section 5.3.3).

6.7 SELF-ORGANIZED TYPE-I QUANTUM DOTS

A first observation of three-dimensional nucleation during growth of InAs/GaAs superlattices was reported by Goldstein *et al.* (1985). In Fig. 6.28 the spectra of two InAs/GaAs 'superlattices' are compared. The narrow line shown in spectrum (a) of Fig. 6.28 stems from approximately two monolayers, InAs separated by 1 nm GaAs; growth is still in a two-dimensional mode. Three-dimensional nucleation is visualized for 2.5 ML InAs by TEM (Fig. 2.5). The photoluminescence peak shifts by 280 meV toward lower energy from the expected position for a 2D layer (spectrum (b) in Fig. 6.28). The peak is one order of magnitude broader than the 2 ML peak caused by the distribution of dot sizes. The importance of this discovery of three-dimensional nano-sized islands, i.e. quantum dots, was recognized much later at the beginning of the 1990s (Guha *et al.*, 1990; Tabuchi *et al.*, 1992; Leonard *et al.*, 1993; Ledentsov *et al.*, 1994a, 1994b). Optimization of interface abruptness, flatness, and smoothness was a goal particularly important for electronic device development in the 1980s.

ELECTRONIC AND OPTICAL PROPERTIES 223

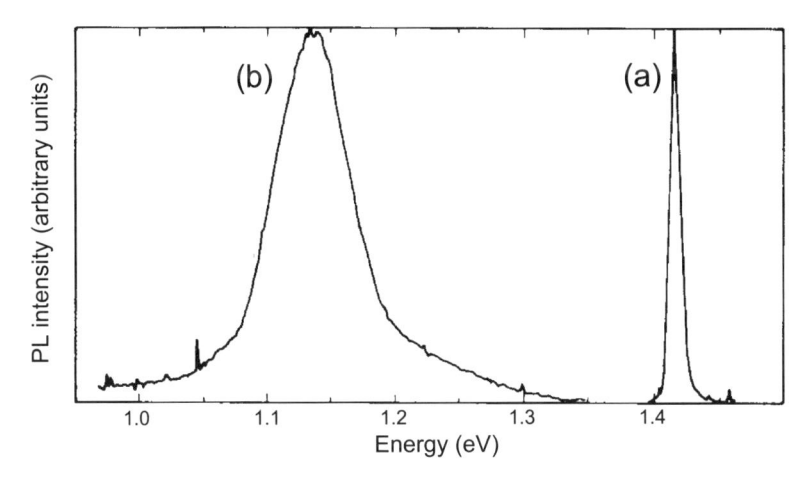

Fig. 6.28 Photoluminescence spectra at 77 K for (a) a two-dimensional superlattice of 1 ML InAs with 5 ML GaAs barriers and (b) a three-dimensionally nucleated layer with islands upon deposition of 2.5 ML InAs. (After Goldstein *et al.*, 1985)

6.7.1 LUMINESCENCE

Figure 6.29 displays a typical low-temperature photoluminescence spectrum at 1.1 eV of a large number of self-organized quantum dots fabricated by deposition of 4 ML InAs on GaAs using MBE (Ledentsov *et al.*, 1995b). MOCVD-grown structures exhibit very similar properties. The localization energy with respect to the

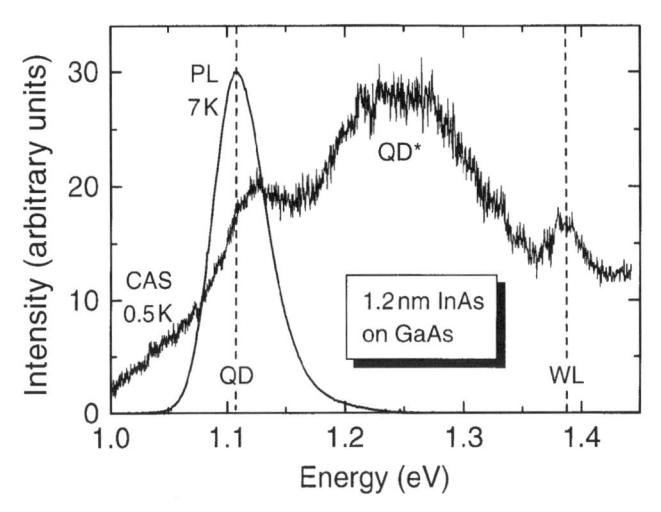

Fig. 6.29 Photoluminescence (PL) and calorimetric absorption spectrum (CAS) of InAs/GaAs quantum dots. (After Grundmann *et al.*, 1995a)

wetting layer (WL) is about $E_{\text{loc}} \approx 300\,\text{meV}$ (Grundmann et al., 1995b). Luminescence from the WL appears only at higher excitation densities. With increasing excitation density, transitions between excited states appear and giant (exciton) sublevel separations of $\Delta E \approx 80\text{--}90\,\text{meV}$ are deduced (Grundmann et al., 1996d). The luminescence efficiency obtained from such quantum dot arrays is similar to that of the best quantum well samples. The predicted primary consequence of a phonon bottleneck effect, a completely quenched luminescence efficiency (Benisty et al., 1991) is not observed. The positions of ground and excited state transitions are practically independent of the excitation density. A relative dot size fluctuation close to $\xi = 0.1$ can be deduced from line-shape fits using a series of Gaussians (for a discussion of size fluctuation $\xi = \sigma_R/R$, see Section 5.4.2).

Similar spectra have been obtained for related material systems, such as InGaAs/GaAs (Fafard et al., 1994; Mirin et al., 1995; Mukai et al., 1996b) or InAlAs/AlGaAs (Fafard et al., 1995; Leon et al., 1995). InP/GaInP (2.4 ML InP deposition) emitting at 1.65 eV (Castrillo et al., 1995) exhibits a similarly large localization energy $E_{\text{loc}} \approx 280\,\text{meV}$. An excited state separation of $\Delta E \approx 23\,\text{meV}$ is reported. Simultaneously, luminescence from smaller, not completely transformed, islands is visible at around 1.8 eV (Fig. 6.30).

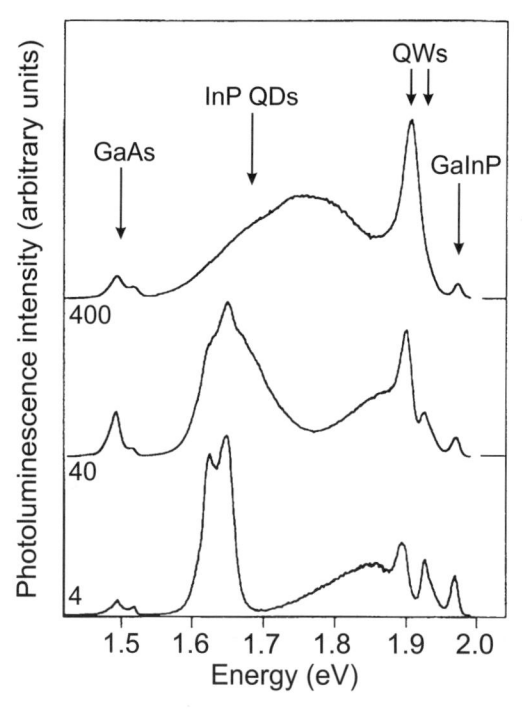

Fig. 6.30 Photoluminescence spectra ($T = 5\,\text{K}$) of InP/GaInP quantum dots at different excitation densities (4, 40, 400 W/cm^2 as indicated). (Reproduced by permission of American Institute of Physics from Castrillo et al., 1995)

ELECTRONIC AND OPTICAL PROPERTIES

Mukai et al. (1996a, 1996b) investigated in detail InGaAs/GaAs quantum dots deposited using multiple InAs/GaAs cycles. With increasing cycle number the dot diameter increases, as observed by plan-view TEM. Accordingly the ground state energy and the sublevel separation decrease due to the reduction of confinement effects.

Theory predicts (Section 5.4.1) the recombination spectrum of a quantum dot to consist of a series of narrow lines independent of temperature. A true quantum dot has sufficiently large sublevel separation that populated excited states give rise to discrete transitions well separated from the ground state.

Indeed, in Marzin et al. (1994) it could be confirmed that the luminescence from ensembles of only a few quantum dots, prepared by etching small (several 100 nm diameter) mesas into a sample containing a layer of self-organized InAs/GaAs quantum dots, consists of a series of sharp lines (Fig. 6.31). The narrowest lines have

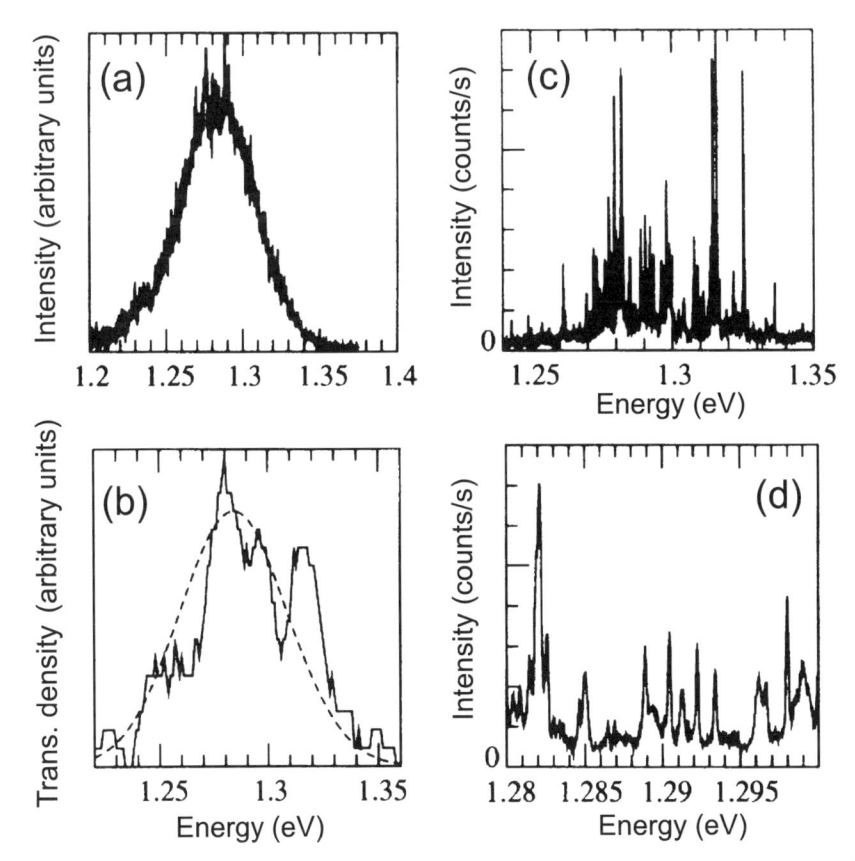

Fig. 6.31 Photoluminescence spectra ($T = 10$ K) of etched mesas with InAs/GaAs self-organized quantum dots: (a) large (5 μm) mesa, (c) 500 nm mesa, (d) same as (c) with different energy scale, (b) number of peaks per unit energy in spectrum (c), compared to Gaussian fit of spectrum (a). (After Marzin et al., 1994)

a width of 0.1 meV limited by the experimental resolution. Model calculations (Marzin and Bastard, 1994; Marzin et al., 1994) result in an energy separation between the ground and first excited transitions of more than 80 meV. Thus all sharp lines within an energy interval of 80 meV around the peak of the ground state luminescence stem from ground states of different single quantum dots. The number of peaks per energy interval resembles the large-area photoluminescence spectrum (Fig. 6.31b).

Using spatially resolved cathodoluminescence, clearly resolved single-line spectra (experimental resolution of 0.15 meV) were found in Grundmann et al. (1995a) and Ledentsov et al. (1995b) for spot focus excitation (Fig. 6.32a). The lines do not change their width with increasing temperature (Fig. 6.32b) up to at least 60 K, where $k_B T$ is an order of magnitude larger than the observed spectral linewidth. This absence of spectral broadening proves the zero-dimensionality of the density of states and yields an upper limit for thermal broadening. Each quantum dot contains here about $(2-3) \times 10^3$ InAs molecules, which leads to a theoretically expected change of transition energy of ≈ 0.06 meV for the addition of one single InAs molecule. Thus the minimal detectable energy separation for sharp lines in that experiment corresponds to a dot size variation of a single InAs molecule. The typical separation of the most prominent lines in Fig. 6.32a corresponds to about sixteen InAs molecules. Monochromatic cathodoluminescence images recorded on single quantum dot recombination lines reveal the local origin of these peaks (Fig. 6.33); the virtual size of the quantum dot in the CL images is determined by carrier diffusion. Obviously, quantum dots of identical ground state energy (within the spectral resolution of the image, 0.3 meV) have an area density of less than 1 μm^{-2}. Conversely, local spectra at the position of particular quantum dots are dominated by a sharp line at the detection wavelength. Minute changes of the recording wavelength lead to completely different intensity distributions because completely different quantum dots come under observation.

In single InP/GaInP quantum dots (Georgsson et al., 1995) of much larger size than in the InAs/GaAs system (approximately $45 \times 60 \times 15$ nm^3) the ground state luminescence appears as three discrete sharp lines which are separated by about 20 meV and persist up to lowest excitation densities and to high temperatures (Hessmann et al., 1996). This triplet ground state is attributed to distinct localization sites for electrons and holes *within* one dot, separated by a considerable energy barrier, suppressing energy relaxation between the sites (Pryor et al., 1997).

The luminescence spectrum from 30 nm InGaAs quantum disks formed during spontaneous reorganization of a sequence of AlGaAs and strained In$_{0.4}$Ga$_{0.6}$As epitaxial films on a GaAs (311)B substrate (Nötzel et al., 1994a) is depicted in Fig. 6.34, together with the spectrum of a reference quantum well of nominally identical composition grown on (001) GaAs. A small blue shift is observed in contrast to the expected red shift, which may hint at indium segregation during the disk formation. The photoluminescence excitation signals for quantum well and disks are rather similar and continuous in both cases. Room temperature photoluminescence of an array of coupled disks of 200 nm diameter reveals a linewidth of 13 meV ($<k_B T$),

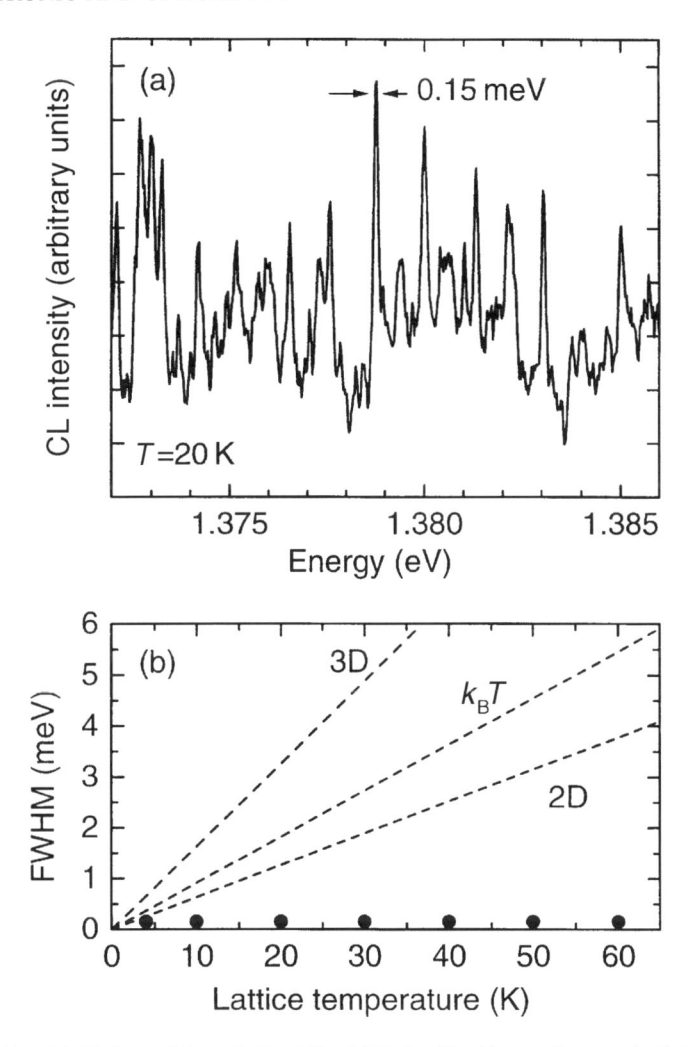

Fig. 6.32 (a) High spatial resolution ($U = 3$ kV, $I = 60$ pA) spot focus cathodoluminescence spectrum of InAs/GaAs quantum dots at $T = 20$ K. (b) Spectral width of individual dot spectra as a function of temperature. Dashed lines are thermal energy and theoretical half-widths of ideal bulk (3D) and quantum well (2D) material. (After Grundmann *et al.*, 1995a)

while the reference quantum well has a spectral width of 27 meV. The broadening of the quantum well consists of the inhomogeneous broadening at low temperature (≈ 11 meV) and the thermal broadening at room temperature for an ideal 2D system ($k_B T \ln 2 \approx 18$ meV). The linewidth below the width for thermal broadening is attributed to efficient lateral localization in the quantum disks. However, inhomogeneous broadening due to size fluctuations is small due to the large lateral disk diameter.

228   QUANTUM DOT HETEROSTRUCTURES

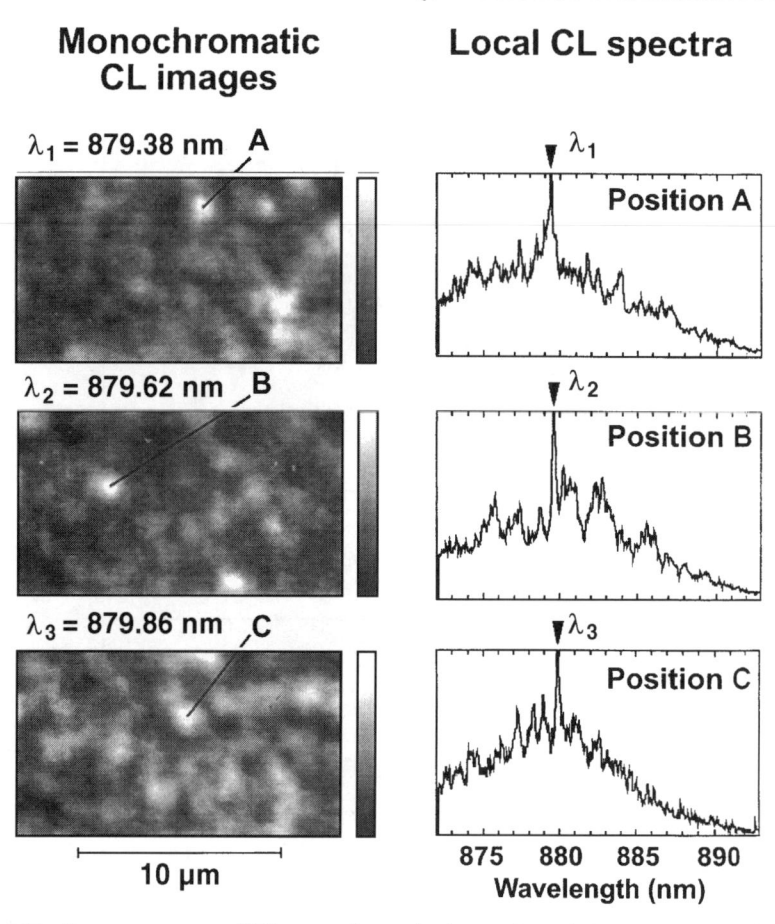

Fig. 6.33 Low-temperature (5 K) monochromatic (spectral resolution <0.2 nm) cathodoluminescence images of InAs/GaAs quantum dots for three detection wavelengths, differing by 0.38 meV. On the right side are local CL spectra recorded with the electron beam at the positions A, B, and C indicated in the intensity images. Arrows on the spectra denote detection wavelengths of the intensity image to the left. (After Grundmann et al., 1995a)

The purple emission of InGaN-based lasers has been recently attributed to excitons localized in three-dimensional In-rich deep traps (Narukawa et al., 1997). Thus quantum dot luminescence has been observed for a variety of systems covering the entire spectral range from infrared to deep blue.

6.7.1.1 Broadening of single dot lines

It was found (Grundmann et al., 1995a; Grundmann, 1996) that single dot luminescence from InAs/GaAs quantum dots has constant FWHM up to at least 60 K. The measured linewidth was limited there by the spectrometer resolution

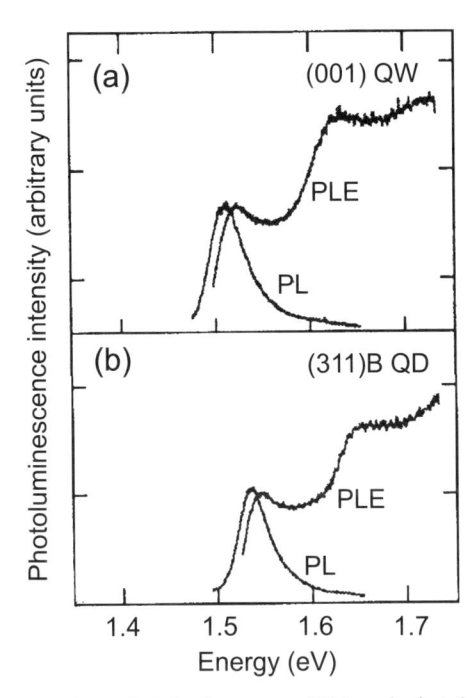

Fig. 6.34 Room temperature photoluminescence (PL) and photoluminescence excitation (PLE) spectra of (a) the reference quantum well and (b) isolated In$_{0.4}$Ga$_{0.6}$As/AlGaAs quantum disks with 30 nm diameter. (Reproduced by permission of American Institute of Physics from Nötzel *et al.*, 1994a)

⩽150 µeV. Gammon *et al.* (1996b) observed via µ-PL that the recombination from excitons in one single quantum dot formed by thickness fluctuations in a quantum well can be as narrow as 23 µeV. The temperature dependence of the linewidth was studied up to 50 K (Fig. 6.35) where it is still below 150 µeV. The observed increase of the homogeneous linewidth of the ground state is explained by enhanced acoustic phonon scattering into excited states. The scattering rate for the µth exciton state, $\Gamma_\mu(T)$, can be written (in perturbation theory) as

$$\Gamma_\mu(T) = \Gamma_\mu + \sum_{\nu > \mu} \gamma_{\mu\nu} n(E_{\mu\nu}, T) + \sum_{\nu < \mu} \gamma_{\mu\nu} [n(E_{\mu\nu}, T) + 1] \quad (6.1)$$

where $n(E_{\mu\nu}, T) = [\exp(E_{\mu\nu}/kT) - 1]^{-1}$ is the number of acoustic phonons with energies equal to the transition energy $E_{\mu\nu}$, given by the Bose function. $\gamma_{\mu\nu} = \gamma_{\nu\mu}$ contains the electron–phonon matrix element between the µth and νth state. Γ_μ is the broadening due to radiative transitions of the µth state. The second term in Eq. (6.1) is due to phonon absorption into higher states and goes to zero for $T \to 0$. The third term describes the phonon emission from a higher state; it is zero for the ground state at all temperatures. In Fig. 6.35 a good fit to the experimentally determined homogeneous linewidth of the ground state is obtained by assuming that

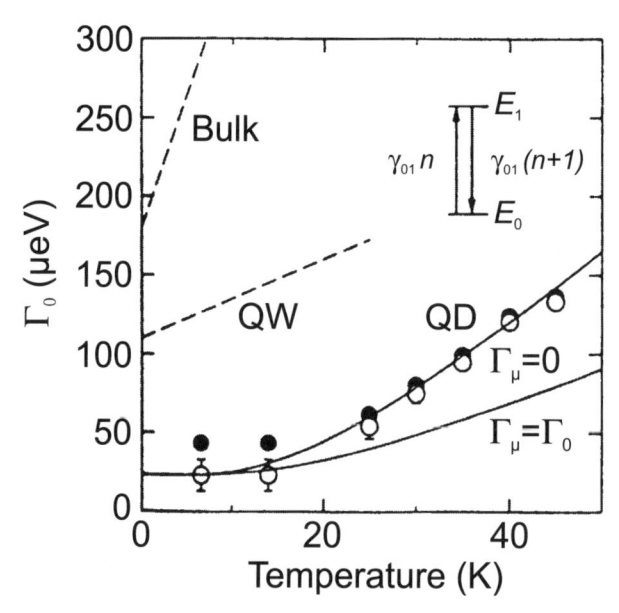

Fig. 6.35 Temperature dependence of the homogeneous linewidth Γ_0 of the quantum dot ground state. Filled (open) symbols denote the width before (after) deconvolution of the spectrometer response. The solid curves are fit with the phonon scattering model discussed in the text (Eq. 6.1) and schematically shown in the inset. Dashed curves are previously measured homogeneous linewidths of a 12 nm wide GaAs/AlGaAs quantum well and a bulk-like 194 nm GaAs layer Kuhl et al., 1989). (After Gammon et al., 1996b)

$\Gamma_\mu = 0$ for all excited states. This is consistent with the lack of luminescence from excited states in the experiment.

The homogeneous width of excited states (as probed by excitation spectroscopy) increases with their energetic separation from the ground state (right inset of Fig. 6.22) (Gammon et al., 1996b), most likely due to an increase of the phonon emission rate. For the first excited state a (absorption) linewidth of 35 μeV is measured which yields a scattering rate of $(19\,\text{ps})^{-1}$, which is in good agreement with the phonon emission rate calculated by Bockelmann (1993) for a GaAs dot with the same level separation. Thus for such dots the phonon bottleneck is indeed almost absent, which is again consistent with the lack of luminescence from excited states in the experiment.

With increasing excitation power (for nonresonant optical excitation in the barrier) for etched InGaAs/GaAs dots (Steffen et al., 1996a) and self-organized InGaAs dots on GaAs (311)B (Kamada et al., 1996b) it is observed that the lines increasingly broaden. This is most likely caused by the increased dephasing probability when the dots are filled with a larger number of carriers. Resonant excitation of the ground state without the presence of additional carriers on excited states is observed to lead to no broadening of the sharp recombination line up into the saturation regime (Kamada et al., 1996b).

ELECTRONIC AND OPTICAL PROPERTIES   231

6.7.1.2 State filling and temperature dependence

With increasing excitation density excited states are increasingly populated. A typical spectrum of an ensemble of self-organized InAs/GaAs quantum dots at different excitation densities is shown in Fig. 6.36. The MBE-grown quantum dots have been identified as pyramids with about 12-nm base length (Bimberg *et al.*, 1995; Grundmann *et al.*, 1995b; Ruvimov *et al.*, 1995). Around an excitation density of 50 W/cm^2, which corresponds to a steady state excitation of two photoexcited electron–hole pairs per dot, the ground state transition saturates. At higher excitation (approximately 200 W/cm^2) the first higher energy transition saturates. This line was originally attributed to the transition of ground state electrons and excited holes on the basis of effective mass calculations (Grundmann *et al.*, 1995c). It has become clear in the meantime that several electron states can exist in QDs of such ground state transition energy and size (e.g. Stier *et al.*, 1998). Therefore the higher energy transition at 1.18 eV is now assigned to the transition between excited electrons and excited holes (Fig. 5.21). Actually two close-lying transitions e(010)–h(010) and e(100)–h(001) contribute, involving the *p*-like electrons levels which are split by the piezolectric field. The transition still at energy of 1.26 eV is thought to involve a higher excited electron state, e(200)–h(020).

The spectral shape is well fitted with three Gaussians. The inhomogeneous broadening of the ground state is FWHM$_0$ = 50 meV. The theoretical change of

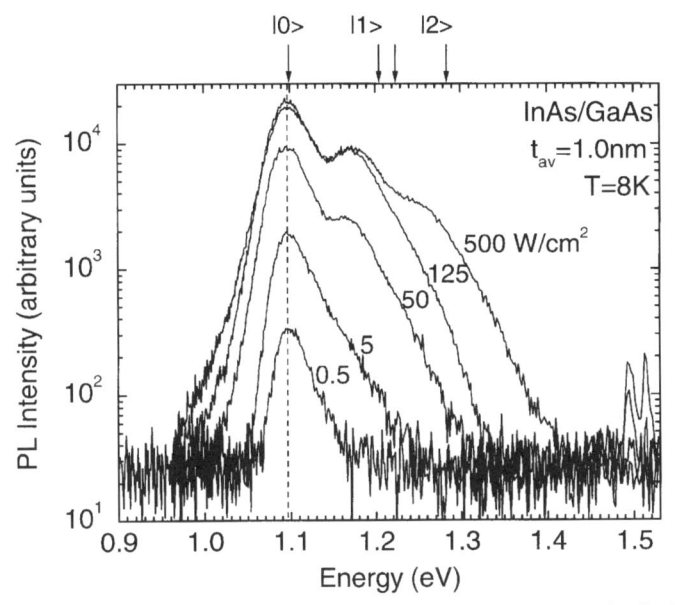

Fig. 6.36 PL spectra (*T* = 8 K) at different excitation densities from nominally 1.0 nm InAs/GaAs quantum dots grown with MBE. Arrows denote different transitions between zero-dimensional states calculated with eight band k·p theory (Stier *et al.*, 1998) (Fig. 5.21). The three peaks are labeled |0⟩, |1⟩, and |2⟩ for further references in the text

the ground state energy with pyramid base length is about $\partial E_0/\partial b \approx -30$ meV/nm at $b = 12$ nm. Thus the relative size fluctuation is $\xi \approx 6\%$, consistent with the evaluation of TEM images ($\xi < 9\%$) (Ruvimov et al., 1995).

Mukai et al. (1996a, 1996b) reported that InGaAs/GaAs quantum dots have a disk-like shape (typical diameter of 20–30 nm). The spectra of these dots consist of at least five almost equidistant lines and are well modeled with an in-plane two-dimensional harmonic oscillator potential. All lines are due to allowed transitions between levels with the same quantum numbers and have a degeneracy (including spin) of $2n$, $n = 1, 2, \ldots, 5$. Since the broadening of the peaks is rather constant for all transitions it is attributed to island height (or composition) fluctuations and not to variations of the disk radius. Similar spectra of almost equidistant peaks of similar width were detected from strain-induced quantum dots (Lipsanen et al., 1995) (Fig. 6.12). With increasing injection current or excitation density higher transitions appear in the spectra with *higher* intensity than the ground state transition due to the higher degeneracy of excited states. The first excited transition appears before the ground state transition is saturated. As discussed in Section 5.5.2.2 this fact alone does not hint at the presence of slowed-down energy relaxation. In the spectra of Lipsanen et al. (1995) and Mukai et al. (1996b), however, the third and higher excited states appear before the ground state is saturated, which indicates some slowing down of energy relaxation.

6.7.1.3 Excitation spectroscopy

In this section the processes governing the capture of carriers and their energy relaxation to the ground state of quantum dots are discussed, as revealed by steady state experiments. Time-resolved studies will be presented in Section 6.7.5. Various energy relaxation mechanisms like scattering by acoustic or optical phonons (Bockelmann, 1993; Heitz et al., 1996a), Coulomb scattering and Auger processes (Bockelmann and Egeler, 1992; Pan, 1994; Vurgaftman and Singh, 1994; Grosse et al., 1997; Uskov et al., 1998), interaction with nearby deep levels (Sercel, 1996), or (infrared) optical inter-level transitions (Sauvage et al., 1997) can contribute.

In self-organized InAs/GaAs quantum dots with a typical lateral size of 12–14 nm the hole level separation is several 10 meV. As initially reported in Ledentsov et al. (1995b) and Heitz et al. (1996a), photoluminescence excitation spectra of such dots do not exhibit peaks due to excited electronic states; instead a series of resonances appears with energies matching 1, 2, 3, or more LO phonon energies. In Fig. 6.37 the excitation spectrum of a typical dot sample is shown for a detection energy in the maximum of the photoluminescence spectrum. The observed resonances are due to intra-dot energy relaxation. Experiments by Steer et al. (1996) corroborate these results. Resonances at phonon energies have additionally been observed for InAs/GaAs quantum dots by Schmidt et al. (1996a) and for AlInAs/AlGaAs quantum dots by Fafard et al. (1995).

ELECTRONIC AND OPTICAL PROPERTIES

As discussed in Section 5.4.2, dots with the same ground state energy (as probed by the chosen photon energy of detection) have different excited states when shape fluctuations are present. Thus the excitation spectrum for a particular detection energy is inhomogeneously broadened for such an ensemble (Fig. 5.36). In order to observe luminescence at the energy of the ground state, however, not only the absorption of the exciting photons by the optical density of excited states but also energy relaxation into the ground state is necessary. The possible energy relaxation processes act as a filter and lead to the modulation of the excitation spectrum. As confirmed by time-resolved measurements (Heitz *et al.*, 1997b), the observed LO resonances are associated with relaxation channels for excited levels which favor fast energy relaxation. The absolute intensity of the LO resonances follows the inhomogeneous quantum dot density represented by the quantum dot photoluminescence spectrum upon excitation in the GaAs barrier (see the inset of Fig. 6.38). The relative intensity of the LO resonances is independent of the detection energy (i.e. quantum dot size). The dominating 3 LO resonance is close to the |1⟩ excited transition found in PL (Fig. 6.36). Further insight is gained by selective excitation of luminescence (Fig. 6.39). For sufficiently high excess energy the luminescence spectrum is not modulated. With decreasing excitation energy first the 3 LO, than the 2 LO, and finally the 1 LO replica becomes dominant, reflecting the density of quantum dots with a matching ground state energy.

Higher resolution excitation spectra reveal a threefold substructure for the 2 and 3 LO resonances (Fig. 6.40). Lineshape fits reveal three different LO phonon energies. The dominating relaxation channels are multiple LO phonon emissions of phonons having the *same* energy. The phonon energy $E_{LO}^{WL} = 29.6 \pm 0.5$ meV, corresponding

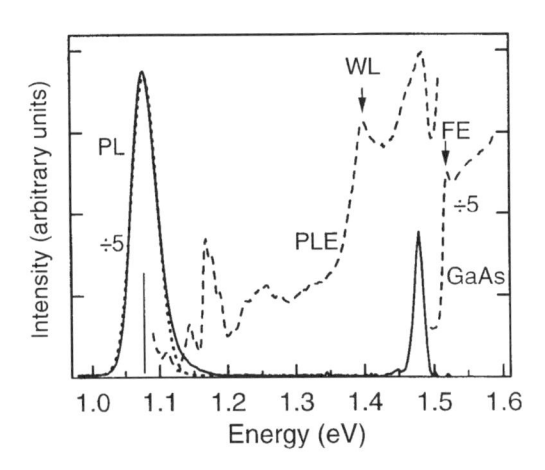

Fig. 6.37 Photoluminescence and PL excitation spectra ($T = 6$ K) for self-organized InAs/GaAs quantum dots ($d_{InAs} = 1.2$ nm) grown at 480 °C. The PL is excited at 1.67 eV, the PLE spectra are recorded at the maximum of the PL spectrum (1.08 eV). (After Heitz *et al.*, 1997b)

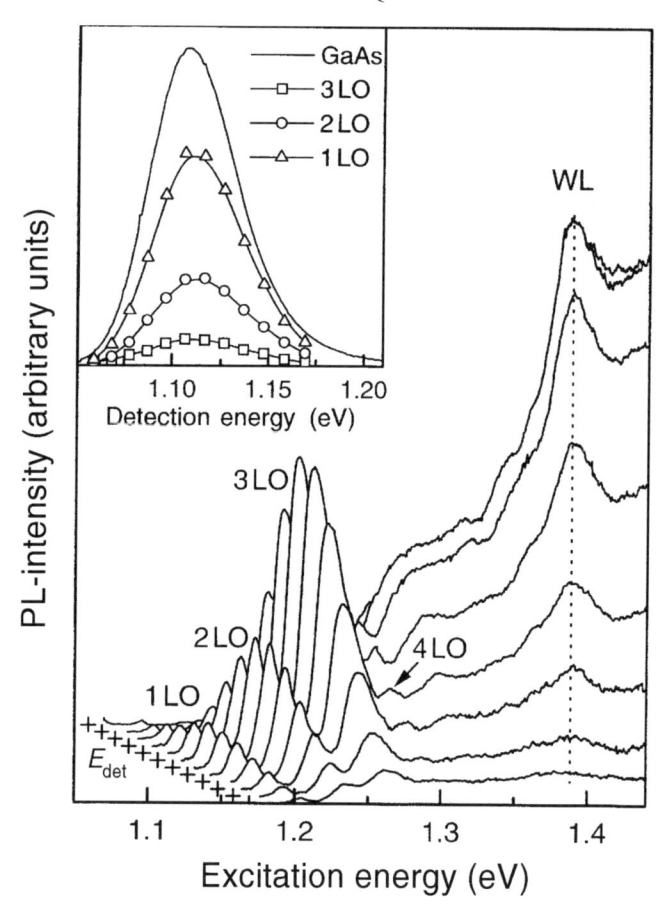

Fig. 6.38 PLE spectra of self-organized InAs/GaAs quantum dots for different detection energies E_{det} marked with crosses. The spectra are shifted in the y direction for clarity. The inset compares the PL intensity for excitation in the 1, 2, and 3 LO phonon resonances and above the GaAs band gap. (After Heitz *et al.*, 1997b)

to the InAs bulk LO phonon energy, is attributed to the InAs wetting layer, for which strain-induced shift and phonon confinement cancel. The mode at $E_{\text{LO}}^{\text{QD}} = 32.6 \pm 0.5$ meV is attributed to the InAs quantum dot LO phonon. This value agrees very well with the value 32.1 meV predicted for strain calculations by Grundmann *et al.* (1995c) and thus confirms the coherent nature of the dots. The third energy of $E_{\text{LO}}^{\text{barr}} = 37.6$ meV is attributed to the GaAs barrier; the increase as compared to GaAs bulk reflects the strain in the vicinity of the dot. A fourth phonon at 35 meV was found in selective excitation spectra reported in Heitz *et al.* (1996a) and was attributed to an interface mode.

The wetting layer and GaAs LO phonon resonances have an FWHM of about 7 meV, with the quantum dot phonon 3 LO resonance even about 14 meV. The large

ELECTRONIC AND OPTICAL PROPERTIES

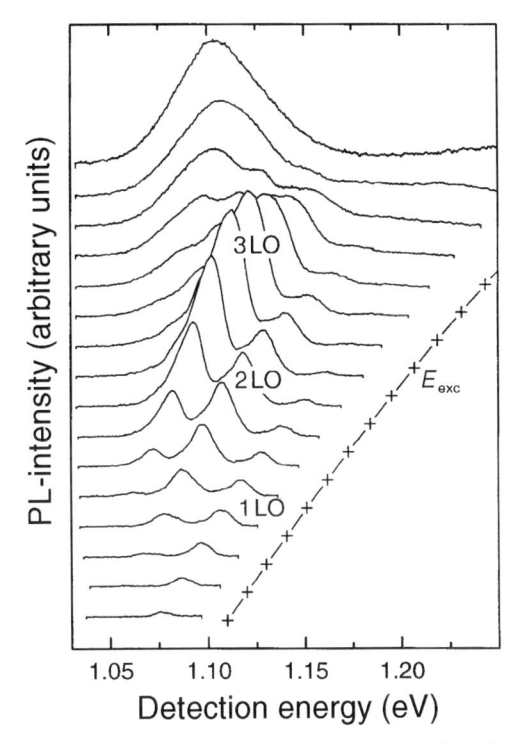

Fig. 6.39 PL spectra of self-organized InAs/GaAs quantum dots for different energies of selective excitation energies E_{exc} marked with crosses. The spectra are shifted in the y direction for clarity. (After Heitz et al., 1997b)

widths might be the result of weakened k-selection rules, strain inhomogeneities within single dots and within the ensemble, or higher-order processes involving LA phonons (Inoshita and Sakaki, 1992). The dominating phonon is the quantum dot LO phonon. For the 3 LO resonance the wetting layer and GaAs phonon replica become stronger as compared to the 1 and 2 LO resonances. The coupling to those modes is enhanced with increasing energy separation from the ground state and consequently decreasing carrier localization in the excited state.

6.7.1.4 Magnetoluminescence

The effects of a static external magnetic field on the energy levels of quantum dot excitons are twofold (and have been theoretically discussed in Section 5.6.2):

(a) diamagnetic shifts of the various states, dependent on the extension of the wavefunction,
(b) Zeeman splitting of states due to orbital and spin degeneracy.

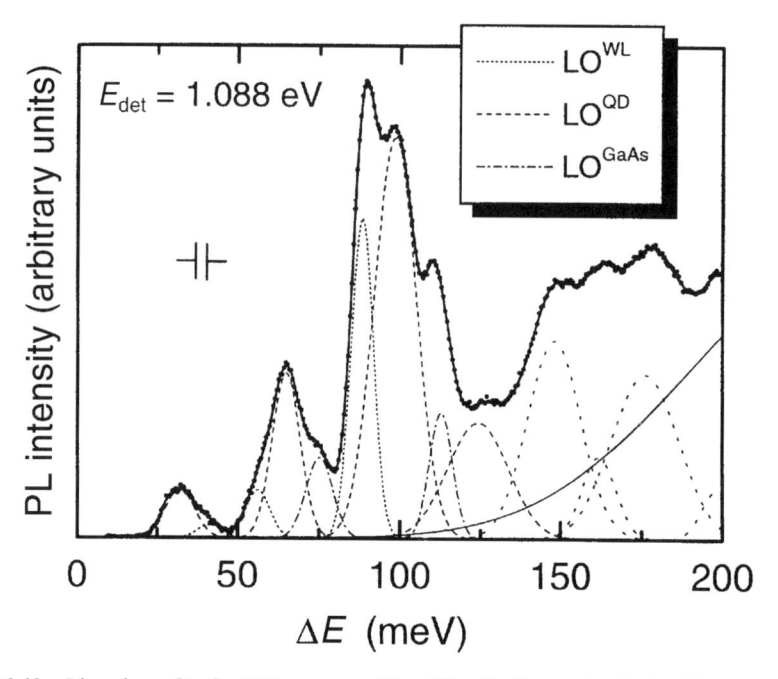

Fig. 6.40 Line-shape fit of a PLE spectrum ($T = 6$ K) of self-organized InAs/GaAs quantum dots at a low excitation density ($D < 0.02$ W/cm^2). The combined experimental resolution of excitation and detection indicated in the figure is 3.6 meV. (After Heitz et al., 1997b)

Both effects have been experimentally investigated for various types of quantum dots. The evolution of the recombination lines of strain-induced quantum dots (Lipsanen et al., 1995) in magnetic fields parallel and perpendicular to the surface, the (xy) plane, has been investigated by Rinaldi et al. (1996). An in-plane field does not lead to observable peak shifts because z is the direction of very strong confinement. In addition no splitting is found. The two-dimensional in-plane harmonic oscillator like states are, however, strongly affected by a magnetic field parallel to the z direction, as shown in Fig. 6.41. The fan chart which approximately describes the experimental results has been calculated using a harmonic in-plane confinement. Transitions between states with $m_j \neq 0$ split, as expected, linearly with increasing field; transitions between $m_j = 0$ states exhibit only a weak diamagnetic shift.

The diamagnetic shift of small self-organized InAs/GaAs quantum dots has been investigated by Itskevich et al. (1997). For magnetic field direction parallel to the growth direction the shift of the photoluminescence line (weighted line center) is quadratic in B for fields up to 23 T (Fig. 6.42). From the magnitude of the shift a spatial extension of the wavefunction in the (xy) plane of 6 nm is obtained. The diamagnetic shift is strongly anisotropic: for fields parallel to the sample surface a *decrease* of the energetic position of the line center with increasing field is observed

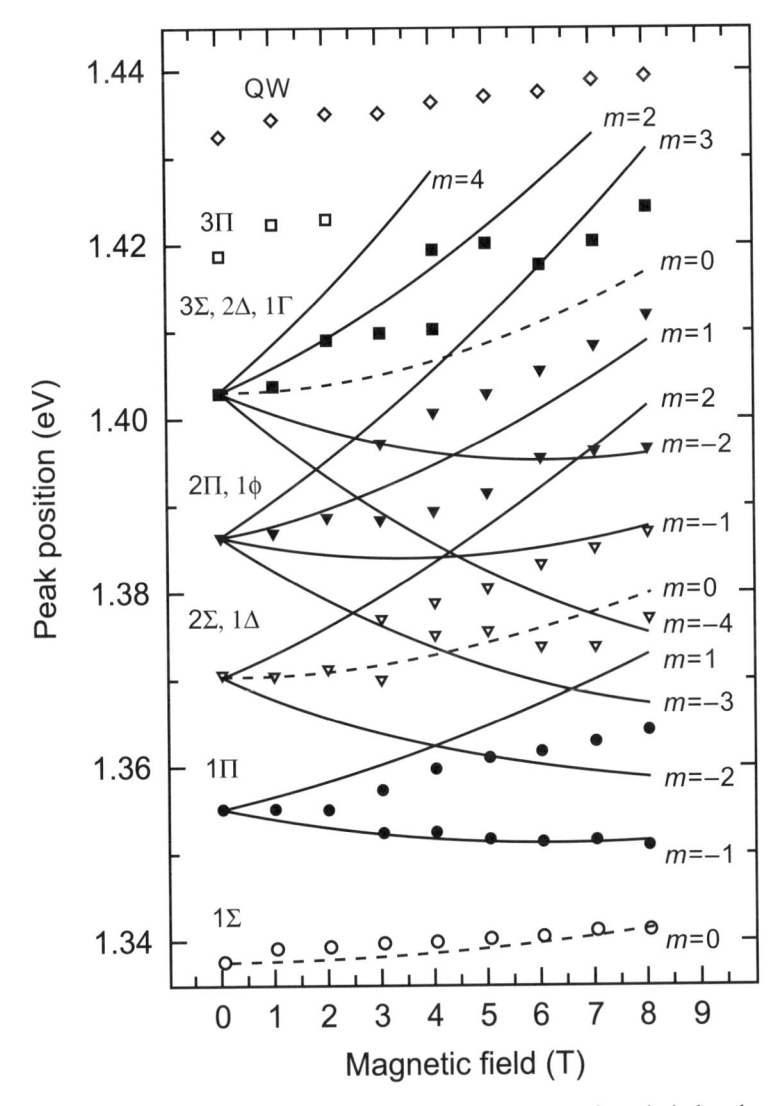

Fig. 6.41 Fan plot of peaks in the magnetoluminescence of strain-induced quantum dots. The magnetic field is along the growth axis, excitation density is 75 W/cm^2. Curves are theoretical calculations; dashed lines indicate nonsplitting transitions ($|m|=0$), solid lines indicate split transition with nonvanishing magnetic momenta ($|m|\neq 0$). Diamonds denote a shift of the QW transition. (After Rinaldi *et al.*, 1996)

for fields up to about 12 T. The subsequent blue shift is smaller than for the magnetic field along the z direction due to the stronger confinement of carriers in the growth direction and the contribution of the larger hole mass. The red shift by almost 2 meV is not accounted for yet.

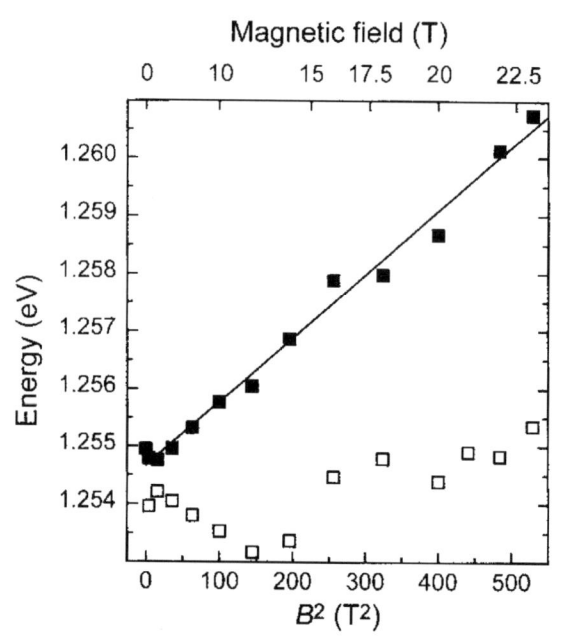

Fig. 6.42 Magnetic field dependence of the energy position (weighted photoluminescence line center) for self-organized InAs/GaAs quantum dots with a ground state energy of about 1.25 eV. Solid (open) squares are for the magnetic field normal (parallel) to the sample plane. Solid line is the least squares fit. (Reproduced by permission of American Institute of Physics from Itskevich et al., 1997)

The energy splitting between states of angular momenta +1 and −1 in the magnetic field has been investigated by Bayer et al. (1996) for cylindrical, wet chemically etched $In_{0.53}Ga_{0.47}As/InP$ dots of different size. As expected from theory, the Zeeman splitting $\Delta E = \hbar eB/\mu$, μ being the reduced mass, depends linearly on the magnetic field and is independent of the dot diameter (Fig. 6.43).

An enhancement of spin splitting with decreasing quantum dot size has been reported in Bayer et al. (1995). Deep etched $In_{0.1}Ga_{0.9}As/GaAs$ quantum dots of cylindrical disk shape were investigated. With increasing magnetic field the difference in energy between the σ^+ and σ^- polarized lines increases linearly (Fig. 6.44a). The energy difference (at a fixed magnetic field) is shown in Fig. 6.44b as a function of the dot diameter d. The experimental data points are best described with a curve $\sim d^{-2}$, which has theoretically been explained with the enhancement of the hole g factor due to the admixture of higher dot levels to the ground state by the spin–orbit interaction (Bayer et al., 1995). Near-field magneto-optical data from self-organized InAs/GaAs QDs reported by Toda et al. (1998) show that the spin splitting scatters for different dots having the same or similar ground state transition energy. No clear trend for the splitting for QDs with ground state energies between 1.25 and 1.37 eV is observable. The g factor varies between 1.21 and 1.73.

ELECTRONIC AND OPTICAL PROPERTIES

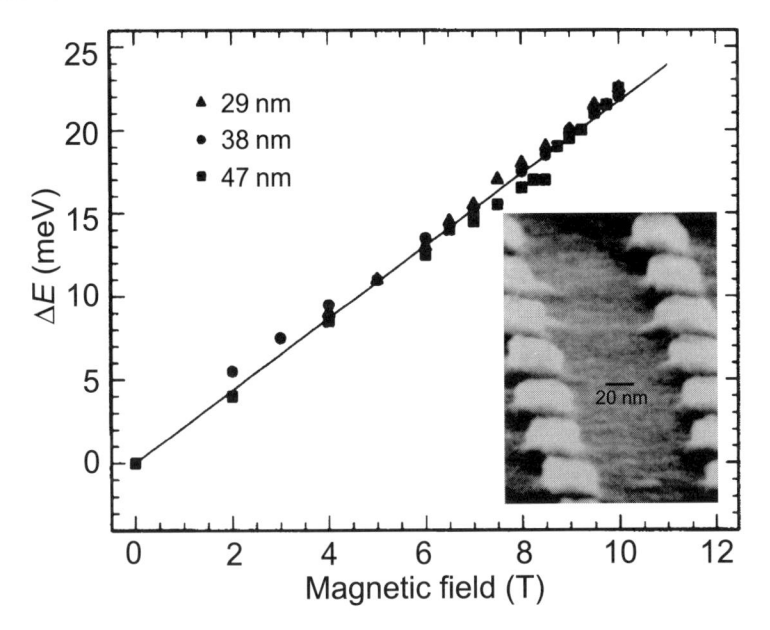

Fig. 6.43 Energy splitting between the transitions involving angular momenta +1 and −1 in cylindrical $In_{0.53}Ga_{0.45}As/InP$ dots as a function of the magnetic field for three different dot diameters $d = 29$ (▲), 38 (●), and 47 (■) nm. (After Bayer et al., 1996)

$GaAs/InAs/Al_{0.36}GA_{0.64}As$ QDs (an asymmetric barrier!) are found to exhibit a smaller spin splitting, which is caused by the wavefunction being pushed out of the QD, predicted earlier by Grundmann et al. (1995c).

The spin splitting in single InAs self-assembled QDs (embedded in AlAs) has been observed by Thornton et al. (1998) using magneto-tunneling spectroscopy. g factors between +0.52 and +1.6 have been found (see Section 7.3).

The magneto-optical properties of excitons localized at quantum well inhomogeneities (see Section 6.5) have been reported in Heller and Bockelmann (1997), with the result that spin is largely conserved during relaxation processes from excited states into the ground state. Spin relaxation of the ground state at zero field is fast in the range of some tens of picoseconds. With an increasing field and subsequent energy splitting of the ground state the spin relaxation is reduced because energy exchange with the lattice is required.

6.7.2 ABSORPTION

In true quantum dots the oscillator strength of the material making up the dots is concentrated in the zero-dimensional states. Accordingly these states give rise to a reasonable absorption signal. A suitable technique is calorimetric absorption spectroscopy (CAS) (Bimberg et al., 1991). In Fig. 6.29 the CAS signal and

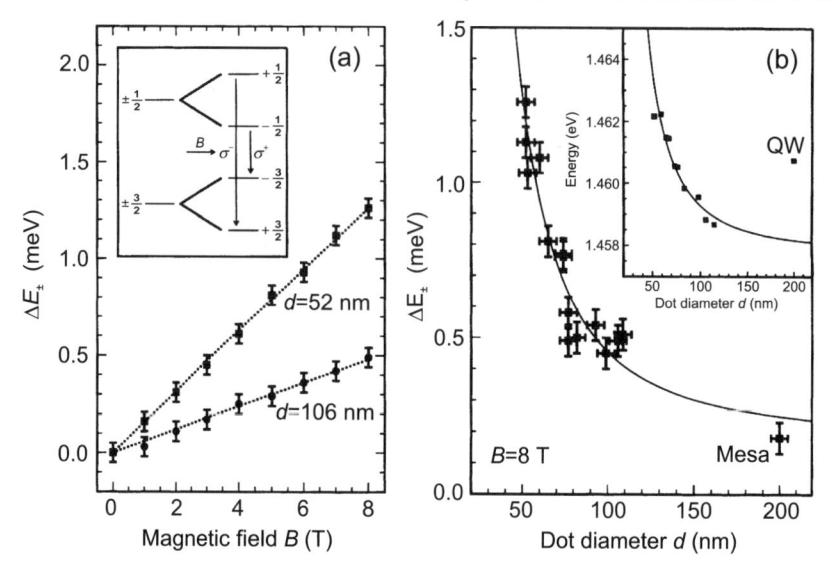

Fig. 6.44 (a) Energetic separation $\delta E_\pm$ between the σ^+ and σ^- polarized lines of $In_{0.1}Ga_{0.9}As/GaAs$ deep-etched quantum dots with two different diameters d ($d = 52$ and 106 nm) as a function of the magnetic field. Inset shows the schematic scheme of investigated levels and transitions. (b) $\delta E_\pm$ at $B = 8$ T as a function of dot diameter d. Solid line is proportional to d^{-2}. Inset shows increase of energy of ground state luminescence ($B = 0$ T) as a function of dot size due to confinement. (After Bayer et al., 1995)

photoluminescence spectra of a typical MBE-grown InAs/GaAs quantum dot sheet for nominal 4 ML InAs deposition were compared (Grundmann et al., 1995a). The ground state absorption and luminescence are in close resonance. Additionally, absorption due to excited levels (QD*) and the wetting layer appears. Their transition energies correspond to luminescence lines found at high-excitation photoluminescence (Grundmann et al., 1996d) (see also Section 6.7.1.2). An almost identical situation is observed for MOCVD-grown quantum dots. The absorption signal from a threefold stack (Fig. 6.45) of InAs/GaAs quantum dots also shows the same transitions as the photoluminescence spectra. Transmission spectra of MBE grown self-organized QDs have been measured using Fourier spectroscopy by Warburton et al. (1997). Several interband transitions were detected. In a field effect structure additionally the population of the QDs with electrons could be controlled, enabling the determination of transition energies of charged excitons in good agreement with results from perturbation theory.

The important question of whether there is resonance between absorption and emission can also be probed by excitation spectroscopy. When the excitation wavelength is tuned into the ground state, the luminescence is emitted at the same wavelength. It can usually only be separated from the unavoidable stray light of the excitation source by time-resolved techniques (Lowisch et al., 1996) (see Section 6.7.5). We note that the resonance between emission and absorption typical for self-

ELECTRONIC AND OPTICAL PROPERTIES 241

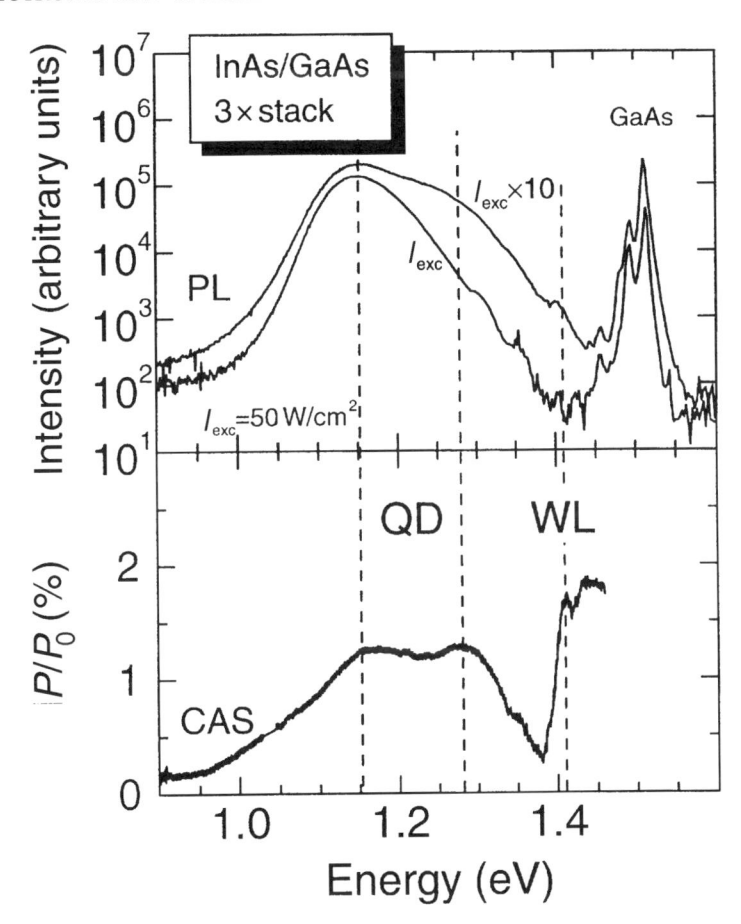

Fig. 6.45 Photoluminescence ($T = 8$ K) and calorimetric absorption spectra of a triple stack of InAs/GaAs quantum dots grown with MOCVD. The excited state at 1.3 eV apparent in the absorption spectrum becomes populated at the higher PL excitation density

organized III–V quantum dots is different from typical absorption and luminescence spectra of II–VI quantum dots in a glass matrix, as shown in Fig. 6.46 (de Oliveira *et al.*, 1995). In this system a large Stokes shift between luminescence and absorption is usually observed, which occurs because absorption takes place in the II–VI dot material, but the luminescence stems from interface traps to which the carriers localize (Roussignol *et al.*, 1989; Bawendi *et al.*, 1990).

6.7.3 ELECTROREFLECTANCE

The contactless electroreflectance (CER) response from vertically coupled InAs/GaAs quantum dots (tenfold stack) has been reported in Aigouy *et al.* (1997a). The

242　　　　　　　　　　　　　　　　　　QUANTUM DOT HETEROSTRUCTURES

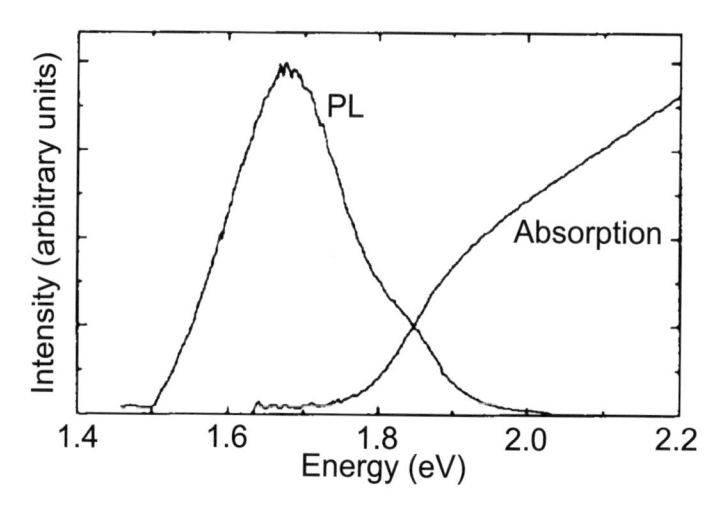

Fig. 6.46 Photoluminescence (PL) and absorption (ABS) spectra of CdTe quantum dots. The PL has been measured at 2 K. The room ABS spectrum has been shifted by 174 meV to account for the temperature change of the bandgap. (Reproduced by permission of American Institute of Physics from de Oliveira et al., 1995)

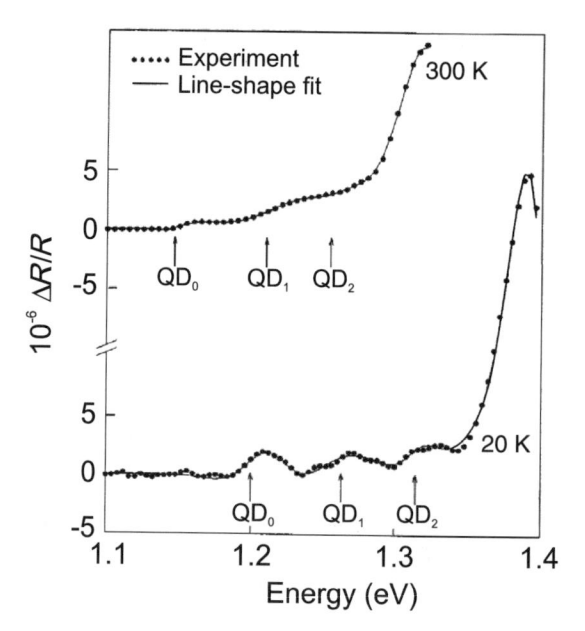

Fig. 6.47 CER spectra (dots) with line-shape fit (solid line) of the sample with a tenfold stack of InAs/GaAs quantum dots at two different temperatures, 300 and 20 K. The transitions due to zero-dimensionally confined states in the QDs are labeled QD_0, QD_1 and QD_2. (After Aigouy et al., 1997a)

energy positions of three QD transitions obtained from peaks in the CER spectra taken at 20 K (Fig. 6.47) are in agreement with measurements using CAS (at 0.5 K) (Ledentsov et al., 1996d). In addition, response is obtained from the wetting layer and the GaAs barrier. Evidence for these states is also found in CER spectra recorded at room temperature. All transitions have also been observed in surface photovoltage spectroscopy (SPS) by Aigouy et al. (1997b).

6.7.4 FOURIER SPECTROSCOPY OF INTER-SUBLEVEL TRANSITIONS

With far- or mid-infrared radiation the transitions between different sublevels in quantum dots can be probed. Depending on the quantum dot size the energy of the corresponding transitions is in the range of a few millielectronvolts below the Reststrahlen regime or for small quantum dots much larger than typical phonon energies.

Sauvage et al. (1997) investigated the infrared absorption of small InAs/GaAs self-organized quantum dots in the energy range 90–250 meV. The samples are illuminated with a pump light source to optically populate the quantum dot ground state. Then the infrared transmission is probed (Fig. 6.48). The transition between the ground and excited state is observed at an energy of 115 meV for a photoluminescence spectrum centered at 1.15 eV; an additional broad feature at 200 meV is attributed to the transition of the confined electron to the continuum (wetting layer or barrier). The two transitions possess different temperature dependences. With increasing temperature smaller quantum dots, having smaller localization energy, are preferentially depopulated. Thus the infrared absorption signal is increasingly caused by larger dots having larger electron localization energy. Therefore the electron confined-to-continuum transition exhibits a blue shift with increasing temperature. On the contrary, the confined transition exhibits a small red shift because the sublevel separation decreases with increasing dot size. It is found that the transition between the confined levels saturates with increasing excitation density when all final states are populated. Absorption peaks at 70 and 88 meV have been observed in Si-doped InAs/GaAs and InAs/Al$_{0.15}$Ga$_{0.85}$As quantum dots, respectively (Phillips et al., 1997b). Normal incidence absorption of a 20-period InGaAs/GaAs quantum dot superlattice at 14–15 μm ($\approx$ 86 meV) with a linewidth of about 11 meV has been reported by Pan et al. (1996). Mid-infrared photoconductivity (300–400 meV) has been found for InAs QDs by Berryman et al. (1997). FIR photoconductivity with peak response at about 70 meV has been reported by Phillips et al. (1998a) for Si-doped InAs/GaAs Qd's for normal incidence.

The far-infrared response of self-organized lens-shaped InGaAs quantum dots in a GaAs/AlAs field-effect-type heterostructure, charged with electrons, has been probed by Drexler et al. (1994). A transition between the electron ground state and first excited state was observed. Under increasing bias voltage the ground state can be occupied by two electrons and a complex pattern of inter-sublevel transitions in a magnetic field was observed (Fricke et al., 1996) (Fig. 6.49b). The splitting of the p-like excited electron state could be followed (Fig. 6.49a). The cause of the

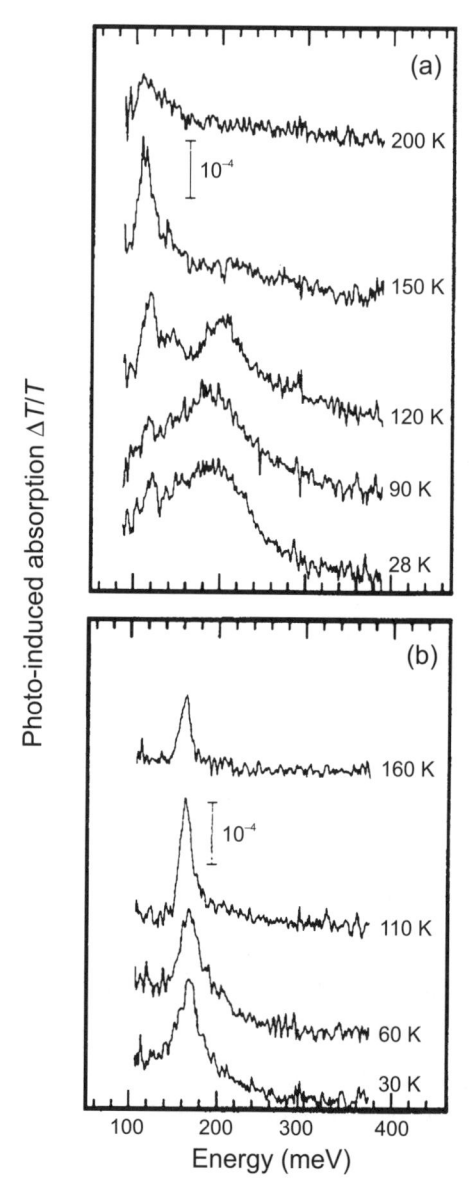

Fig. 6.48 Photo-induced infrared absorption of self-organized InAs/GaAs quantum dots for different temperatures. (a) For a sample with broad distribution of ground state energies (approximately 1.05–1.35 eV), (b) for a sample with narrow energy distribution (approximately 1.13–1.25 eV). (Reproduced by permission of American Institute of Physics from Sauvage *et al.*, 1997)

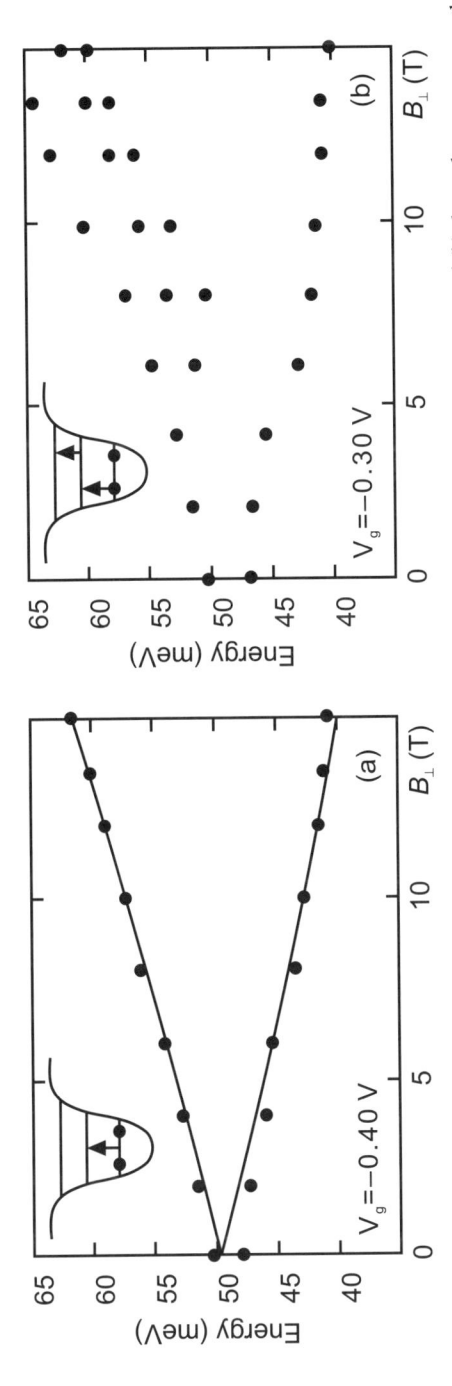

Fig. 6.49 Positions of the far-infrared resonances of lens-shaped InGaAs/GaAs quantum dots filled with (a) two and (b) three electrons versus the magnetic field. The partial filling of the p-like states in (b) gives rise to additional resonances. Insets display schematically the observed resonances in a simplified single-particle picture. (After Fricke *et al.*, 1996)

measured zero magnetic field splitting can be manifold (see Section 5.2.4). Spontaneous emission of far infrared radiation in an InGaAs/AlGaAs quantum dot laser has been reported by Vorob'ev *et al.* (1998).

6.7.5 CARRIER DYNAMICS

In this section the dynamics of carrier capture, energy relaxation and radiative recombination is discussed. An attempt is made to separate these three mechanisms, although all three typically appear in the same time-resolved luminescence experiment. Especially capture and energy relaxation are intrinsically connected. In a restricted sense energy relaxation will be understood here as relaxation of energy from carriers on confined levels within the quantum dot.

6.7.5.1 Carrier capture and relaxation

For device applications, generally, fast carrier capture is desirable. The capture process consists of two essential parts:

(a) the *transport* of carriers towards the quantum dots from the barriers and within the wetting layer plane;
(b) the actual quantum mechanical capture process of a carrier near the quantum dot, involving real space transfer and energy relaxation, potentially into the lowest energetic state.

For structures with a low density of quantum dots ($\approx 2 \times 10^9 \, \text{cm}^{-2}$) (Wang *et al.*, 1996a), the small volume filling factor is believed to be responsible for the observed large capture time of $\tau_c = 350 \, \text{ps}$ due to the dominance of transport. Such limitation is not expected in dense arrays of self-organized quantum dots with typical lateral densities of 4×10^{10} to $2 \times 10^{11} \, \text{cm}^{-2}$.

Time-resolved studies of the onset of ground state luminescence upon resonant and nonresonant excitation in self-organized MBE-grown 4 ML InAs/GaAs quantum dots reveal a time constant of $28 \pm 5 \, \text{ps}$, independent of excitation energy in a wide energy range $>400 \, \text{meV}$ (Fig. 6.50) (Heitz *et al.*, 1997b). These times are of the order of relaxation times predicted for LO + LA scattering (Inoshita and Sakaki, 1992), indicating that the multiphonon emission probability does not drastically decrease for higher-order LO processes. The rise time of the excited state photoluminescence is much faster ($<10 \, \text{ps}$). Thus the observed rise time of the ground state luminescence is mainly given by the relaxation time constant from the excited electronic state into the ground state. The carrier capture itself occurs on a $<10 \, \text{ps}$ time scale.

In somewhat larger MOCVD-grown InAs/GaAs quantum dots reported in Adler *et al.* (1996), a much larger rise time of 156 ps for the ground state luminescence at 1.036 eV was found. This is attributed to slow energy relaxation within the dot; higher states show much smaller rise times in the 20 ps range, also indicating fast capture from the barrier into the quantum dots. Ohnesorge *et al.* (1996) reported the dependence of the rise time on the excitation density (Fig. 6.51) for self-organized

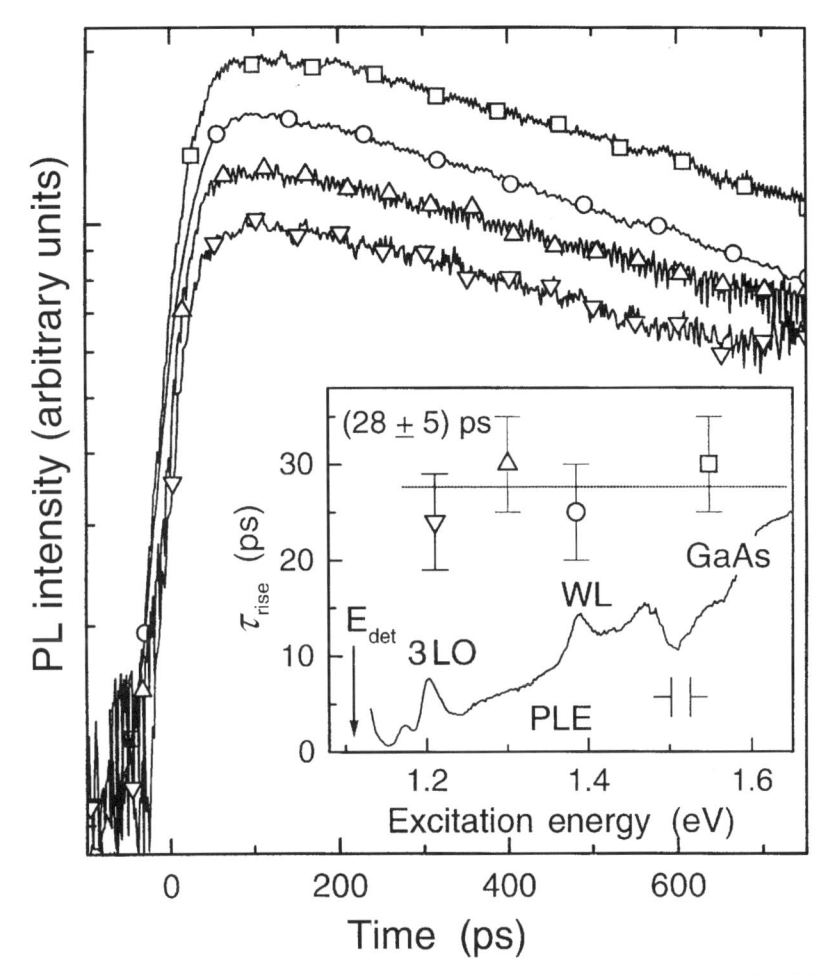

Fig. 6.50 Transients of the PL of self-organized InAs/GaAs quantum dots for different excitation energies. The inset displays the PL rise time which depends on the excitation energy together with a PL excitation spectrum. Spectral resolution is 20 meV. (After Heitz *et al.*, 1997b)

MOCVD-grown InGaAs/GaAs quantum dots with ground state luminescence at around 1.34 eV. While below an excitation density of about 4 W/cm² the rise time is basically constant (90 ps), it exhibits a drop for higher excitation (to 35 ps for 100 W/cm²). For the low excitation density regime the rise time depends on temperature, ruled by the Bose distribution function

$$\tau^{-1} \propto 1 + \left[\exp\left(\frac{E}{kT}\right) - 1\right]^{-1} \tag{6.2}$$

248 QUANTUM DOT HETEROSTRUCTURES

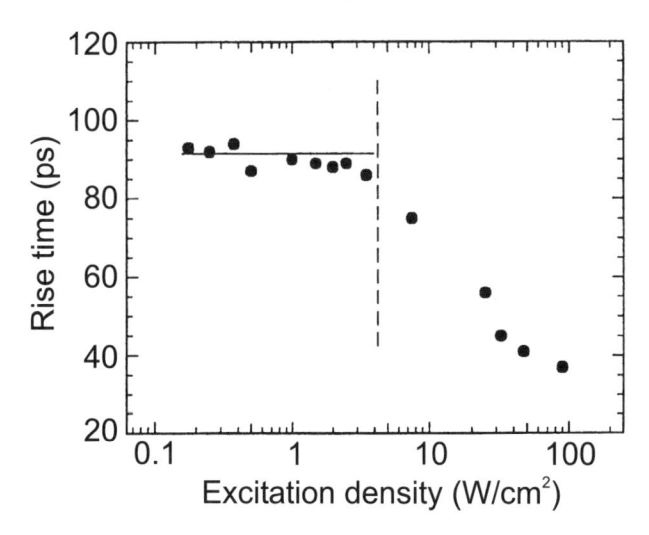

Fig. 6.51 Rise time of luminescence of InGaAs/GaAs quantum dots as a function of excitation density. (After Ohnesorge *et al.*, 1996)

while at high excitation density it does not. The two regimes are thus attributed to multiphonon processes of the type $n \times \text{LO} + \text{LA}$ at low excitation density (<one eh pair per dot) and Auger processes at high excitation densities (Ohnesorge *et al.*, 1996). For the latter process, capture and relaxation processes cannot be strictly separated.

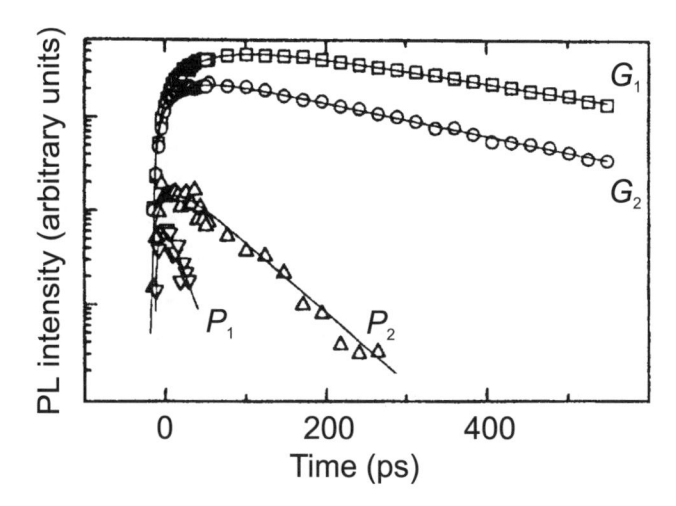

Fig. 6.52 Temporal dynamics of resonantly excited self-organized InP/InGaP quantum dots. P_1 (P_2) denotes the 1 (2) LO phonon replica; G_1 and G_2 denote the ground state PL. (After Vollmer *et al.*, 1996)

ELECTRONIC AND OPTICAL PROPERTIES

Time-resolved studies (Vollmer et al., 1996) after selective excitation of 5 nm high InP islands with a lateral size of 30–50 nm embedded in an $In_{0.48}Ga_{0.52}P$ matrix yield fast rise times (below the time resolution of 10 ps) of luminescence at integer multiples of LO phonon energy below the excitation energy (Fig. 6.52). The 1 and 2 LO phonon resonances decay with time constants of 17 and 58 ps, respectively. Once an exciton state is reached that cannot relax further via LO phonon emission acoustic phonon scattering sets in with an emission time of several tens of picoseconds. The LO phonon replica is thus explained in a 'hot exciton luminescence' picture, as suggested by Wang et al. (1996b). The rise times of the ground states (in an ensemble with two different types of quantum dots with different sizes) apart from the LO resonances are 59 and 30 ps, respectively, also indicating the contribution of acoustic phonon emission.

6.7.5.2 Carrier recombination

The lifetime of the ground state of a quantum dot can only be determined unambiguously when no carriers are refilled from upper states. In Fig. 6.53 transients at different excitation densities are shown for self-organized 4 ML InAs/GaAs quantum dots (Heitz et al., 1997b). The detection energies correspond to the ground state, the $|1\rangle$ and $|2\rangle$ excited states, and the wetting layer (Fig. 6.36). The wetting layer shows a fast decay for all excitation densities. With increasing excitation density the upper levels of the QD are more and more populated. Consequently, the ground state is refilled for some time and the start of the decay of the ground state transient becomes delayed. Finally, for long delay times at all excitation densities the ground state decays with the lifetime $\tau_r = 1.07 \pm 0.1$ ns.

In Fig. 6.54 the transient for the first excited state luminescence is shown together with two fits. A fit of the transient for nonresonant excitation using a capture time of $\tau_0 = 30$ ps yields a perfect fit of the transients using Master equations for the microstates (see Section 5.5.2.4) ($\tau_{|0\rangle} = 0.85$ ns and $\tau_{|1\rangle} = 0.24$ ns, $\tau_0/\tau_{|0\rangle} \approx 0.04$). We note that fits for any value of $\tau_0 = 0$ ps to $\tau_0 < 100$ ps are practically identical. Thus the decay of nonresonantly excited transients is *insensitive* to the inter-level relaxation constant if it is small compared to τ_r; in this case the decay constant of the excited state is given by

$$\tau_{decay} = \left(\frac{2}{\tau_{|0\rangle}} + \frac{1}{\tau_{|1\rangle}} \right)^{-1} \tag{6.3}$$

(compare with Eq. 5.108) and therefore unsuited for the determination of τ_0. If instead of Master equations for the microstates a conventional (but incorrect) rate equation system for the average level populations is applied, nonexponential transients result in contrast to experiment and a value of about $\tau_0 = 240$ ps is necessary in order to fit roughly the experimental data for the excited state. For the correct value of $\tau_0 = 30$ ps the decay of the excited state transient is too fast (Fig. 6.54).

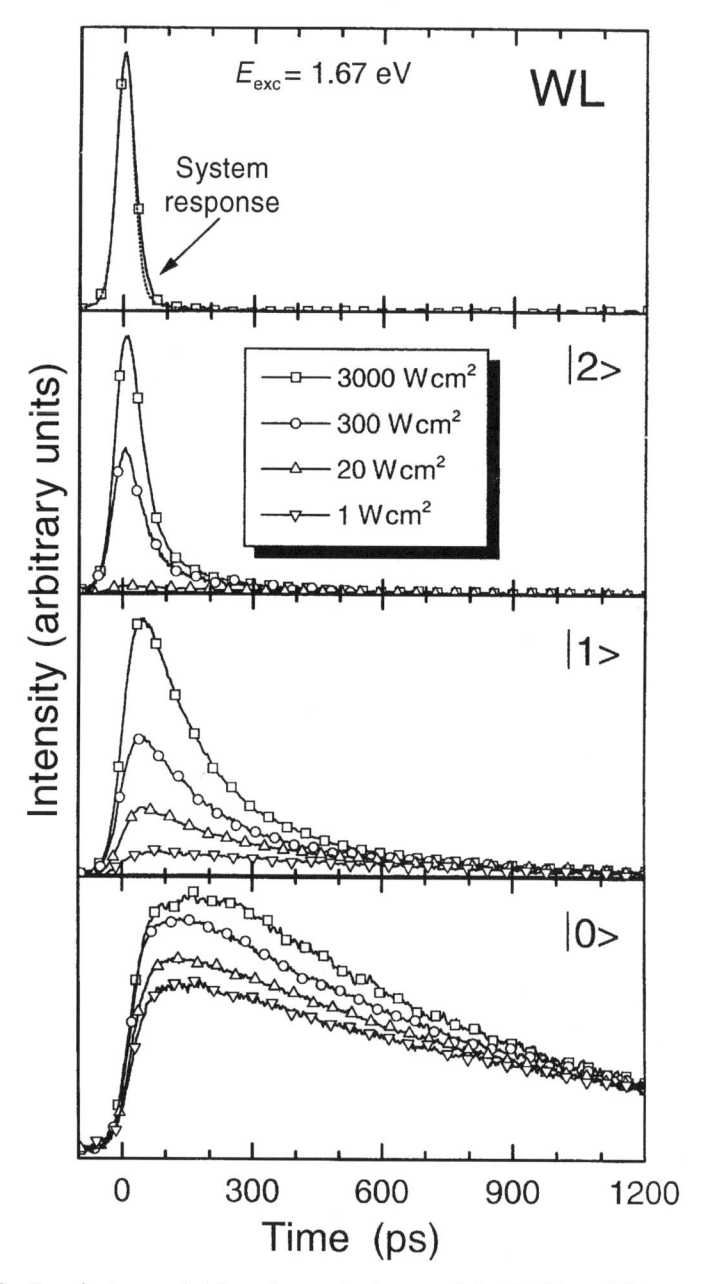

Fig. 6.53 Transients recorded from the wetting layer, and the $|0\rangle$, $|1\rangle$, and $|2\rangle$ states (see Fig. 6.36) of self-organized InAs/GaAs quantum dots for an excitation energy of 1.67 eV in the GaAs barrier at various excitation densities $D = 1$, 20, 300, and 3000 W/cm^2. Transients are scaled for the longest decay component. (After Heitz *et al.*, 1997b)

ELECTRONIC AND OPTICAL PROPERTIES 251

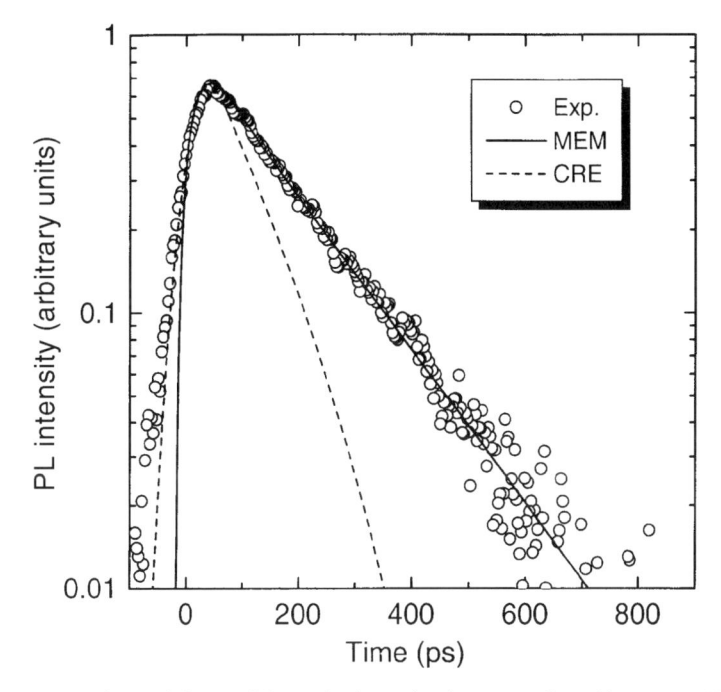

Fig. 6.54 Experimental decay of the excited state luminescence from 12 nm InAs/GaAs self-ordered quantum dots. Lines are fits with master equations for the microstates (MEM) and conventional rate equations (CRE), both for an inter-sublevel scattering time $\tau_0 = 30$ ps

The decay time constants of luminescence from the ground $|0\rangle$ and excited $|1\rangle$ states reported in Kamath et al. (1997a) are (at $T = 13$ K) $\tau_{|0\rangle} = 700$ ps and $\tau_{|1\rangle} = 250$ ps, respectively. They closely fulfill the relation $\tau_{|1\rangle} = \tau_{|0\rangle}/3$ expected for the case where the recombination time constants of electron–hole pairs in both states are identical (Eq. 5.108).

The lifetime of the ground state of InAs/GaAs quantum dots depends on the quantum dot size (ground state energy). For quantum dots of base size 12–14 nm $\hbar\omega = 1.1$ eV) the lifetime is around 1 ns. It is found to drop with decreasing size (Heitz et al., 1997b), becoming 340 ps for smaller dots ($\hbar\omega = 1.37$ eV) (Grundmann et al., 1995b). This can be due to nonradiative contributions or can have an intrinsic reason. Following Sugawara et al. (1995), the increase of oscillator strength (decrease of lifetime) is connected with a reduction of confinement effects (Section 5.3.1.5), similar to the base of bound excitons discussed by Rashba (1975). In the model of Sugawara et al. (1995), the lifetime converges against the limit (Eq. 5.63) for small-sized quantum dots when infinite barriers are used. In actual quantum dots with finite confinement potentials, the lifetime will decrease again for smaller quantum dots because the wavefunction penetrates into the barrier and the strong confinement is lost.

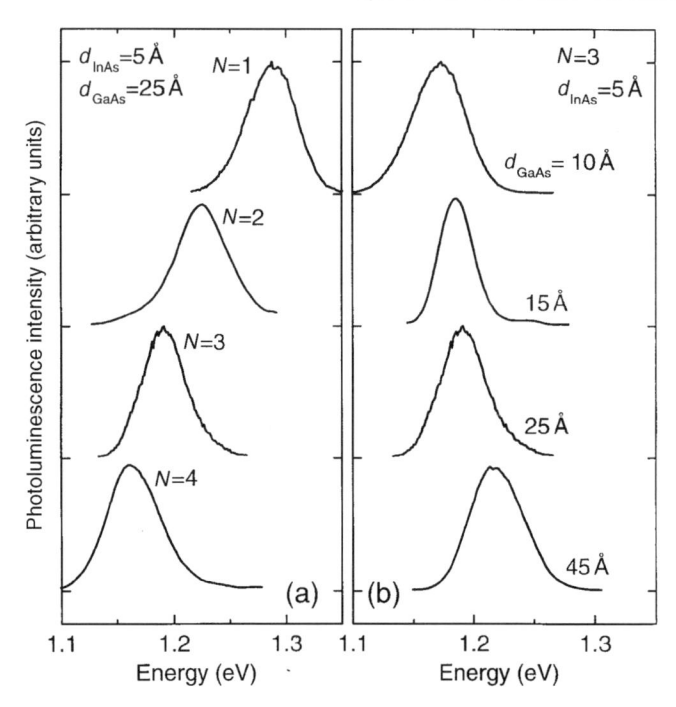

Fig. 6.55 Photoluminescence spectra at 77 K of vertically stacked InAs quantum dots for 0.5 nm InAs deposition per layer. (a) Spectra for different numbers $n = 1$, 2, 3, and 4 layers with a constant GaAs separation layer thickness $d = 2.5$ nm; (b) spectra for a threefold stack with different GaAs separation layer thicknesses $d = 1$, 1.5, 2.5, and 4.5 nm. (After Ledentsov et al., 1996d)

6.7.6 VERTICALLY STACKED QUANTUM DOTS

In general, electronic coupling of vertically stacked quantum dots will lead to a low energy shift of the ground state, as observed by Grundmann et al. (1996b), Ledentsov et al. (1996d), and Solomon et al. (1996). Photoluminescence spectra indeed exhibit a red shift with an increasing number of layers and decreasing separation layer thickness (Fig. 6.55). While the general trend is clear, a quantitative analysis of PL spectra is made difficult because dots in different layers of the stack can have a different geometry (see, for example, Fig. 4.27) and therefore different ground state energy, modifying the observable energy change of the electronically coupled system. However, in combination with absorption spectra the formation of coupled states can be verified (Ledentsov et al., 1996d). Also, the inhomogeneous broadening is reduced in vertical stacks (Grundmann et al., 1996b).

For stacked dots with sufficiently large spacing, e.g. 10 nm for InAs/GaAs, electronic coupling is negligible and thus the properties of individual quantum dots are not altered. However, ensemble properties are affected, e.g. the total absorption

ELECTRONIC AND OPTICAL PROPERTIES

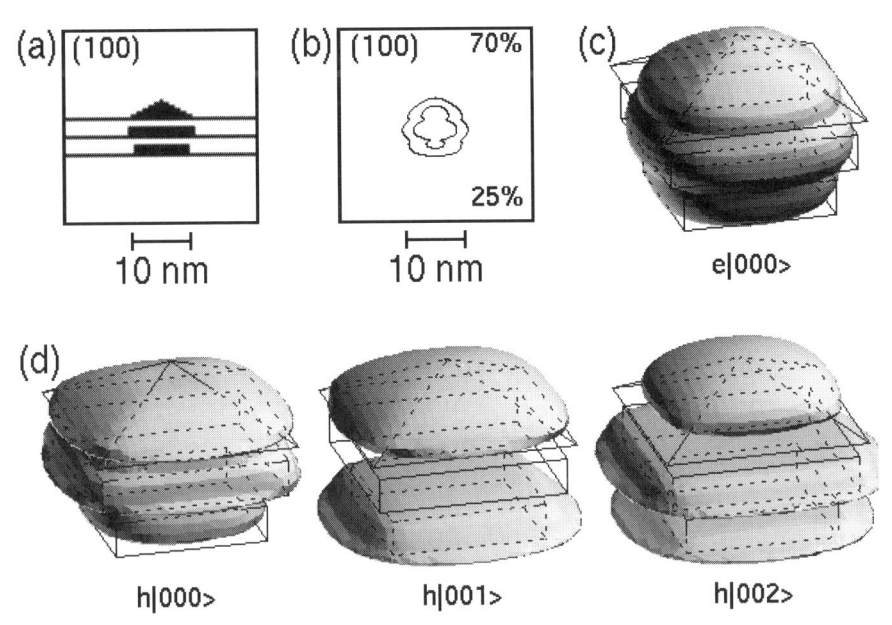

Fig. 6.56 Numerical simulation of vertically stacked and electronically coupled quantum dots. (a) Geometry of the structure, i.e. indium concentration; (b) (100) cross section of the contour plot for the 25 and 70% the ground state electron wavefunction orbital; (c) three-dimensional plot of the electron ground state wavefunction; and (d) three-dimensional view of the ground and |001⟩ and |002⟩ excited hole states. (After Ledentsov *et al.*, 1996d)

cross section increases linearly with the total number of quantum dots and the ground state luminescence saturates at a higher excitation density (Grundmann and Bimberg, 1997c).

For small barrier thickness in between the dots the strong coupling regime is reached (Grundmann *et al.*, 1996b; Ledentsov *et al.*, 1996d), where the electron ground state is coherent over the entire stack (Fig. 6.56c) and the next excited state with a node along the stack axis has a higher energy than the in-plane *p*-like state. For the holes, having larger mass, the lowest energy states are those with nodes in the *z* direction, |000⟩, |001⟩, and |002⟩ states (Fig. 6.56d). The separation of the first excited hole state from the hole ground state is predicted to be smaller than the LO phonon energy.

In addition to the decrease of the ground state transition energy with increasing numbers of layers, the coupled wetting layer transition energy decreases to some lesser extent. Thus the localization energy of carriers in coupled dots is effectively increased. Simultaneously, the inhomogeneous broadening of the ensemble of coupled dots is reduced and is smaller than that of a sheet of single dots at the same transition energy. Sugiyama *et al.* (1997) reported a FWHM of 24 meV and sublevel separation of 52.2 meV at 77 K has been reported for a fivefold stack of InAs/GaAs quantum dots with a small vertical spacing of $d = 2$ nm.

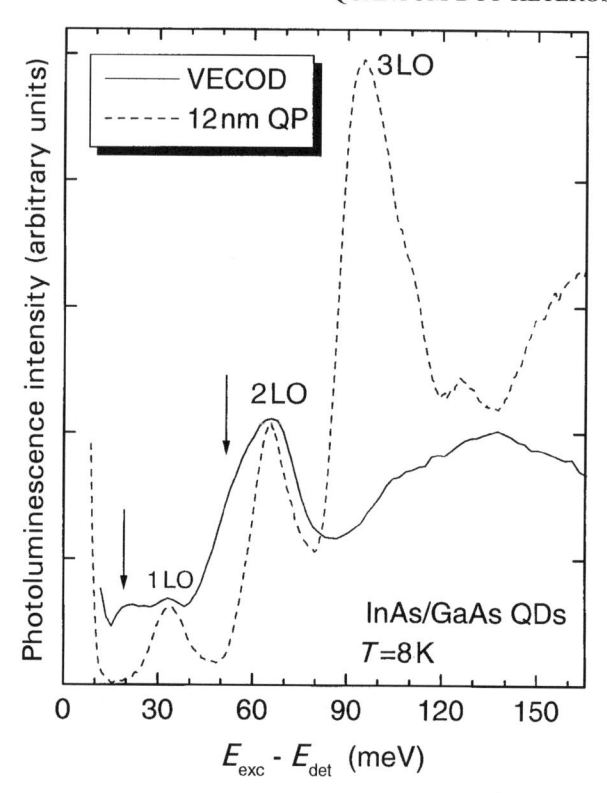

Fig. 6.57 Photoluminescence excitation spectra of a 12 nm InAs pyramid and threefold vertically coupled InAs/GaAs stack. Arrows denote the response due to low-lying electronic levels. (After Grundmann *et al.*, 1996c)

Besides the LO phonon resonances known from single dots (Section 6.7.1.3) excitation spectra of coupled dots exhibit an additional peak at about 20 meV excess energy (Fig. 6.57), which is attributed to the |001⟩ lowest excited hole level (Grundmann *et al.*, 1996b).

Excitation spectroscopy on a fivefold stack of quantum dots grown with migration enhanced epitaxy, which ensures particularly high material quality of the low-temperature GaAs top layers, reveals a characteristic dependence of resonances on the ground state energy (Fig. 6.58) (Heitz *et al.*, 1996b). The most prominent resonance is due to the |001⟩ hole level. When the carriers absorbed into the excited state are allowed more time to relax (limited by the radiative lifetime of the excited state) into the ground state, the phonon selection criterion is no longer dominating the shape of the excitation spectrum (Heitz *et al.*, 1998). With an increasing number of layers the photoluminescence excitation intensity concentrates in the excited electronic state resonance and its dependence on the detection energy becomes stronger. This effect is attributed to a reduction in inhomogeneous broadening in the

ELECTRONIC AND OPTICAL PROPERTIES 255

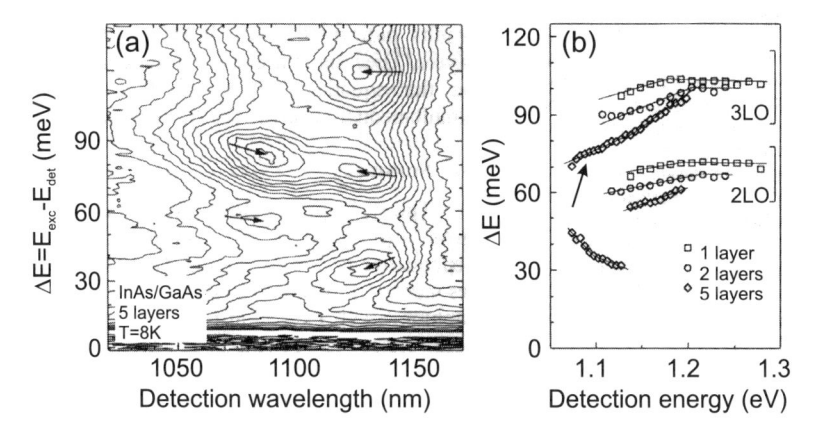

Fig. 6.58 Low-temperature PL excitation spectroscopy of stacked self-organized InAs/GaAs quantum dots. (a) Dependence of PL intensity on the detection wavelength and excess excitation energy δE for a fivefold stack; (b) PLE resonances for different excitation energies (i.e. dot sizes) for samples with 1, 2, and 5 stacked layers with a layer separation of 10 nm Arrow indicates PLE response due to electronic level. (After Heitz et al., 1996b)

vertically aligned quantum dots, causing a stronger correlation of ground state energy and excited state separation in the ensemble.

Stacked quantum dots can be realized with varying degrees of coupling, which modifies the oscillator strength. The decrease of spontaneous recombination lifetime with increasing coupling (decreasing barrier width between stacks) is demonstrated in Ledentsov et al. (1996d). Lifetimes are found to be independent of excitation density between 1 and 100 W/cm^2 (Bimberg et al., 1997).

6.7.7 ANNEALING OF QUANTUM DOTS

One major problem for the growth of QD laser structures is a consequence of the difference in growth temperatures for the deposition of the different layers constituting the complete device. The optimum growth temperature T_G for barriers, waveguides and surrounding superlattices in the AlGaAs/GaAs system is very high, 700 °C, as compared to T_G of 475–530 °C, which is typically used for the deposition of high-density QD arrays. Therefore, after growing the underlying layers at high temperature, T_G is decreased for the deposition of the dot layer and is afterwards increased again for the growth of barriers, waveguide and contact layers on top of the QDs.

For quantum wells the influence of thermally activated interdiffusion is well understood (Deppe et al., 1988; Kuttler et al., 1996). Thermal annealing experiments on QDs have been reported in Kosogov et al. (1996), Leon et al. (1996), Malik et al. (1997) and Heinrichsdorff et al. (1997a). An increase in the effective QD size together with a reduction of the In content in the QDs due to thermally activated

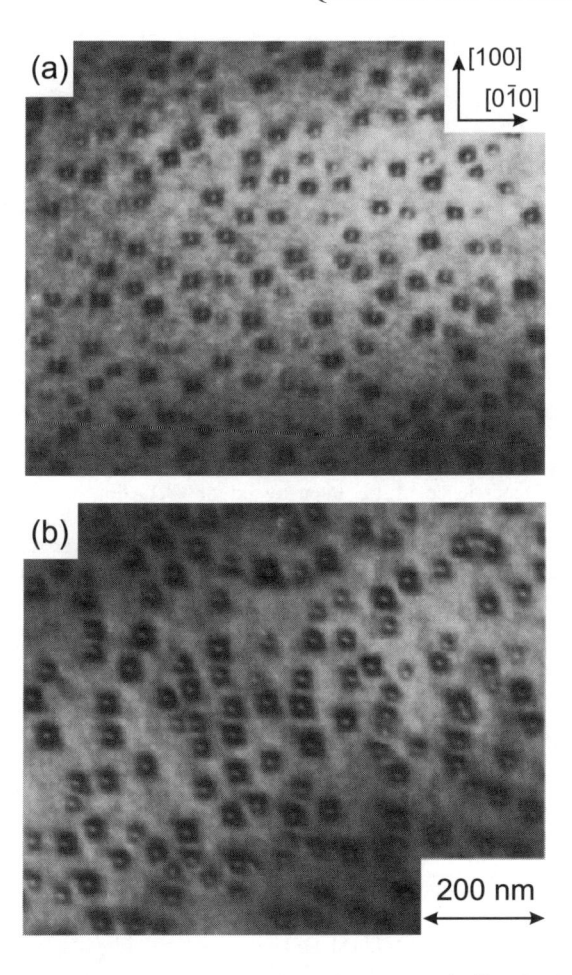

Fig. 6.59 TEM plan-view image taken under [001] zone axis illumination of a threefold stack of InAs/GaAs (0.5 nm/4.5 nm) quantum dots, annealed at $T_A = 700\,°C$ for (a) 10 min and (b) 30 min. (After Kosogov et al., 1996)

intermixing of QD and barrier material has been reported (Kosogov et al., 1996; Leon et al., 1996) and can be directly visualized by TEM images (Fig. 6.59). In these reports relatively high annealing temperatures of $T_A = 700–950\,°C$ have been applied, leading to a strong blue shift of the QD emission energy together with a reduction of the luminescence peak width (Leon et al., 1995; Malik et al., 1997; Lobo et al., 1998).

Heinrichsdorff et al. (1997a) investigated the influence of indium interdiffusion for annealing temperatures between 580 and 700 °C. It was found that the inhomogeneous broadening of ground state and excited state energies of the QD ensemble increases for small diffusion lengths L_D but decreases below that of the

ELECTRONIC AND OPTICAL PROPERTIES

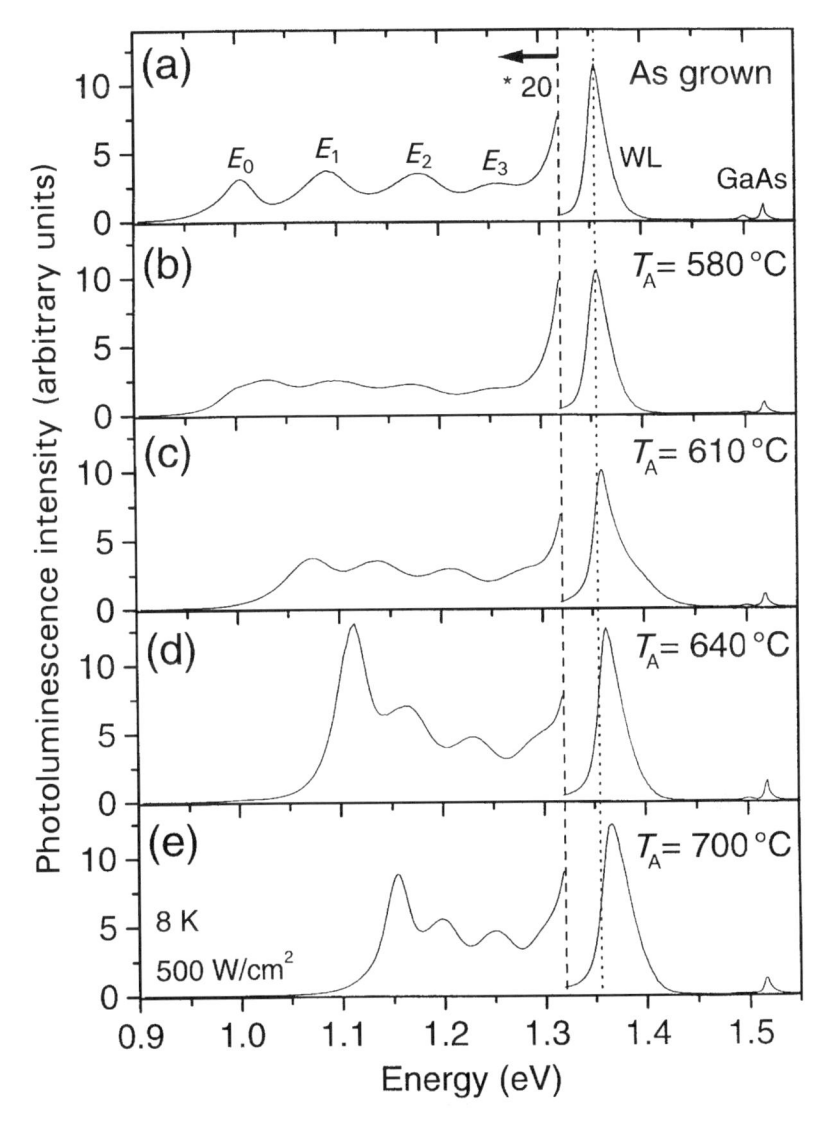

Fig. 6.60 Low-temperature PL spectra of (a) as-grown and (b)–(e) annealed QD samples at high excitation densities (500 W/cm^2). (After Heinrichsdorff *et al.*, 1997a)

as-grown samples for large values of L_D. The results qualitatively agree with calculations for spherical dots assuming Fickian interdiffusion of the dot and barrier. Low-temperature PL spectra of as-grown and annealed samples are shown in Fig. 6.60. For the as-grown sample the dot luminescence consists of four peaks between 1.0 and 1.3 eV (E_0–E_3) due to recombination of excitons in the QD ground and excited states. Additionally, the wetting layer luminescence line at 1.35 eV and the

GaAs related luminescence at 1.52 eV can be seen. Obviously, the peak width is smallest for the ground state luminescence and increases for the higher energy QD peaks, as expected for a size distribution induced inhomogeneous sublevel broadening (see Section 5.4.2). The PL spectra of the annealed samples show a continuous shift of the QD luminescence lines to higher energy, together with a reduction of the peak separations. The high-energy QD peak (E_3) is shifted toward the wetting layer energy; it can no longer be resolved for $T_A = 700\,°C$. At the same time the wetting layer peak is only slightly shifted from 1.355 eV for the as-grown sample to 1.367 eV for $T_A = 700\,°C$.

The width of the peaks E_0–E_3 strongly depends on the thermal annealing temperature. For T_A being as low as 580–610 °C the peak width *increases* as compared to the as-grown sample. A further increase of T_A to 640 °C and eventually 700 °C, however, leads to a *decrease* of the sublevel broadening, even below the half-width of the as-grown sample. The evolution of the peak positions and widths for the first three sublevels as deduced from curve fits using multiple Gaussians are summarized in Fig. 6.61. The width of the wetting layer line, partly due to band filling effects, increases from 20.8 eV for the as-grown sample to 30.3 meV for $T_A = 700\,°C$.

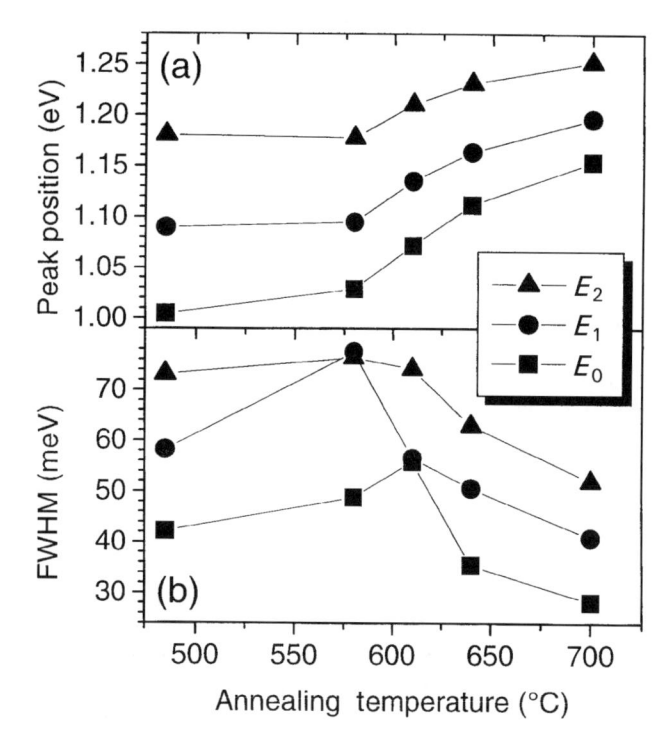

Fig. 6.61 Peak positions and broadening for the three low-energy QD PL peaks (E_0–E_2). Values are deduced from curve fits using multiple Gaussians. $T_A = 485\,°C$ represents the as-grown sample. (After Heinrichsdorff *et al.*, 1997a)

For a qualitative simulation of the interdiffusion of QDs with the surrounding barrier a simple model is used with a spherical attractive potential which drastically reduces the computational efforts as compared to the realistic (e.g. pyramidal) geometry (Grundmann et al., 1995c). Strain and piezoelectric effects are not included, as would be necessary for a more quantitative consideration. Fickian interdiffusion of the QD and barrier material is assumed with linear dependence of the potential on the concentration of the QD material. The initial potential depth is chosen to be 0.5 eV, with an effective carrier mass of 0.1 m_0 and initial radii $r_0 = 2-8$ nm. The influence of Fickian diffusion on an arbitrary spherical distribution of indium is described by the heat conduction equation. For comparison, eigenstate energies for two-dimensional (quantum wells) and one-dimensional (cylindrical quantum wires) potentials are also calculated to find the dependence of the intermixing-induced changes of the eigenstates on the dimensionality (Heinrichsdorff et al., 1997a). Starting from an initially steep (step-like) potential, the new potential for the given diffusion length L_D was calculated on a grid of $(5-6) \times 10^3$ mesh points using the Crank–Nicholson scheme. The eigenenergies in the resulting potential were calculated by discretizing and solving the single-particle Hamiltonian on the same grid, yielding an absolute accuracy of better than 0.5 meV for the eigenenergies. For dots, wires, and wells the initial diameter and well thickness, respectively, were varied from 2 to 16 nm and the diffusion length was varied from 0 and 6 nm, resulting in 4200 calculated data points for each case.

It is found that the influence of interdiffusion on the number n of bound eigenstates strongly depends on the dimensionality. For the quantum well the intermixing leads to an increase of n, whereas n decreases upon interdiffusion of quantum dots. For the quantum wire case, the value of n is mostly unaffected by variation of L_D, except for small values of L_D where eigenstates with higher radial quantum numbers show a different dependence on L_D (Heinrichsdorff et al., 1997a).

The dependence of the QD ground state energies and their ensemble inhomogeneous broadening $\partial E / \partial r$ on the diffusion length is depicted in Fig. 6.62. The intermixing of QDs with the barrier results in the expected continuous blue shift of the QD energy. In contrast, the inhomogeneous broadening of the QD ensemble shown in Fig. 6.62b (the vertical derivative of Fig. 6.62a) shows a more complex behavior. Initially, the value of $\partial E/\partial r$ increases for all QD sizes and reaches a maximum as L_D increases. When L_D is further increased, $\partial E/\partial r$ decreases and drops below the initial value when the QD potential becomes more shallow. It should be noted that similar results are obtained for quantum wells and quantum wires (not shown here), but the sharp maximum of $\partial E/\partial r$ is most pronounced for the quantum dot case. The ground state inhomogeneous broadening becomes maximum when the potential minimum starts to rise, i.e. when the QD potential well starts to become shallow. For eigenstates with higher angular quantum numbers the maximum of $\partial E/\partial r$ is reached for smaller values of L_D, but the increase of $\partial E/\partial r$ is not as pronounced as for the ground state. These results are in qualitative agreement with the experimental findings shown in Fig. 6.61.

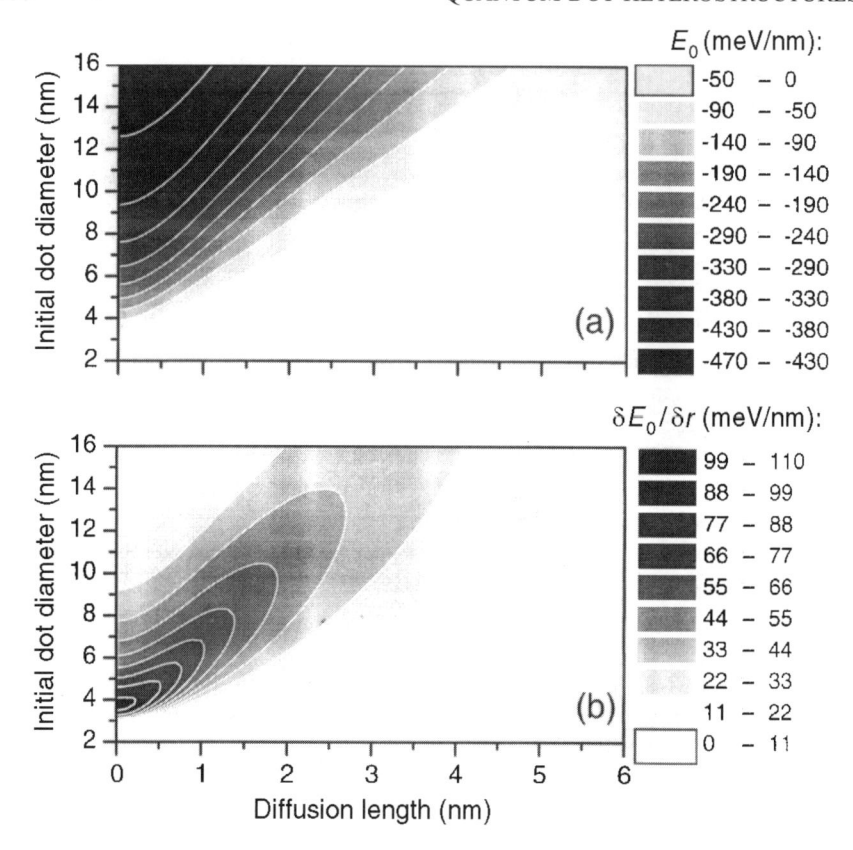

Fig. 6.62 Dependence of (a) ground state energies and (b) inhomogeneous broadening of spherical quantum dots on intermixing. (After Heinrichsdorff et al., 1997a)

6.8 SELF-ORGANIZED TYPE-II QUANTUM DOTS

The particular band alignment of a type-II system has been discussed in Section 5.3.2.

6.8.1 CW PROPERTIES

The optical recombination spectrum of a type-II GaSb/GaAs quantum dot system has been reported in Hatami et al. (1995). For this material system the holes are bound in the QD whereas the electrons are localized outside in the vicinity. In Fig. 6.63 the photoluminescence spectra of a series of samples within increasing GaSb quantum well thickness on GaAs are plotted. Above a critical thickness of about 3 ML, an additional peak due to GaSb dots with 22 nm base width occurs below the recombination from the two-dimensional wetting layer. The positions of the wetting layer and the quantum dot luminescence line shifts to higher energy with increasing

ELECTRONIC AND OPTICAL PROPERTIES

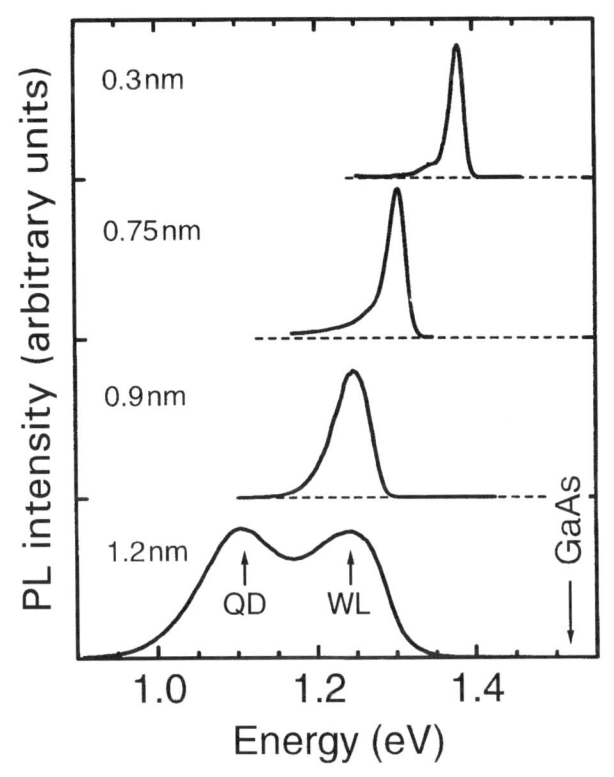

Fig. 6.63 Low temperature ($T = 2$ K) photoluminescence spectra of GaSb insertions in the GaAs matrix for four different GaSb layer thicknesses. $D = 5$ W/cm^2. (After Hatami et al., 1995)

excitation density D due to dipole layer formation with a typical $\Delta E \sim D^{1/3}$ behavior (Ledentsov et al., 1995a) (Fig. 6.64). For the quantum dot at higher excitation density the shift is larger due to state filling in the zero-dimensional hole density of states, but the ground state transition obtained from a line-shape fit still follows the $D^{1/3}$ line (Hatami et al., 1998). Such shift due to state filling was also reported by Hogg et al. (1998).

A similar line position of 1.15 eV for GaSb/GaAs quantum dots has been reported by Glaser et al. (1996), who also report about luminescence from the type-II quantum dot systems InSb/GaAs and AlSb/GaAs (see Fig. 6.65). The use of an Al$_{0.1}$Ga$_{0.9}$As barrier instead of GaAs leads to a blue shift of the GaSb dot luminescence by about 80 meV, which is expected from the transitivity rule for the band line-up (Christensen, 1988). The peaks of all these type-II quantum dot systems exhibit the typical blue shift with increasing excitation density (Glaser et al., 1996).

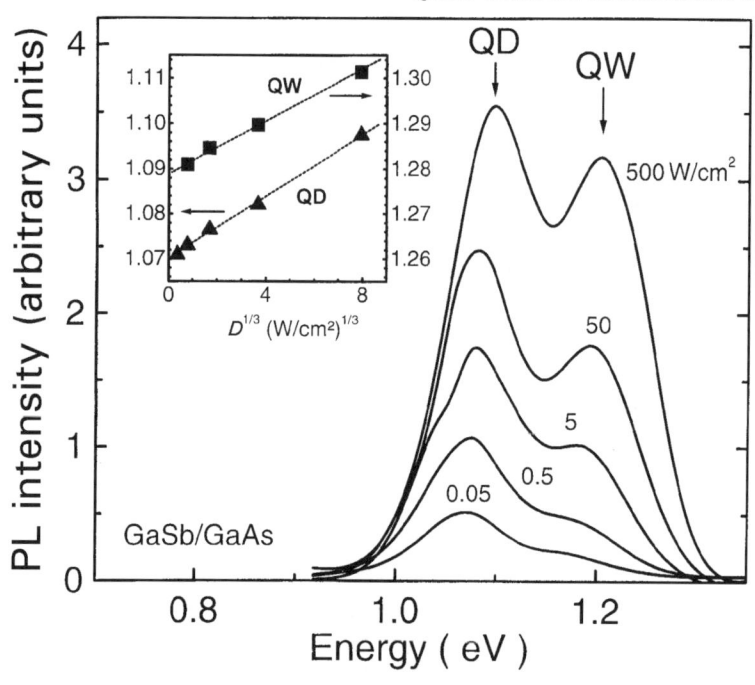

Fig. 6.64 Dependence of the photoluminescence spectrum of a 1.2 nm GaSb/GaAs quantum dot structure on the excitation density D. Low-energy line due to quantum dots dominates the spectrum at a low excitation density. Inset: shift of peak maximum energy with excitation density for 0.75 and 1.2 nm GaSb in GaAs. Dashed lines are a guide to the eye $\sim D^{1/3}$ line. (After Hatami *et al.*, 1995)

6.8.2 TIME-RESOLVED EXPERIMENTS

For type-II GaSb/GaAs quantum dots the (initial) recombination time constant τ_1 is very similar to that of the wetting layer (~ 5 ns) (Hatami *et al.*, 1998) (Fig. 6.66) and is independent of the temperature up to 65 K. Above this temperature the intensity starts to drop, probably due to exciton dissociation. Using a spherical quantum dot model (Laheld *et al.*, 1995), a practically vanishing wavefunction overlap $<10^{-3}$ is predicted for the material parameters given. However, the geometric situation in a typical sample geometry with self-organized quantum dots is very different from this model:

(a) When dots are charged with holes and additional holes are in the wetting layer, the electrons are attracted by a two-dimensional charge distribution.
(b) Along the growth direction the GaSb/GaAs layer is surrounded by AlGaAs barriers, confining the electrons close to the dot plane.

At moderate or high excitation density ($\geqslant$ one eh pair per dot), it is thus understandable that the wavefunction overlap between electrons and holes in the wetting layer and the quantum dots is very similar and quite high. Using a typical

ELECTRONIC AND OPTICAL PROPERTIES 263

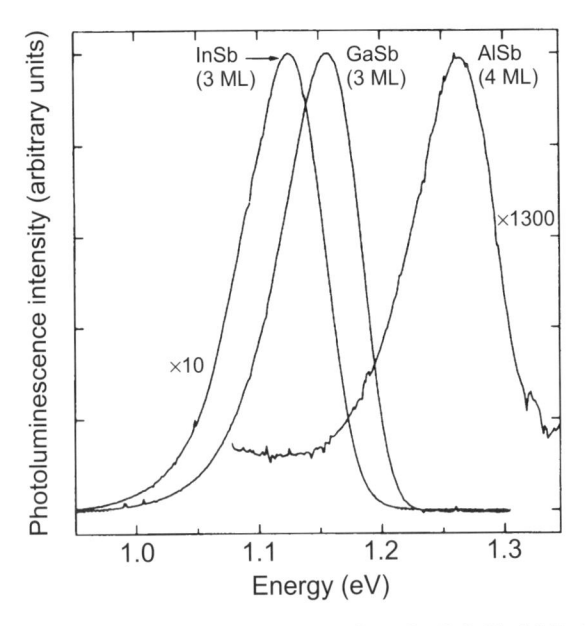

Fig. 6.65 Photoluminescence spectra at $T = 1.6$ K from 3 ML InSb, 3 ML GaSb, and 4 ML AlSb in GaAs quantum dots. Excitation density was 2.8 W/cm^2. (Reproduced by permission of American Institute of Physics from Glaser *et al.*, 1996)

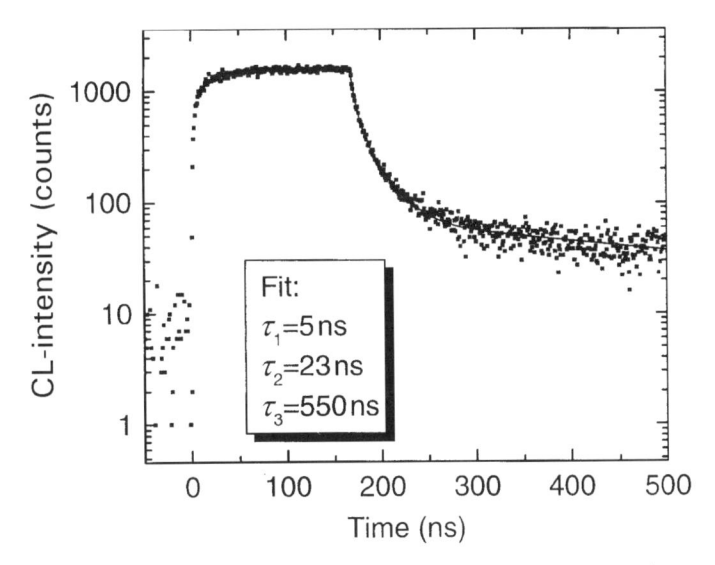

Fig. 6.66 Cathodoluminescence transient of GaSb quantum dots ($T = 5$ K) at a detection energy of 1.16 eV with fit using the sum of three exponentials to simulate the nonexponential decay. The initial lifetime τ_1 and the lifetime at large delay τ_3 are discussed in the text

exciton lifetime of $\tau_{\text{type-I}} \approx 1$ ns in type-I quantum dots for nearly 100% wavefunction overlap, the wavefunction overlap $f_{\text{II}} \approx (\tau_{\text{type-I}}/\tau_{\text{type-II}})^{1/2}$, determined from the initial fast decay time constant $\tau_1 = 5$ ns, is around 50% for the investigated type-II dots. The luminescence decay is strongly nonexponential because the electron–hole overlap decreases with increasing delay time; however, a fit with three exponentials yields a satisfactory fit. Finally, when the carrier density is very low, independent single electron–hole pairs at widely separated quantum dots decay. The observed time constant at a large delay time, τ_3, is approaching 1 μs, indicating indeed a small wavefunction overlap for single excitons at isolated type-II quantum dots.

7 Electrical Properties

Transport of charge occurs *always* in quanta of the elementary charge e. This discreteness has in general no consequences for the current flow in *bulk* material, except for the underlying microscopic scattering mechanisms. The number of electrons involved in the operation of present-day transport devices is so large that single-electron effects are not observed. Today's memory chips operate, for example, with storage charges of about 10^4–10^5 electrons per bit. In order to lower the power consumption of electronic circuits it is, however, highly desirable to decrease the number of carriers involved in a switching or storage operation. The goal is to reduce this number down to *one* single electron representing a bit using quantum confinement effects (single-electron transistor).

The foremost modification of charge transport by the presence of a quantum dot in the current path is that single electron effects become essential. When the change of the energy of a quantum dot by adding a single electron is larger than $k_B T$ and the quantum dot is connected to reservoirs with appropriate Fermi levels, transport through the dot can be altogether forbidden and no current may flow (Coulomb blockade; Section 5.3.4). Excellent collections of papers cover this topic for quantum dots induced by lithographically defined structures (e.g. see Grabert and Horner, 1991; Fukuyama and Ando, 1992; Grabert and Devoret, 1992; Kastner, 1993; Kouwenhoven *et al.*, 1997). In these structures, however, lateral confinement energies are small and quantum effects are limited to the low-temperature regime, $T \leqslant 4\,\text{K}$. Ongoing research in this field is beyond the scope of this book. We will focus here on the new field of transport via growth-induced self-organized quantum dots. Experiments on charging of quantum dots will be discussed in Section 7.1 and time-resolved capacitance spectroscopy (DLTS) in Section 7.2 Tunneling experiments are reviewed in section 7.3. In Section 7.4 lateral transport is discussed.

7.1 CV SPECTROSCOPY

The charging of quantum dots with electrons or holes can be probed with capacitance voltage (CV) spectroscopy. By moving the Fermi level E_F through the states of the quantum dot via the bias voltage, the differential capacitance is probed with an additional (small) a.c. signal. (Note that the temperature, potential depth, and measurement frequency are such that a thermal equilibrium is established.) The sheet-like density of electrons in the quantum dots (for a n-doped structure) is

$$n_{\text{QD}} = N_D \sum_{n=1}^{M} \left[1 + \exp\left(\frac{E_n - E_F}{kT}\right) \right]^{-1} \qquad (7.1)$$

where E_n denotes the energy level of a quantum dot filled with n electrons, N_D is the area dot density, and M is the maximum number of electron levels in the quantum dot. Taking into account the finite spread $N(E_n)$ of energies within an inhomogeneous ensemble, Eq. (7.1) has to be extended to

$$n_{QD} = \sum_{n=1}^{M} \int N(E_n) \left[1 + \exp\left(\frac{E_n - E_F}{kT}\right)\right]^{-1} dE_n \qquad (7.2)$$

with $N_D = \int N(E_n) \, dE_n$ for all n, and $N(E_n)$ being typically a Gaussian distribution. Assuming that a quantum dot is charged with one electron in the s-like ground state, a second electron can fill the ground state but experiences an additional charging energy $E_{21} = E_2 - E_1$ of the order e^2/C_0, where C_0 is the capacity of the dot (see Section 5.3.4). Thus the Fermi level has to be increased in order to fill another electron into the dot. This phenomenon is called Coulomb blockade and is discussed in detail in, for example, Grabert and Horner (1991), Kastner (1993), and Kouwenhoven et al. (1997). Coulomb blockade is basically observable at temperatures for which $E_{21} \gg k_B T$.

The double-peak structure of an s-like ground state due to the charging energy can be washed out in experiments by the inhomogeneous broadening, high temperatures and the magnitude of the a.c. signal used to probe the capacitance. Further electrons have to be filled into excited states (if there are more bound states at all). The energy difference $E_{32} = E_3 - E_2$ consists, apart from the next charging energy, also of the energy difference between the single-particle ground and excited states.

The scenario described above has been experimentally investigated for MBE-grown lens-shaped InGaAs self-organized quantum dots (Drexler et al., 1994; Fricke et al., 1996). In Fig. 7.1 the CV characteristics of such dots is shown for different magnetic fields, which furthermore lead to a splitting of the p-like excited state. A clear separation of the lowest two peaks is reported in Fricke et al. (1996) (Fig. 7.2). Miller et al. (1997) observed a fine structure in the CV peaks for small gate electrode areas. The CV spectrum consists of a superposition of similar subcurves with an offset of integer multiples of 7 meV. The origin of this effect remains speculative; discrete island size variation by monolayer height fluctuation or Coulomb interaction between nearest-neighbor dots could cause shifts of the observed magnitude.

Brunkov et al. (1996a, 1996b) investigated the charging of self-organized InAs/GaAs quantum dots (Fig. 7.3). The experimental data are fitted by a model based on the numerical solution of the one-dimensional Poisson equation. The localization energy E_{loc} of the electron level, measured from the GaAs conduction band, is used as a fit parameter. The capacitance is calculated in the quasi-static approximation (Brunkov et al., 1996b). Inhomogeneous broadening of E_{loc} is considered, but finite charging energy is neglected. The best fit is clearly obtained for $E_{loc} = 100$ meV, confirming the predictions for the magnitude of the electron localization and the existence of only one bound state for electrons in small InAs dots (Grundmann et al., 1995c). At low temperatures n_{QD} is found to be close to twice the density of quantum dots, each dot being filled with two electrons. With

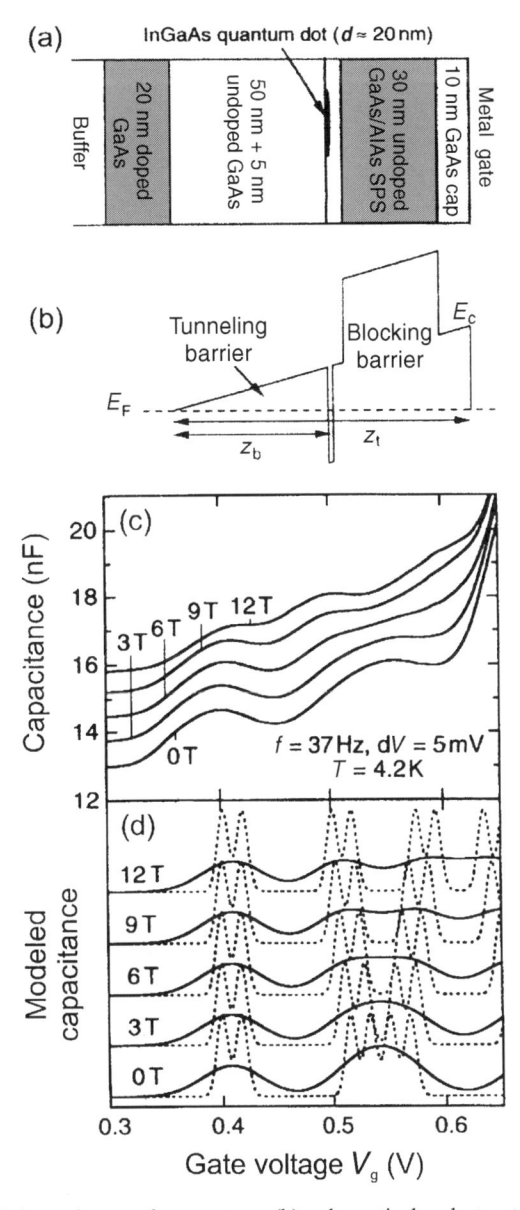

Fig. 7.1 (a) Schematic sample structure, (b) schematic band structure, and (c) measured capacitance of the sample with self-organized InGaAs/GaAs quantum dots as a function of gate voltage V_g for different magnetic fields B perpendicular to the sample surface. Vertical scale refers to the bottom trace ($B = 0$ T); to the other traces constant offsets were added for clarity. (d) Calculated peak positions using a capacitance $C_0 = 9.2$ aF for a single quantum dot, and $\hbar\omega_0 = 41$ meV and $m^* = 0.07$ for the parabolic confinement used to model the magnetic field dependence. Inhomogeneous Gaussian broadening with $\sigma_E = 5$ (25) meV is assumed for the dashed (solid) lines. (After Drexler et al., 1994)

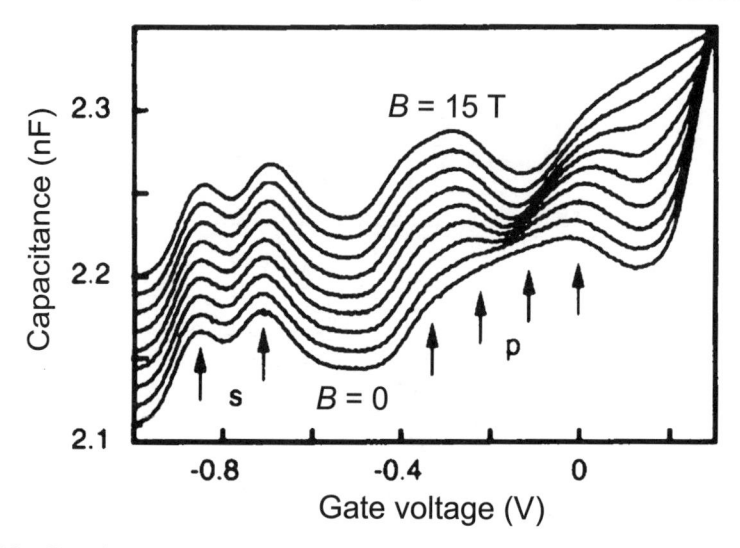

Fig. 7.2 Capacitance–voltage characteristics of lens-shaped InGaAs/GaAs quantum dots at different magnetic fields. (After Fricke *et al.*, 1996)

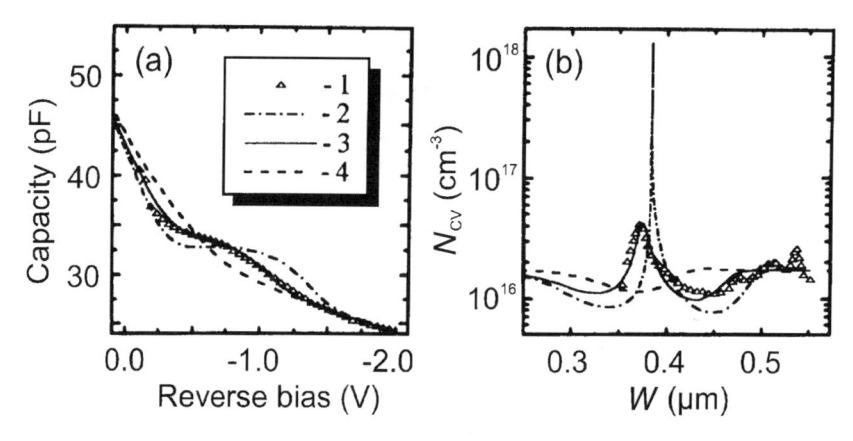

Fig. 7.3 (a) Capacitance voltage and (b) carrier number versus depth characteristics of the structure with self-organized InAs/GaAs quantum dots at $T = 200$ K. The experimental data (1) are compared to model calculations with quantum dot electron localization energy $E_{\text{loc}} = 150$, 100, and 75 meV for curves 2, 3, and 4, respectively, and fixed density of quantum dots $N_D = 6 \times 10^{10}$ cm^{-2} as determined from TEM, and inhomogeneous broadening $\sigma_E = 25$ meV. (After Brunkov *et al.*, 1996a)

increasing temperature the electron occupation of the dots decreases (Brunkov *et al.*, 1996a). Such evaporation of carriers from the dots has been observed in luminescence on similar (undoped) structures (Grundmann *et al.*, 1996a; Xu *et al.*, 1996).

A model taking into account the inter- and intra-dot Coulomb interaction by using a Monte Carlo simulation of CV experiments has been reported in Medeiros-Ribeiro *et al.* (1997). From the dependence of the capacitance peak splitting and width (for charging the *s*-like state) as a function of the dot–gate distance it has been deduced that dot–dot interaction is important for the quantitative interpretation of the experimental data.

Besides the charging of electronic levels, the charging of hole levels in InGaAs/GaAs quantum dots has been demonstrated in Medeiros-Ribeiro *et al.* (1995) using a *p*-doped structure. However, only the hole ground state could be revealed. Photocurrent and photocapacitance measurements revealing absorption in QDs have been reported by Pettersson *et al.*, (1996).

7.2 DLTS

Quantum dots can be considered as traps for charge carriers. Applying time-resolved capacitance spectroscopy methods using data analysis methods known from deep level transient spectroscopy (DLTS) (Lang *et al.*, 1974; Blood and Orton, 1992) allows investigation not only of the electronic level structure of the QDs, but also of the dynamics of emission and capture processes. The QDs are usually placed close to the space charge region of a pn-junction or Schottky contact at zero bias. First, the junction is reverse-biased such that the QDs are inside the depletion region and empty (Fig. 7.4a). Then a forward filling pulse is applied and carriers are captured into the QDs (Fig. 7.4b). After re-establishing the reverse bias condition the carriers are emitted from the QDs (Fig. 7.4c). Since the depletion width and therefore the junction capacitance depends on the amount of charge inside the space charge region, it is possible to measure the amount of charge in the QDs directly and its change in time leads to capacitance transients which can be recorded. The DLTS signal is taken as the change ΔC of the capacitance at a fixed reference emission rate (boxcar method) for varying temperature.

Emission of electrons from self-organized InP QDs in a GaInP matrix was observed using DLTS by Anand *et al.*, (1995). In Fig. 7.5 the DLTS signal from an ensemble of three layers of vertically coupled InAs QDs embedded in GaAs grown by MOCVD is shown (Kapetyn *et al.*, 1998). The peak at about 40 K is due to thermal emission of electrons from the QDs, whereas the constant contribution towards lower temperatures is due to tunnel emission from the QD ground state into the conduction band of the barrier material. The activation energy of the thermal peak as obtained from an Arrhenius plot (inset of Fig. 7.5) is 94 meV and represents the energy difference of the QD ground state and excited QD states. From the tunnel emission rate an energy barrier height of $\Delta E = 162$ meV is directly obtained, in good agreement with the expected ground state electron localization energy obtained from numerical simulations using eight-band k·p theory including strain effects (Stier *et al.*, 1998.

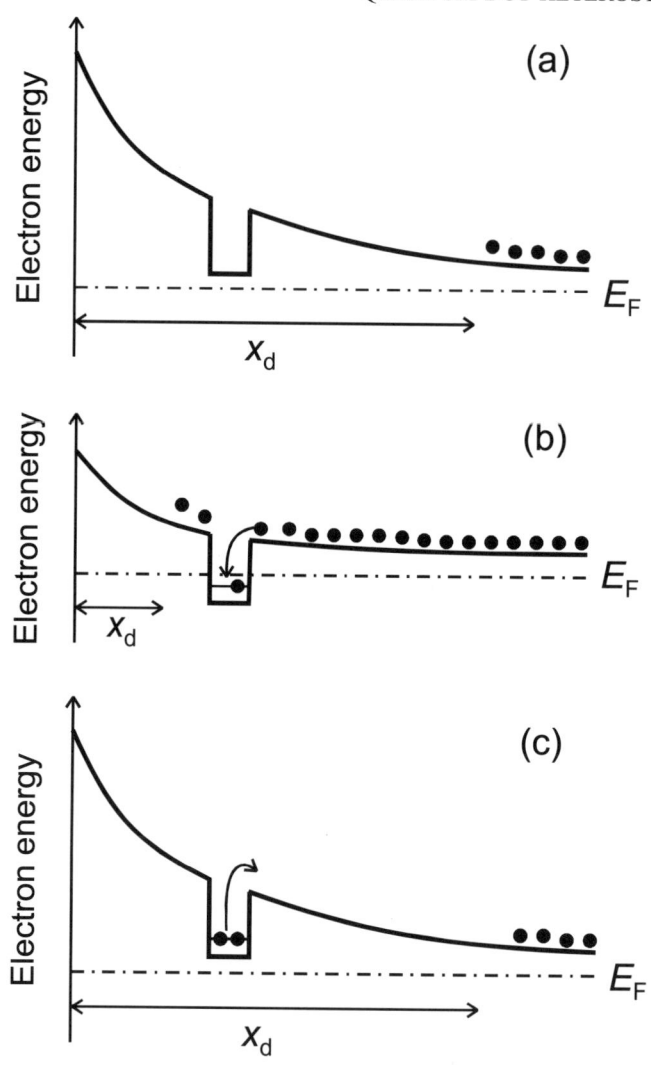

Fig. 7.4 Principle of DLTS measurement, where x_d denotes the width of the depletion region of a pn junction or Schottky contact. (a) Reverse detection bias applied, (b) during filling pulse, (c) after filling pulse, for observation of carrier reemission

7.3 VERTICAL TUNNELING

When a quantum dot is sandwiched between two barriers current can flow (tunnel) through the dot only at a bias voltage for which an occupied emitter state is resonant with a quantum dot level. Resonant tunneling diodes containing self-organized InAs/GaAs quantum dots have been investigated by Itskevich *et al.* (1996) and Narihiro *et al.* (1997).

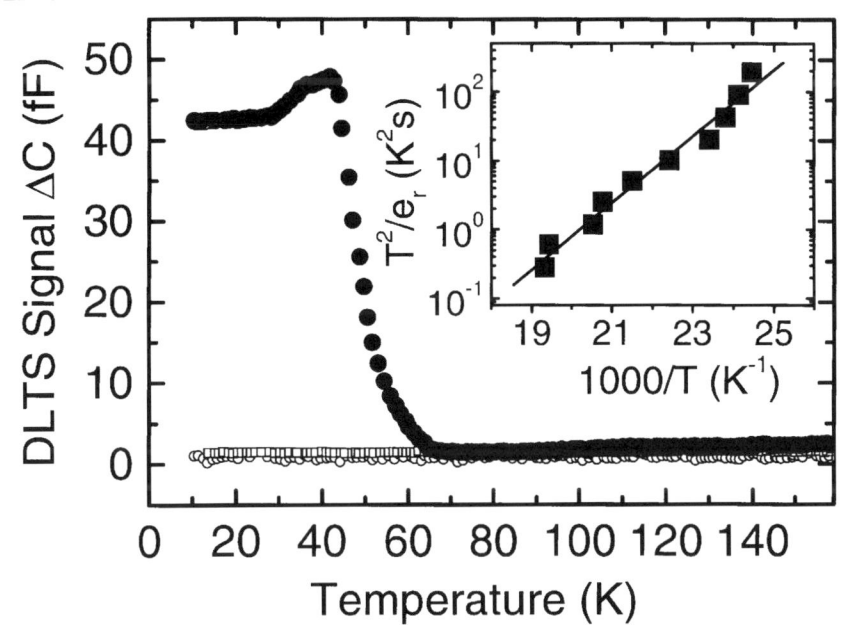

Fig. 7.5 DLTS signal of a strongly coupled threefold QD stack sample (●) and reference samples with WL only and without InAs (○) at an emission rate of $e_r = 47.6\,\text{s}^{-1}$. The diodes were biased $-3.5\,\text{V}/-2.5\,\text{V}$ during transient recording and the 10 μs pulse, respectively. The constant contribution in the DLTS signal below 30 K is due to tunnel emission; the peak at about 40 K is due to thermal activation. Inset: Arrhenius plot of the thermal emission peak of the QD sample; from the linear fit an activation energy of 94 meV is obtained. (After Kapteyn et al., 1998)

A schematic band line-up (Itskevich et al., 1996) is depicted in Fig. 7.6a. The $I(V)$ characteristics are shown in Fig. 7.6b. The control sample without quantum dots exhibits only a continuously rising background current only (curve I). In the sample containing quantum dots within the tunneling barrier additional features due to tunneling through zero-dimensional quantum dot states are observed. In forward bias, when electrons flow from the substrate to the mesa contact (diameter 100 μm), a number of peaks (with pA height) appears (curve IV). In reverse bias a set of distinct steps is observed at 350 mK (curve III), which is almost completely smeared out at 4.2 K (curve II).

This asymmetry is attributed to the asymmetric geometry of the tunneling barrier. Although the quantum dots are nominally inserted into the center of a 10 nm AlAs barrier, the pyramidal-like shape of the dots causes the barrier directly on top of the dots to be thinner. Thus for reverse bias (electrons tunnel into the top of the dots) the rate for tunneling into the dot, which may be dominated by inelastic processes, is much larger than for tunneling out. As each dot moves below the emitter Fermi level, the $I(V)$ curve exhibits another step as an additional tunneling path is opened (Itskevich et al., 1996).

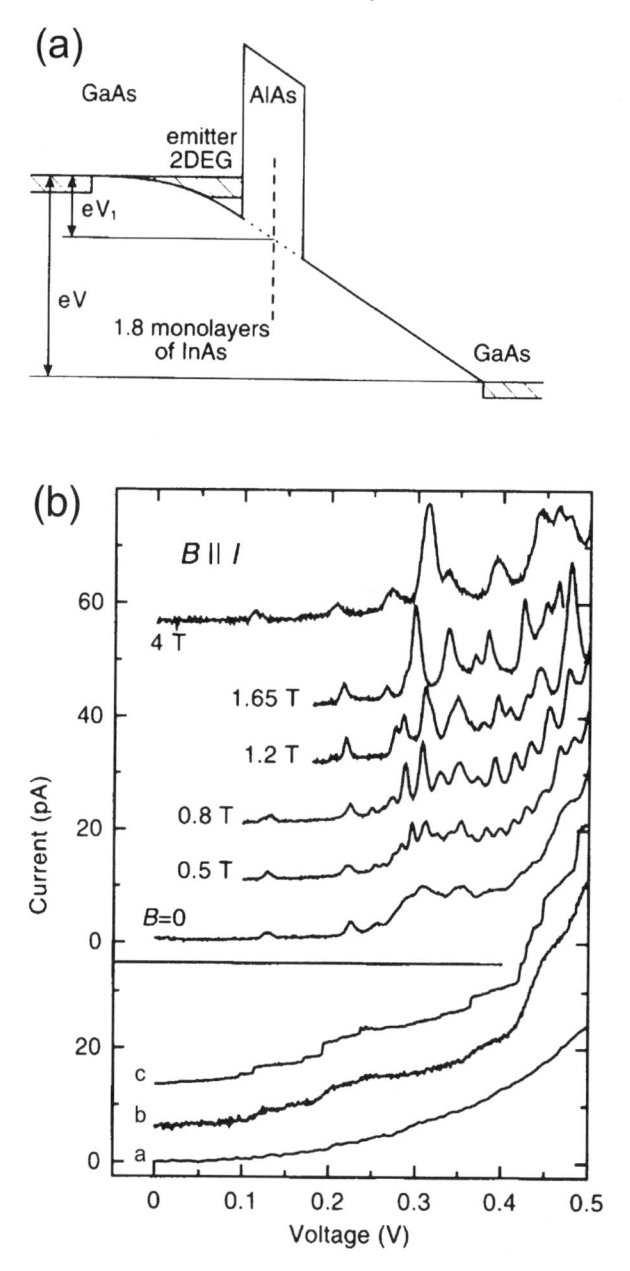

Fig. 7.6 (a) Schematic diagram of the conduction band edge of a quantum dot tunnel structure under an applied (reverse) voltage V. (b) $I(V)$ characteristic of 100 μm diameter mesa. Curve I: control sample without dots; curves II (III) for $B = 0$ T at 4.2 K (350 mK) for *reverse* bias. Curves IV are for *forward* bias at 4.2 K at various magnetic fields (parallel to the current). Offsets have been added for clarity. (After Itskevich *et al.*, 1996)

ELECTRICAL PROPERTIES

For forward bias the tunneling rate into the dots is smaller than out of the dots, and the current is a voltage-tunable probe of the single-particle levels in the emitter 2DEG. The peak width depends accordingly on the temperature (Itskevich et al., 1997). The peak positions depend on the applied magnetic field and a fan plot (Fig. 7.7) exhibits sets of Landau levels of the 2DEG in the emitter which come into resonance with states from different (independent) quantum dots contributing to the total $I(V)$ characteristic of the mesa (Itskevich et al., 1996). Since there are about 10^7 quantum dots under the mesa, the observed signal must stem from only a few dots in

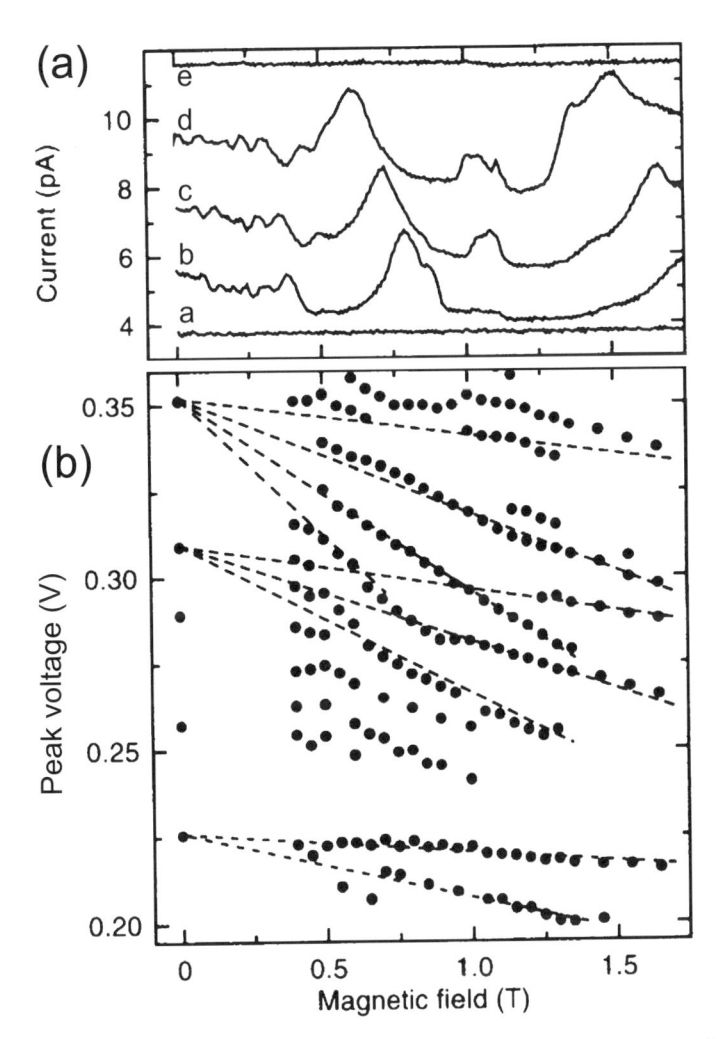

Fig. 7.7 (a) $I(B)$ characteristic at various forward bias V: curve a, 105 mV; curve b, 114 mV; curve c, 115 mV; curve d, 116 mV; and curve e, 130 mV; curves are offset. (b) Fan plot of the peaks in the $I(V)$ characteristic versus the magnetic field strength. (After Itskevich et al., 1996)

the distribution (presumably with the largest ground state energy). The electron spin splitting in a single InAs self-organised QDs (in AlAs) has been observed by Thornton et al., (1998) using magneto-tunneling spectroscopy. g factors between $+0.52$ and $+1.6$ have been found, having the opposite sign than the value for InAs bulk (-14.8) (See also Section 6.7.1). The difference was explained within a simple three band $k \cdot p$ calculation.

Resonant tunneling at 300 K with a peak-to-valley ratio of 10 has been measured for a single self-assembled Si/SiO_2 dot (Fukuda et al., 1997).

7.4 LATERAL TRANSPORT

The presence of quantum dots in a layer or channel largely modifies its transport characteristics. Sakaki et al. (1995) found that the presence of a sheet of InAs quantum dots close to a two-dimensional electron gas (2DEG) at an AlGaAs/GaAs heterointerface decreases the mobility of electrons in the channel drastically, which is attributed to the scattering potential introduced by the dots. In Fig. 7.8 the mobility and electron sheet concentration are plotted versus the spacing between the 2DEG and the InAs dots. The decrease of mobility by almost a factor of 100 when the spacing is about 15 nm is attributed to a dramatic increase of the scattering potential. Via their strain field, which induces potential variations, the quantum dots can interact over a larger distance with free carriers than expected from pure carrier scattering and tunneling arguments.

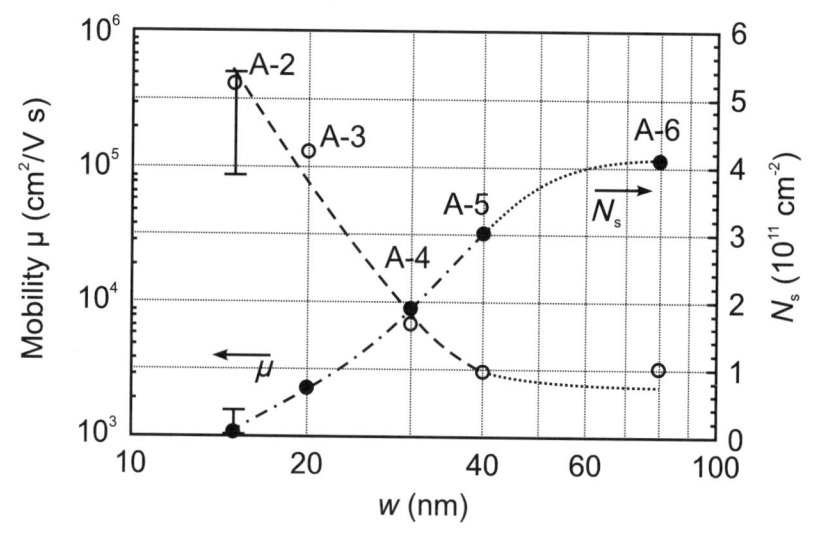

Fig. 7.8 Electron mobility and sheet density at 77 K for a two-dimensional electron gas close to a layer of InAs quantum dots as a function of the separation layer width w. (Reproduced by permission of American Institute of Physics from Sakaki et al., 1995)

ELECTRICAL PROPERTIES

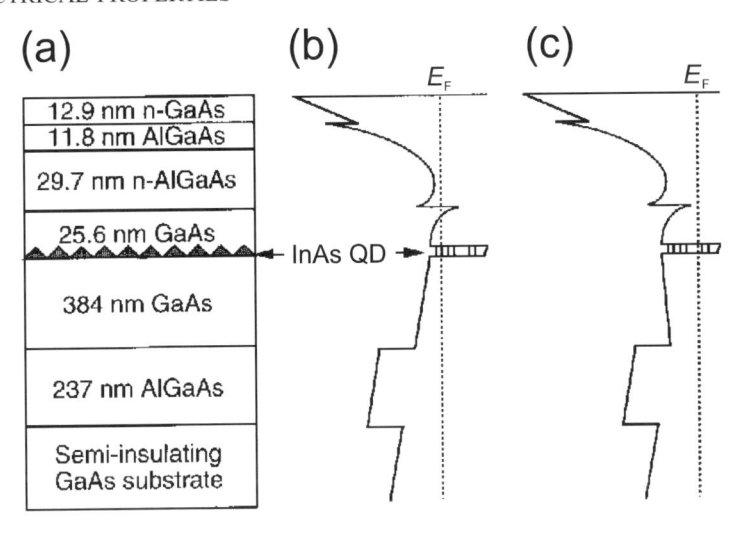

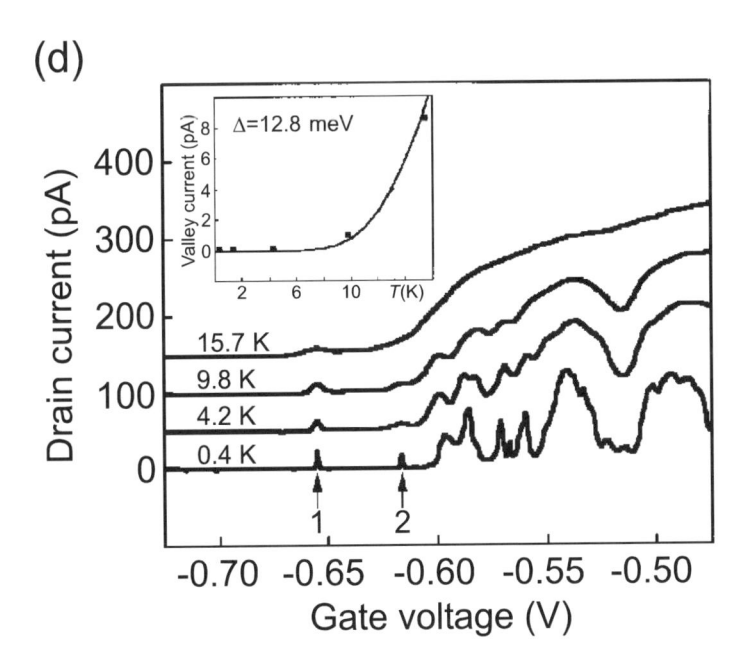

Fig. 7.9 (a) Schematic layer sequence of epitaxial structure comprising an n-AlGaAs/GaAs heterointerface with a two-dimensional electron gas and a layer of InAs/GaAs quantum dots. (b) and (c) Corresponding band diagrams with no gate bias and gate voltage below the critical value, respectively. (d) Experimental dependence of drain current on gate voltage in a split gate structure at a drain source voltage of $10\,\mu V$. Inset: dependence of valley current on temperature (squares) with fit (see text). (Reproduced by permission of American Institute of Physics from Horiguchi *et al.*, 1997)

The transport through single dots in a similar structure (Fig. 7.9a to c) has been investigated by Horiguchi *et al.* (1997). The number of quantum dots contributing to transport has been largely reduced by defining a split gate structure with a depletion zone such that only a few or possibly one dot is located in the channel. Up to 20 K the drain source current exhibits a peak due to tunneling through the ground state of a single quantum dot at a particular gate voltage (Fig. 7.9d). Below 10 K another peak is observable at a lower gate voltage, which in most samples is separated from the first one by a charging energy of about 14 meV. A similar value of 12.8 meV is obtained from fitting the valley current (inset in Fig. 7.9d). In a similar InGaAs/AlGaAs modulation doped field effect transistor distinct steps and a region of negative differential resistance were observed by Phillips *et al.* (1998b) in the current-voltage characteristics (up to room temperature).

8 Photonic devices

8.1 PHOTO-CURRENT DEVICES

A wavelength domain multiplication optical memory device based on self-organized InAs/GaAs quantum dots has been reported in Imamura *et al.* (1995), similar to the concept presented in Muto (1995) and Jimenez *et al.* (1997). The device consists of an n-GaAs buffer and an asymmetric i-GaAs layer in which an ensemble of InAs quantum dots with size inhomogeneity is embedded. Above the dots an AlAs layer is incorporated. The structure is covered by a metal forming a Schottky diode (Fig. 8.1a). When the device is illuminated under negative bias with light of a certain energy E_2 ('write'), electron–hole pairs are created in dots that have a matching ground state energy. Larger or smaller dots will be excited only for light with different energies E_1 and E_3, respectively, $E_1 < E_2 < E_3$, enabling wavelength domain multiplication. The electrons will tunnel from the quantum dots through the i-GaAs, moving towards the n-GaAs, resulting in a photo-current. Due to the large AlAs barrier the holes cannot tunnel toward the metal contact and remain in the dots. There they bleach the absorption, decreasing the photo-current for the next light pulse ('read') with energy E_2. However, the read pulse will fill the dots with holes if they were empty, thus creating filled dots regardless of the stored information. Decrease of light power for the read pulse may allow read–write cycles to increase, while clearing of information can be achieved by applying a positive bias to the Schottky diode.

The difference between write and read photo-currents decreases with a time constant of 0.48 ms at 300 K (Fig. 8.1b). This decrease has been attributed in Imamura *et al.* (1995) to the recombination of stored holes with electrons thermally excited from the n-type region. The retention time becomes larger when the dot electrons have a larger localization energy. Another mechanism contributing to the decrease of photo-current response for light with a particular energy could be tunneling of stored holes into neighboring dots of different ground state energy. The inter-dot tunneling is expected to decrease when the dot hole level has a larger localization energy.

A field effect transistor structure sensitive to the charge state of quantum dots embedded near the channel has been investigated by Sakaki *et al.* (1995) and Yusa and Sakaki (1996, 1997). Yusa and Sakaki (1996) reported that the gate voltage dependence of the electron concentration is varied by trapping and de-trapping of one or two electrons in the dots. Photogenerated carriers are also trapped, making the device characteristics sensitive to the illumination conditions (Yusa and Sakaki, 1997).

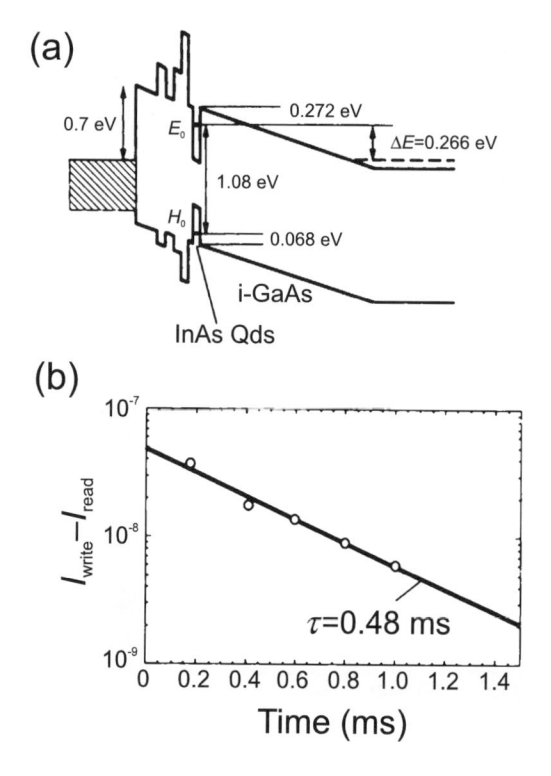

Fig. 8.1 (a) Estimated band alignment for InAs quantum dots embedded in a Schottky diode structure under negative bias. (b) Time dependence of photocurrent difference. (Reproduced by permission of American Institute of Physics from Imamura *et al.*, 1995)

Electrical detection of charge storage in self-assembled QDs was reported by Finley *et al.* (1998a, 1998b). Charge storage times of 22 min (8 h) were found at $T = 200$ K (140 K).

A photodiode with a QD absorbing region has been reported in Campbell *et al.* (1997). Since the absorption coefficient ($\alpha d \approx 3-4 \times 10^{-3}$) of the QD material is quite low, a resonant cavity design was employed. Within a spectral bandwidth of 1.2 nm around 1267 nm, for a top mirror consisting of 4 MgF/ZnSe pairs, a peak maximum external quantum efficiency of 49% has been achieved.

Nie *et al.* (1998) reported a QD-based photodetector operating at 1.06 μm. The absorption region consists of two five-layer stacks of 6 ML $In_{0.5}Ga_{0.5}As/GaAs$ dots in GaAs placed at the antinodes of a resonant cavity. The peak efficiency was 57% within a spectral bandwidth of 1.3 nm.

A far-infrared response due to inter-sublevel transitions has been demonstrated for a quantum dot absorber (Phillips *et al.*, 1997a). A tenfold stack of InAs/GaAs quantum dots has been placed in an intrinsic GaAs region between n-doped contact layers. A photocurrent response is found for normal incidence in the range of 10–21 µm with its peak at 17 µm (Phillips *et al.*, 1998). A mid-infrared detector based on a tenfold stack of self-organized InGaAs QDs with InGaP barriers on GaAs substrate was reported by Kim *et al.* (1998b). The 16 nm large dots had an area density of 3×10^{10} cm^{-2}. The photoconductivity signal peaking at 220 meV (with a FWHM of 48 meV) was observed up to 130 K. At 77 K the maximum responsivity was 0.13 A/W. The detectivity for vertical incidence was 4.74×10^7 cm Hz$^{1/2}$/W.

8.2 QUANTUM DOT LASER

A quantum dot laser includes a laser host material; a plurality of quantum dots disposed in the host material; and a pumping source for exciting and inducing a population inversion in the quantum dots (US Patent, 1993).

8.2.1 HISTORY

Semiconductor laser diodes have a long history (Alferov, 1998). The most important milestones after the initial observation of laser emission from GaAs p-n-homojunctions at 4 K were the introduction of the double heterostructure (Alferov and Kazarinov, 1963; Kroemer, 1963; Alferov, 1967) and the separate confinement (Casey and Panish, 1978; Tsang, 1981; Hersee *et al.*, 1982) concepts. The evolution went and still goes toward creating population inversion in increasingly constricted geometric regions and energy ranges (quantum wells, wires, and dots). The lowest threshold current densities (<50 A/cm^2 at room temperature) are currently obtained using compressively strained quantum wells as active medium (Chand *et al.*, 1991). The ultimate step forward for reducing j_{th} and improving other decisive fundamental parameters will be the introduction of equisized quantum dots as active medium, concentrating the oscillator strength in a narrow energy range.

The first attempt to study the lasing properties of three-dimensionally quantized carriers was to investigate quantum well lasers in strong magnetic fields (Arakawa and Sakaki, 1982; Vahala *et al.*, 1987; Berendschet *et al.*, 1989). These experiments demonstrated a reduction in the temperature dependence of j_{th} but did not prove a reduction in j_{th}. Berendschet *et al.* (1989) actually observed an increase in j_{th}.

The first light-emitting diode structure containing structural quantum dots was realized from a MOCVD-grown GaInAsP/InP quantum well. It was twofold patterned with a holographic process, etched and subsequently overgrown by liquid phase epitaxy InP (Miyamoto *et al.*, 1987). More recent approaches using lithographic patterning yield a laser operation, but are still troubled by extremely high threshold current densities (e.g. 7.6 kA/cm^2 at 77 K; see Hirayama *et al.*, 1994), most probably due to the process-introduced damage.

In 1994 the first Fabry–Perot injection laser based on self-organized quantum dots was presented (Kirstaedter *et al.*, 1994). In an AlGaAs/GaAs GRIN-SCH structure a single sheet of InGaAs/GaAs quantum dots was embedded. The laser structure was processed to yield a shallow mesa stripe geometry with as-cleaved facets. At 77 K the threshold current density for a 1 mm cavity length and uncoated facets was 120 A/cm^2. A large characteristic temperature $T_0 = 350$ K between 50 and 120 K was found.

For this laser, j_{th} increased strongly to 950 A/cm^2 at room temperature and the laser emission shifted to the high-energy side of the photoluminescence spectrum, very close to the wetting layer transition. This shift was caused by both evaporation of carriers out of the dots into the barrier and gain saturation (Schmidt et al., 1996b). In the meantime the figures of merit for quantum dot lasers have significantly improved (Bimberg et al., 1997). Presently, room temperature operation on the quantum dot ground state is realized with a record low $j_{th} = 12.7$ A/cm^2 for a 1.5 mm cavity length at 77 K and 110 A/cm^2 at room temperature (Heinrichsdorff et al., 1997d) (Fig. 8.2). For a four-side cleaved geometry a threshold of 62 A/cm^2 at room temperature has been realized (Ledentsov et al., 1996d).

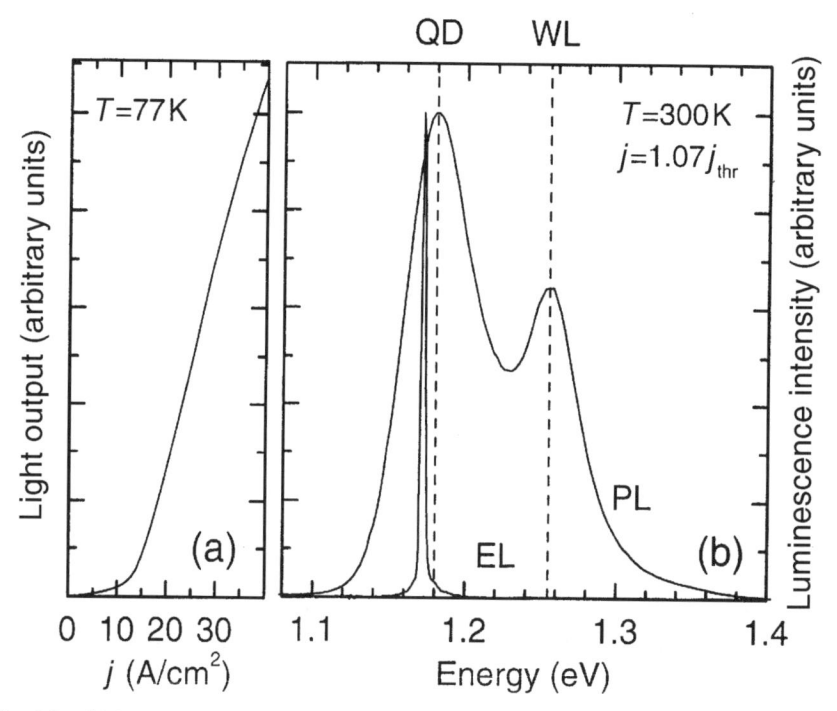

Fig. 8.2 (a) LI curve for an MOCVD-grown single-sheet InAs/GaAs quantum dot laser at 77 K, $j_{th} = 12.7$ A/cm^2. (b) Photoluminescence (PL) and lasing (EL) spectra of an MOCVD-grown threefold stack of InAs/GaAs quantum dots with AlGaAs/GaAs waveguide at room temperature. (After Heinrichsdorff et al., 1997d)

8.2.2 STATIC LASER PROPERTIES

8.2.2.1 Gain and threshold

Laser operation relies on the fact that a quantum dot ensemble exhibits (positive) gain for sufficiently high carrier injection levels. As outlined in Section 5.8.3.2, gain occurs in a simple twofold degenerate energy level system when the ground state population is inverted, i.e. when on average more than one electron and hole resides in the quantum dot ground state. Laser operation starts at a current density j_{th} where the losses are overcome, i.e.

$$g_{mod}(j_{th}) = \Gamma g_{mat}(j_{th}) = \alpha_{int} + \alpha_{mirr} \qquad (8.1)$$

where g_{mod} is the modal gain, Γ the optical confinement factor, and the total loss is separated in the internal loss α_{int} in the waveguide and the mirror loss α_{mirr}. Large material gain is theoretically predicted (Asada *et al.*, 1986), and also for inhomogeneously broadened ensembles (Asryan and Suris, 1996; Grundmann and Bimberg, 1997b).

The gain of InGaAs/GaAs quantum dot structures has been measured (Kirstaedter *et al.*, 1996) using different methods. By converting photoluminescence spectra at different excitation densities into material gain spectra (Fig. 8.3), a saturated material gain of close to 10^5 cm^{-1} is obtained for a single layer of dots. This value is more than an order of magnitude larger than that for comparable quantum well lasers (e.g. see Hu *et al.*, 1994).

Similar spectra of the modal gain are obtained using the 'stripe length method', introduced by Shaklee *et al.* (1985), where single-pass amplified spectra are measured for different excitation lengths (Fig. 8.4). The modal gain is transformed

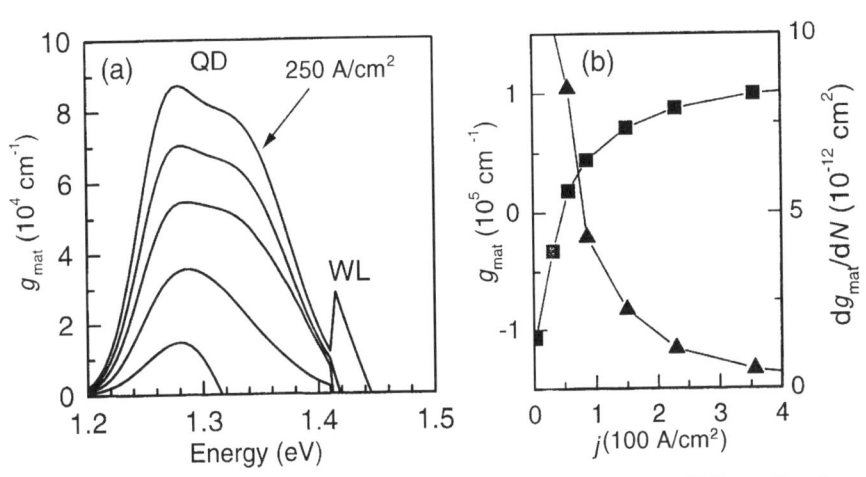

Fig. 8.3 (a) Material gain spectra of InAs/GaAs quantum dot laser, $T = 77$ K as a function of excitation density ranging from 20 to 250 A/cm^2. The WL is assumed to be a ≈ 0.5 nm quantum well. (b) Peak material gain at ≈ 1.28 eV and differential gain as a function of injection current density. (After Kirstaedter *et al.*, 1996)

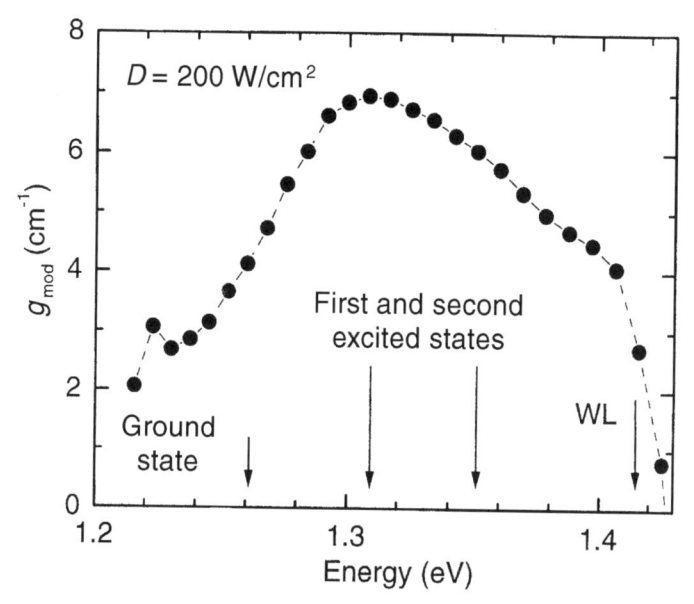

Fig. 8.4 Modal gain spectrum for an InAs/GaAs quantum dot laser, $T = 77$ K derived from single-pass amplification. The transition energies of the QD ground state and excited states and the WL are marked. (After Kirstaedter et al., 1996)

into the material gain using the optical confinement factor $\Gamma = 1.2 \times 10^{-4}$ of the structure. The gain spectra are asymmetric on the high-energy side due to the contributions of excited states. The modal gain of the dot ensemble was also determined from the dependence of the threshold and differential external quantum efficiency (Section 8.2.2.2) on the cavity length. This method yields values of the modal gain close to those obtained with the stripe length method (Kirstaedter et al., 1996).

The increase and saturation of the gain with current density is shown in the inset of Fig. 8.3 (Kirstaedter et al., 1996), together with the differential gain g'_{mat}:

$$g'_{mat} = \frac{d g_{mat}(n)}{dn} \quad (8.2)$$

where n is the carrier density averaged over the SCH. The maximum observed values next to the threshold for the differential gain in the 10^{-12} cm^{-2} range are orders of magnitude larger than those for quantum well lasers (e.g. see Hu et al., 1994). Further discussion of the impact of the differential gain on the modulation properties can be found in Section 8.2.3.1.

For a single sheet of quantum dots the extrapolation of the threshold current for zero mirror loss (Fig. 8.5) yields a very low value around 20 A/cm^2. For high mirror loss in short cavities the threshold increases strongly and the laser no longer operates on the quantum dots but on the wetting layer (Fig. 8.6). This is attributed to gain

PHOTONIC DEVICES

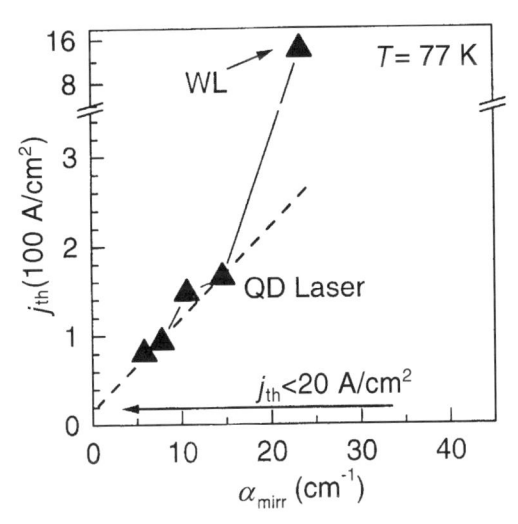

Fig. 8.5 Threshold current density of InAs/GaAs quantum dot lasers with different mirror loss (cavity length). Extrapolation for $\alpha_{mirr} \to 0$ yields a transparency current density of about 20 A/cm^2. At short cavities gain saturation causes lasing on the wetting layer states. (After Kirstaedter et al., 1996)

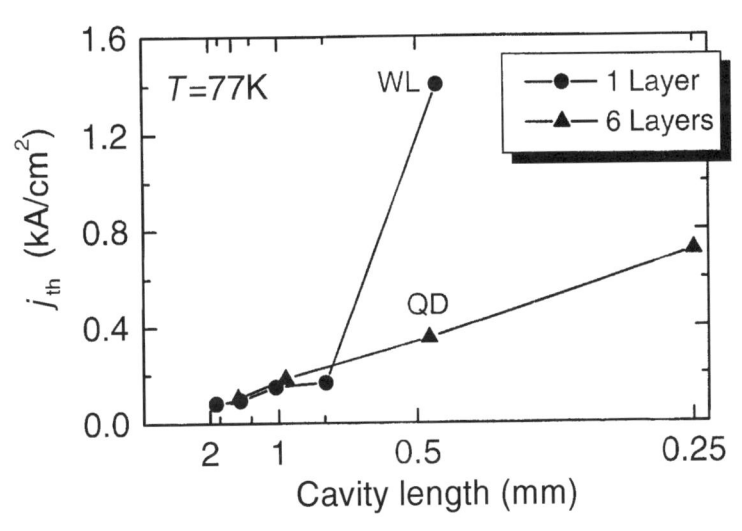

Fig. 8.6 Comparison of lasing characteristics of InAs/GaAs quantum dot lasers with single sheets of dots and a sixfold stack for various cavity lengths, i.e. mirror losses. (After Schmidt et al., 1996b)

saturation. For similar reasons, lasing on the excited transition has been observed in Kamath *et al.* (1996a) for $In_{0.4}Ga_{0.6}As/GaAs$ quantum dots in a 1 mm cavity at room temperature and $j_{th} = 650 \, A/cm^2$ in Mirin *et al.* (1996).

The maximum modal gain g_{mod}^{max} of a single sheet of quantum dots has a finite value of about $10 \, cm^{-1}$. Thus only cavities with sufficient length L

$$\alpha_{mirr} + \alpha_{int} < g_{mod}^{max}, \quad \text{i.e. } L > \frac{g_{mod}^{max} - \alpha_{int}}{\ln(1/\sqrt{R_1 R_2})} \quad (8.3)$$

can show laser operation on the dots. R_1 and R_2 denote the mirror reflectivities. For typical internal loss $\alpha_{int} = 2.5 \, cm^{-1}$ and uncoated as-cleaved facets with $R_1 = R_2 \approx 30\%$ we find $L \geqslant 1.6 \, mm$.

At least two schemes to prevent gain saturation are feasible:

(a) increase of the modal gain by stacking layers of quantum dots (Schmidt *et al.*, 1996b);
(b) decrease of mirror loss by high reflection (HR) coating of the facets (Shoji *et al.*, 1996).

As shown in Fig. 8.6, the linear gain regime is extended and shorter cavities achieve lasing when a sixfold stack of InGaAs/GaAs quantum dots is used as the gain medium (Schmidt *et al.*, 1996b).

With increasing number of layers the threshold current density (at room temperature) of InAs/GaAs and InGaAs/GaAs quantum dot lasers decreases (Fig. 8.7). Numbers given in this figure are obtained for otherwise identical four-side cleaved structures. For stripe lasers ($L = 1.5-2 \, mm$) the threshold is typically 20–40% larger. Current densities of $90 \, A/cm^2$ are achieved for a tenfold $In_{0.5}Ga_{0.5}As/GaAs$ stack and $62 \, A/cm^2$ for a threefold $In_{0.5}Ga_{0.5}As/Al_{0.15}Ga_{0.72}As$ stack (Ledentsov *et al.*, 1996d). For a threefold stack of MOCVD-grown InAs/GaAs quantum dots the dots become transparent for a current density of $40 \, A/cm^2$ at room temperature (Mao *et al.*, 1997).

The carrier distribution function and threshold also depend on barrier thickness between individual layers. An intermediate case between thermal and nonthermal carrier distribution has been found in Schmidt *et al.* (1996b) for slightly coupled layers.

For single-sheet InAs/GaAs dot lasers investigated in Shoji *et al.* (1996) the laser emission at room temperature for as-cleaved facets occurs close to the photon energy of the wetting layer for a 0.9 mm cavity length. A threefold stack emits from the first excited state of the dots. The decrease of mirror loss by using HR coated facets ($R = 95\%$) eventually leads to a shift of the laser emission to the ground state. Alternatively, of course, a decrease in temperature, trapping carriers on the lower states, results in a switching of the laser emission from the excited state back to the ground state for uncoated facets (Shoji *et al.*, 1997).

Optical gain has been also investigated for self-organized InP/GaInP quantum dots (Moritz *et al.*, 1996). As depicted in Fig. 8.8, the quantum dot ensemble exhibits a positive modal gain around 1.75 eV for TE polarization with a maximum

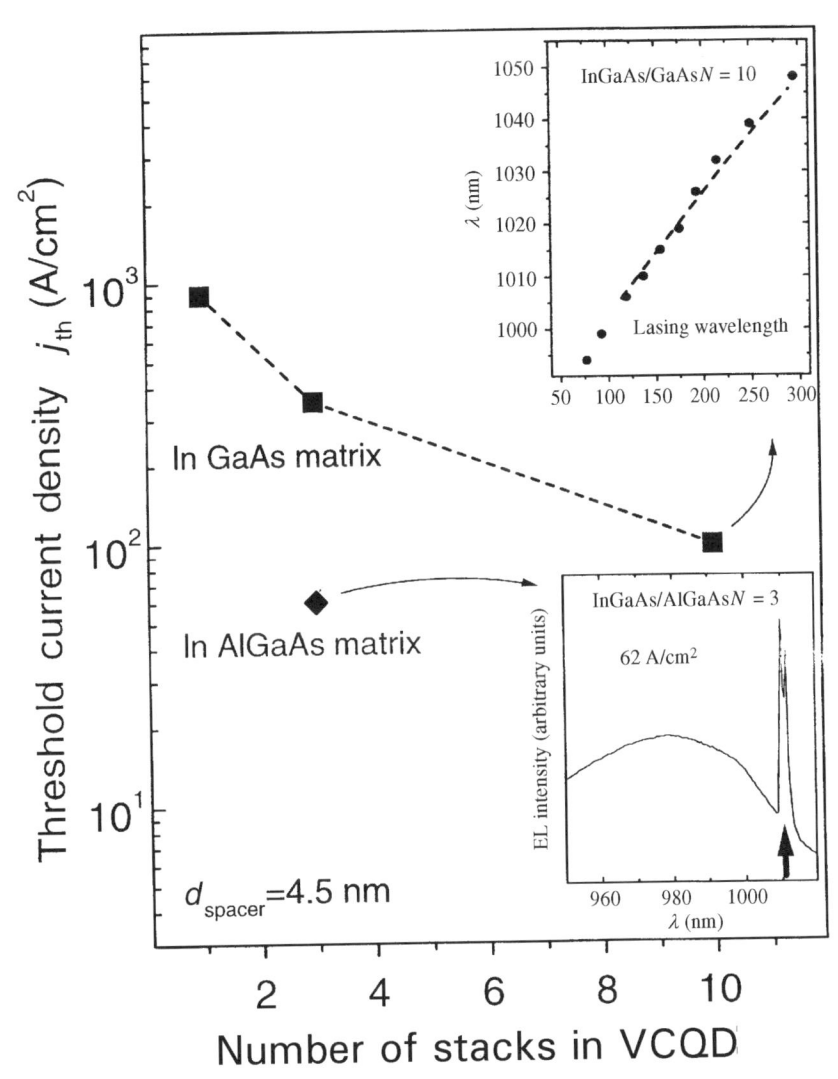

Fig. 8.7 Threshold current density at 300 K for lasers with vertically coupled quantum dots with varying number of periods; spacer layer thickness is 4.5 nm. Upper inset shows the temperature dependence of the lasing wavelength, dashed line represents the shift of GaAs band gap. Lower inset shows emission spectrum of a four-sided cleaved InGaAs/AlGaAs laser. (After Ledentsov et al., 1996d)

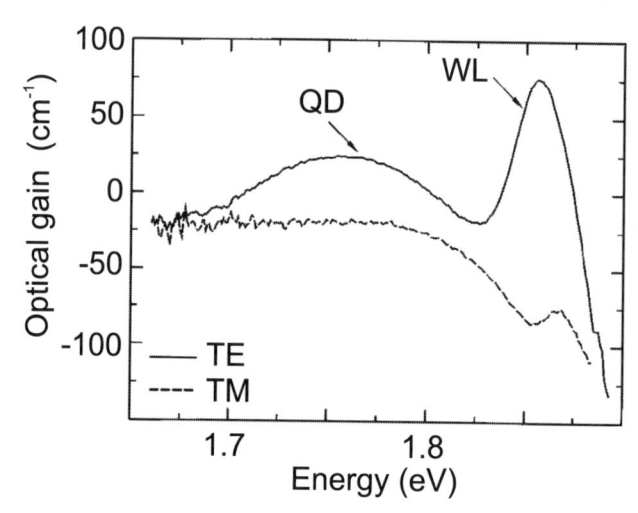

Fig. 8.8 Optical gain (TE and TM polarization) of nominally 3 ML InP quantum dots in GaInP at room temperature; excitation density was 30 kW/cm². (Reproduced by permission of American Institute of Physics from Moritz *et al.*, 1996)

value of 25 cm^{-1}, while for TM polarization only absorption is observed. A separate peak at 1.85 eV is due to a large gain of 75 cm^{-1} in the wetting layer.

8.2.2.2 Quantum efficiency

An important parameter characterizing the laser quality is the internal differential quantum efficiency η_{int}. This value can be derived from the dependence of the external differential quantum efficiency η_{ext} on the mirror loss according to the equation

$$\frac{1}{\eta_{ext}} = \frac{1}{\eta_{int}} \left(\frac{\alpha_{int}}{\alpha_{mirr}} + 1 \right) \tag{8.4}$$

where η_{ext} can be directly determined from the slope of the light output versus current characteristic. In Fig. 8.9 the 300 K inverse external quantum efficiency is shown versus the inverse mirror loss for a tenfold InAs/GaAs QD stack. By extrapolating for $\alpha_{mirr} \to \infty$, i.e. $L \to 0$, the internal quantum efficiency of 50% is obtained (Bimberg *et al.*, 1997; Zaitsev *et al.*, 1997). From the slope of the line the internal loss is determined to 11 cm^{-1}, which is larger than the value of 2.5 cm^{-1} found for a single layer at 77 K (Kirstaedter *et al.*, 1996). Recently, values of 70% (Bimberg *et al.*, 1997), 81% (Mirin *et al.*, 1996), and 91% (Mao *et al.*, 1997) have been obtained for ternary InGaAs/GaAs quantum dot lasers grown by MBE or MOCVD (Mao *et al.*, 1997). These values are not much lower than the recently reported record internal efficiency of 99% for a quantum well laser (Zhang, 1994a).

PHOTONIC DEVICES

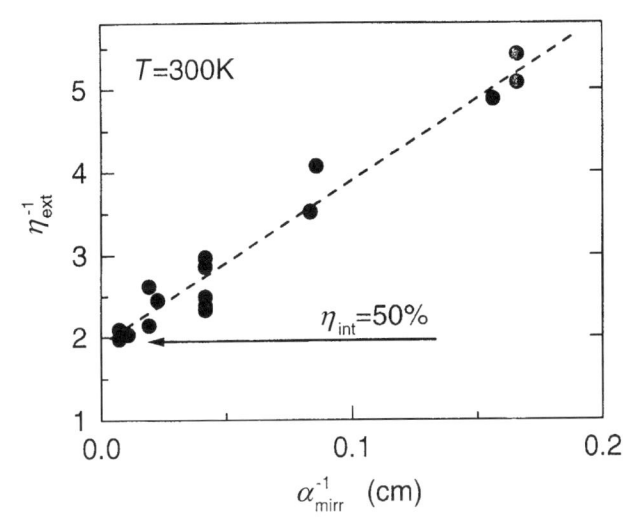

Fig. 8.9 External differential quantum efficiency at room temperature for a tenfold stacked InAs/GaAs QD laser as a function of mirror losses. The internal quantum efficiency is about 50%. (After Bimberg et al., 1997)

8.2.2.3 Temperature dependence

A δ-function density of states or a suppression of an equilibrium distribution of carriers between all dots in a dot ensemble should prohibit thermal broadening of the gain spectrum. Then the peak gain would remain constant as a function of temperature at constant injection level and the threshold current density is temperature independent. The characteristic temperature T_0 (Eq. 5.128) is infinite.

The first observation of an increase of T_0 from 144 to 313 K at 0 °C due to carrier confinement was reported in Arakawa and Sakaki (1982) for a double-heterostructure laser placed in a magnetic field. A dramatic increase of T_0 for a single layer of self-organized quantum dots up to 350 K was reported in Kirstaedter et al. (1994) for operating temperatures between 50 and 120 K. A still larger value of $T_0 = 425$ K is shown in Fig. 8.10a for operating temperatures smaller than 100 K (Bimberg et al., 1996). Above that temperature the breakdown of the nonequilibrium carrier distribution and the temperature-dependent leakage current reduce the value of T_0. In this particular structure the activation energy for electrons to move from the quantum dot into the wetting layer is only ≈ 25 meV, leading also to a decrease in photoluminescence intensity and carrier lifetime (Fig. 8.10b). The quantum dots become thermally depopulated and the injection current has to be increased in order to maintain the threshold gain. In addition, nonradiative recombination in the barrier might lead to current leakage. This leakage current is related to the quality of the barrier (in particular to the part grown at low temperatures) and does not present a principal limitation of the threshold current.

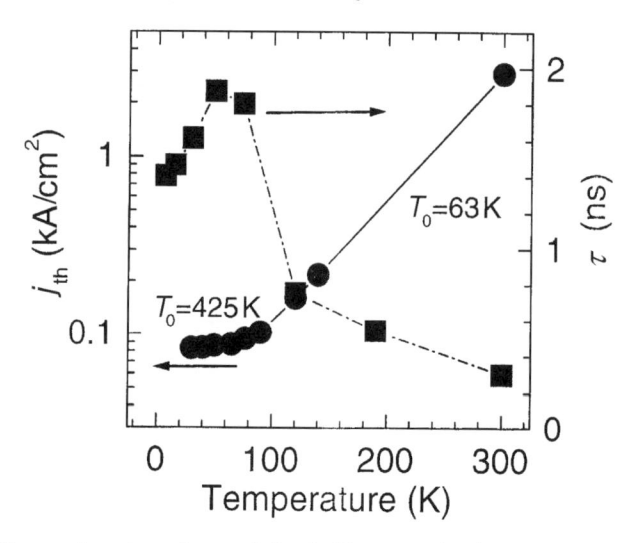

Fig. 8.10 Temperature dependence of threshold current density and carrier decay constant on temperature for single-layer InGaAs/GaAs quantum dots

Higher growth temperatures of the cladding layers and optimization of the dot size with the goal of an increase of the activation energy shifts the onset of current leakage to 220 K with an ultra-high T_0 of 530 K between 80 and 220 K (Alferov et al., 1996). Another way to decrease the leakage current is to place the dot layers in a GaAs/AlGaAs quantum well, thus drastically reducing the escape probability of carriers from the dot. With this approach the onset of current leakage has been increased up to room temperature while maintaining a high $T_0 = 385$ K between 80 and 330 K (Maximov et al., 1997a, 1997b). Large T_0 values of up to 477 K at low temperatures of 60–110 K have been also reported (Shoji et al., 1997). At room temperature, however, $T_0 < 50$ K and a decrease of luminescence efficiency are found.

A negative characteristic temperature in the temperature range of about 60–130 K has been found in Zhukov et al. (1997). It has been attributed to the transition of a nonthermal into a thermal carrier distribution with increasing temperature, as is discussed in Grundmann and Bimberg (1997c) (see Section 5.8.3.2).

8.2.2.4 Spectra

The emission of a single-layer quantum dot laser consisting of a dot ensemble with only one energy state for electrons and holes is expected to occur at the maximum of the photoluminescence emission and should not show any energy shift with injection current. In the case of real structures emission from both the ground state and excited state(s) may be observed. Thus the line shape of the gain spectrum is no longer independent of the injection level, generally causing a slight blue shift of the laser

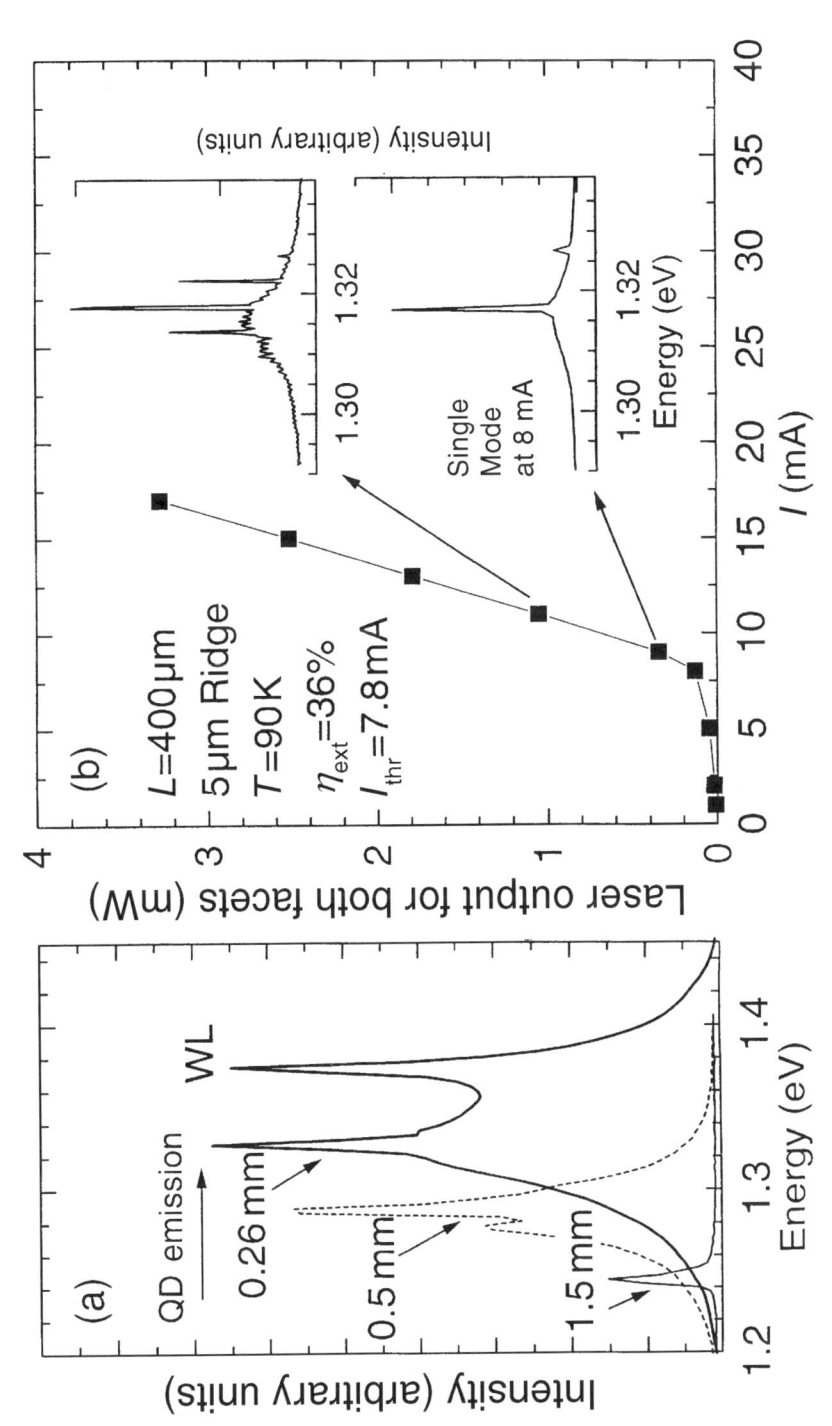

Fig. 8.11 (a) Blue shift of the quantum dot emission due to state filling at 77 K. Each spectrum is marked with the corresponding cavity length. For short cavity lasers the QD gain is completely saturated and the emission jumps to wetting layer states. (b) Light current characteristic of a sixfold quantum dot laser with a 400 μm long and 5 μm wide cavity showing single-mode operation just above the threshold. (After Bimberg et al., 1997)

emission with increasing injection current or increasing loss. Such a blue shift is demonstrated in Fig. 8.11a for different cavity lengths (Bimberg et al., 1997). Obviously excited quantum dot states provide higher modal gain. A similar effect was found for laser structures with different facet reflectivities, i.e. mirror losses (Shoji et al., 1997). Upon complete saturation of quantum dot gain, the laser emission jumps to the wetting layer states, which have a much higher modal gain due to the larger confinement factor.

The width of the ensemble gain spectra of a single layer of quantum dots is in the range of typically FWHM $\approx$ 50 meV. This value is far away from a gain width below 1 meV to fulfill the condition for single-mode operation in a stripe geometry at a typical cavity length $\sim$ 0.5 mm. Nevertheless, single-mode operation just above threshold (Kirstaedter et al., 1994) ($<1.1 j_{thr}$) is normally observed (Fig. 8.11b). For quantum well lasers it is well known that they can switch to single-mode operation at injection levels far above threshold. Two possible mechanisms might be responsible for the single-mode operation of QD lasers. Due to inhomogeneities of the dot distribution the gain spectra have local minima near the gain maximum, increasing the threshold of other longitudinal modes. Consequently, the first suppressed mode will start lasing only at higher injection levels. In addition, different subsets of dots, distinguished, for example, by their size, may have different capture times; thus only rapidly refilled dots contribute to lasing. Threshold emission spectra with series of well-defined groups of longitudinal modes, separated by non-lasing gaps were reported by Harris et al. (1998). Within each group of longitudinal modes a

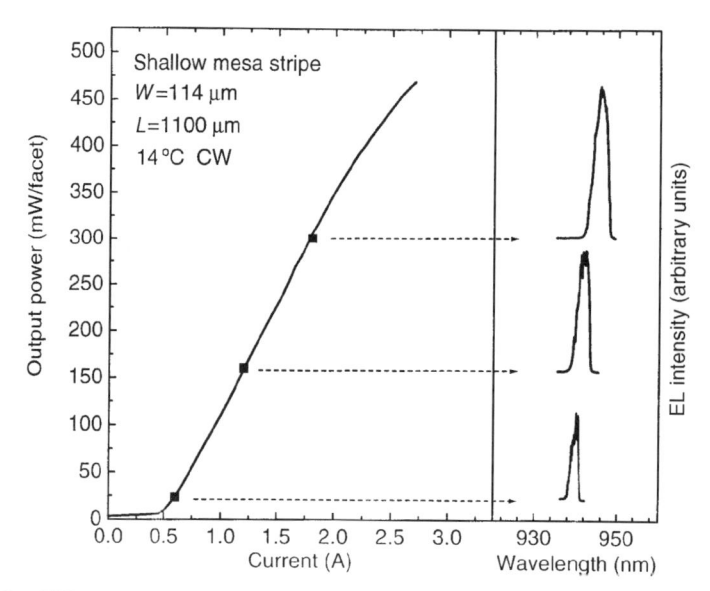

Fig. 8.12 CW output power per facet versus pump current of the laser incorporating a tenfold stack of InGaAs/AlGaAs quantum dots with spectra at different pump currents of 0.6, 1.2, and 1.8 A, respectively. (After Shernyakov et al., 1997)

highly nonuniform intensity distribution is found. The spectrally separated groups of longitudinal modes correspond to different lateral intensity profiles (far-field patterns).

8.2.2.5 High power performance

Shernyakov et al. (1997) reported total cw output power of up to 1 W at 14 °C heat sink temperature for a tenfold stack of $In_{0.5}Ga_{0.5}As$ with inter-layers of $Al_{0.15}Ga_{0.85}As$ (Fig. 8.12). The laser structure was a shallow mesa of width 114 µm. The differential efficiency was 40%; the threshold current density for a 1100 µm long device was 290 A/cm². The maximum temperature of the heat sink for which lasing could be obtained was ∼ 70–75 °C. Advantages of QD lasers for high-power operation are expected in the future due to reduced COD, relaxed cooling requirements due to larger T_0, and slightly improved wall plug efficiency due to the reduced j_{th}.

8.2.3 DYNAMIC LASER PROPERTIES

8.2.3.1 Modulation bandwidth

Part of the attraction of the quantum dot laser is the prediction of high differential gain which might lead to a higher modulation bandwidth (Arakawa and Yariv, 1986). The orders of magnitude of higher differential gain of quantum dot lasers should lead to an increased modulation bandwidth. Small signal analysis of standard rate equations yields an approximation of the modulation bandwidth defined as the −3 dB value of the ratio of photon density S and current density I (Nagarajan et al., 1992):

$$\frac{S(\omega)}{I(\omega)} \propto \left| \left(\frac{1}{1 + i2\pi\omega\tau_{cap}} \right) \frac{1}{4\pi^2 \omega_R^2 - 4\pi^2 \omega^2 + i2\pi\omega\gamma} \right|$$

$$\omega_R^2 \cong \frac{1}{4\pi^2} \frac{v_g \Gamma g' S_0}{\tau_{phot}(1 + \varepsilon S_0)}$$

$$\gamma = K\omega_R^2 + \left[(1 + \chi)\tau_{QD}\right]^{-1}$$

$$K = 4\pi^2 \left[\tau_{phot} + \frac{(1 + \chi)\varepsilon}{v_g g'} \right]$$

(8.5)

ω_R being the relaxation oscillation frequency, γ the damping factor, and K the proportionality factor between γ and ω_R^2; ω represents the modulation frequency, τ_{cap} the quantum dot capture time, τ_{QD} the carrier decay time, τ_{phot} the photon round trip time in the laser cavity, g' the differential material gain, Γ the confinement factor, v_g the group velocity in the laser, ε the gain compression coefficient, S_0 the photon density in the laser cavity at the bias level, and χ the ratio of quantum dot capture and emission time (Kirstaedter et al., 1996).

The material gain depends on the carrier density (state filling), which is clamped during laser operation close to the threshold carrier density, and the gain compression factor ε, which depends on the photon density. Various physical processes like transport and carrier capture (Agrawal, 1990; Nagarajan et al., 1992), carrier heating, spectral hole burning and two-photon absorption (Willatzen et al., 1991) may contribute to gain compression. The form we use here for quantum dots is

$$g_{\mathrm{mat}}(n, S) = \frac{g'(n - n_0)}{1 + \varepsilon S} \quad (8.6)$$

which exactly describes the gain compression for a homogeneously broadened two-level system (Agrawal, 1990). The gain compression describes the fast reduction of inversion by the photons. The coefficient ε is inversely proportional to the saturation photon density of the laser. From Eq. (8.5) it is clear that not only the differential gain but also gain compression influences the high-frequency behavior of the laser. The bandwidth is in any case limited by $\sim \tau_{\mathrm{cap}}^{-1}$.

For determining experimentally the cut-off frequency, relaxation oscillations (see Fig. 8.13) have been measured as a function of the photon density (i.e. the power in the laser cavity) (Bimberg et al., 1997). Using a fit, the differential gain g' and the gain compression coefficient ε are determined and therefrom the K value. In Fig. 8.14 the relaxation oscillation frequency of a typical quantum dot laser (sixfold

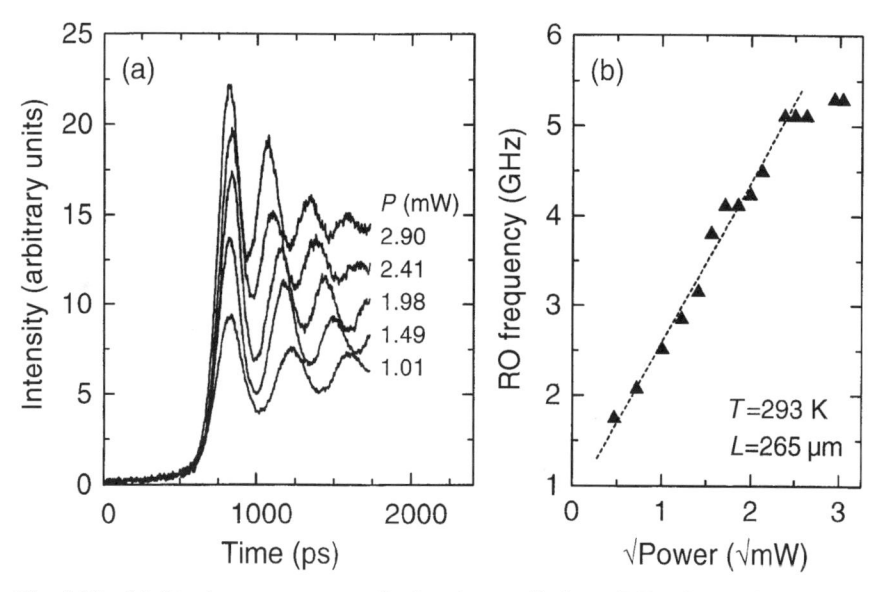

Fig. 8.13 (a) Streak camera traces of relaxation oscillations (RO) of an MOCVD-grown quantum dot laser with a threefold stack of InAs/GaAs quantum dots. (b) RO frequency versus the square root of the output power per facet; dashed line is best fit of linear region with slope 1.8 GHz/mW$^{1/2}$. (After Mao et al., 1997)

PHOTONIC DEVICES

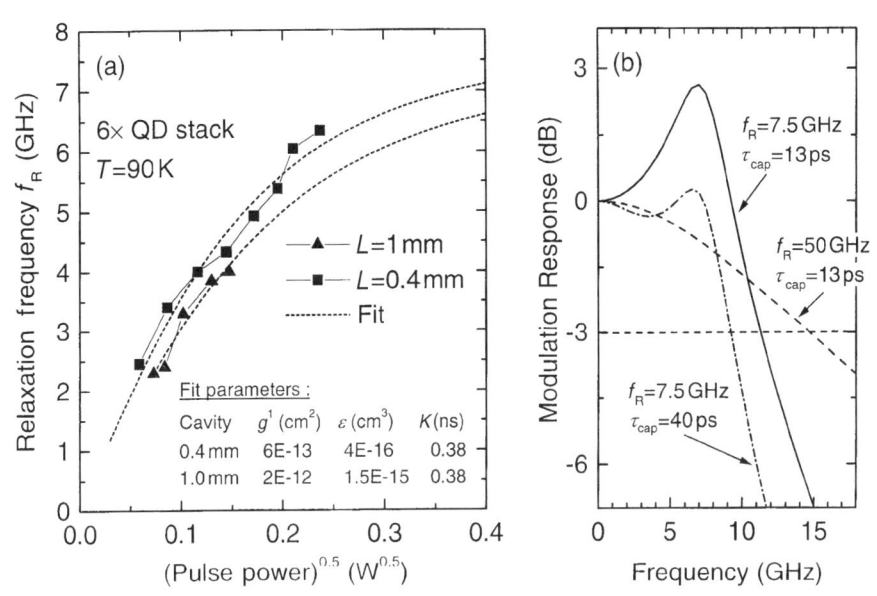

Fig. 8.14 (a) Relaxation oscillation frequency of a ridge waveguide sixfold stacked InAs/GaAs quantum dot laser as a function of the square root of pulse power in the laser cavity for two different cavity lengths. (b) Calculated modulation bandwidth for different relaxation oscillation frequencies and capture times using the parameters from (a). (After Bimberg et al., 1997)

InAs/GaAs stacked dots) is plotted as a function of the square root of the pulse power. It shows a deviation from the expected proportionality to the pulse power $\sim \sqrt{S_0}$, indicating the onset of gain compression at a power level of about 50 mW. The gain compression coefficient has a minimum of 4×10^{-16} cm^3 at 90 K, equivalent to a power saturation coefficient of about 100 mW in the laser cavity. The gain compression coefficient is more than one order of magnitude higher than typical values ($\sim 10^{-17}$ cm^3) in InGaAs/InGaAlAs quantum well lasers (Grabmeier et al., 1991). The differential gain of 2×10^{-12} cm^2 obtained from the fit is close to the value derived from the steady state experiments (Section 8.2.2.1).

Since the gain compression is closely correlated with the capture time of carriers into the dots, the physical origin of the enhanced gain compression in quantum dot laser structures may be connected to the slower relaxation rate of the dots. This mechanism reduces the modulation bandwidth even when the differential gain is high. The maximum modulation bandwidth for the sixfold stacked quantum dot laser investigated by Bimberg et al. (1997) is about 11 GHz. It is not limited by the capture time of about 13 ps, corresponding to a 3 dB frequency of 23 GHz, but by the small confinement factor, limiting the differential modal gain. The measured K factor of 0.38 ns is a factor of 2–3 times higher than the best values reported for InGaAs quantum well lasers (Nagarajan et al., 1992) at 100 K, indicating that the modulation bandwidth in quantum dot lasers is still somewhat lower than in quantum well lasers.

Mao et al. (1997) found the relaxation oscillation frequency of up to 5.3 GHz for threefold stacked MOCVD-grown self-organized quantum dots, corresponding to a 3 dB cut-off frequency of 8.2 GHz (Fig. 8.13). Kamath et al. (1997b) reported a 3 dB bandwidth of 7.5 GHz for $In_{0.4}Ga_{0.6}As/GaAs$ quantum dot lasers grown by MBE. An increase of the bandwidth from 5 GHz at room temperature to 20 GHz at 80 K has been reported by Klotzkin et al. (1998). It was attributed to the higher effectivity of electron-hole scattering (Vurgaftman and Singh, 1994) at lower temperatures.

8.2.3.2 Chirp

A key parameter for the application of quantum dots in directly modulated light sources is the magnitude of the chirp, i.e. the change of emission wavelength during direct current modulation. The physical origin of this shift is related to the coupling of the real and imaginary parts of the complex susceptibility in the laser medium. A variation of gain due to a change of injection current leads to a variation of the

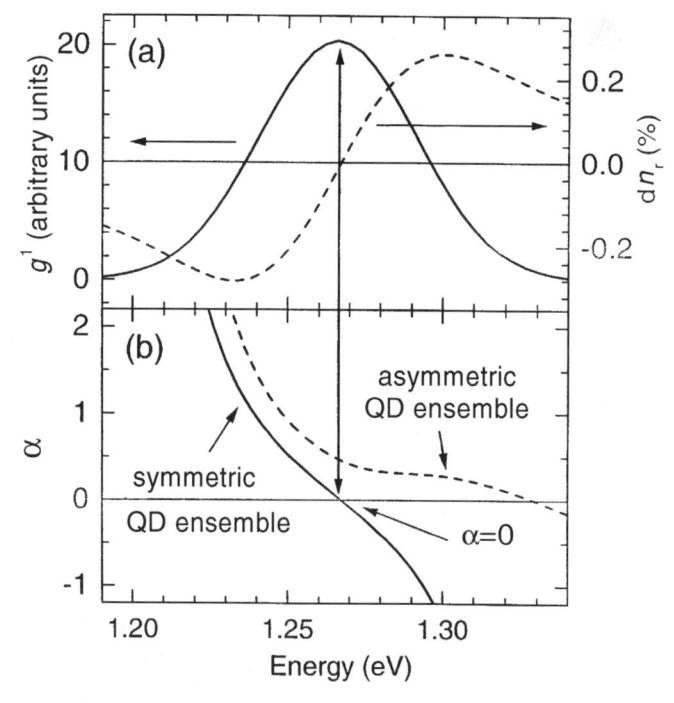

Fig. 8.15 (a) Differential gain g' and differential refractive index n'_r for a quasi-ideal quantum dot ensemble with a symmetric gain curve and only one energy level for electrons and holes. (b) Calculated linewidth enhancement factor α for the ideal QD ensemble (solid line) and the experimentally reported single-layer QD ensemble with the asymmetric gain spectrum shown in Fig. 8.4 (dashed line)

refractive index that modifies the phase of the optical mode in the laser cavity. The coupling strength is defined by the linewidth enhancement factor α defined as

$$\alpha = \frac{\partial \chi_r / \partial N}{\partial \chi_i / \partial N} = -2\frac{\hbar c}{E} \frac{\partial n_r / \partial N}{\partial g_{\text{mat}} / \partial N} = -2\frac{\hbar c}{E} \frac{n'_r}{g'} \quad (8.7)$$

where n_r is the real part of the complex refractive index, c is the light velocity in vacuum, E is the photon energy, and g' is the differential material gain. The linewidth enhancement factor can be calculated from the gain spectrum via the Kramers–Kronig relation. In the case of a dot laser with a dot ensemble showing a perfect Gaussian energy distribution and only one energy level for electrons and holes, the gain spectrum and the differential gain are perfectly symmetric around the peak gain energy. In this case the differential change of refractive index is exactly zero at the lasing energy, i.e. the peak gain position. Thus $\alpha = 0$, and the ideal quantum dot laser operates *chirp-free*, as shown in Fig. 8.15. In the case of real quantum dots the contribution of excited states causes an asymmetric gain curve, which increases the linewidth enhancement factor depending on the excitation level to about 0.5. This value is still considerably lower than values of $\alpha \cong 1-2$ (Raghuraman et al., 1993), which are considered as being very good for quantum well lasers.

8.2.4 VCSEL

In vertical cavity surface emitting lasers (VCSELs) the active medium is embedded in a short λ cavity (e.g. see Iga et al., 1988; Geels and Coldren, 1990). In order to keep the losses low, the mirrors have to be highly reflective and are usually realized with stacks of $\lambda/4$ distributed Bragg reflectors. Lateral mode shaping is achieved by structuring mesas with diameters in the $\sim 10\,\mu\text{m}$ range.

The vertical cavity represents a (one-dimensional) microcavity. It has a stop band of typically several 10 meV width and within the stop band a pass band of about 1 meV width. Using a narrow quantum dot ensemble whose ground state energy is matched to the cavity pass band, optimum photon and electron control would be achieved (Fig. 8.16a). Since the Fabry–Perot mode distance is large, due to the short cavity single-mode operation is automatically realized. The laser wavelength is in any case clamped to the pass band, i.e. if the quantum dot ensemble gain maximum is not well matched to the pass band an increase in the threshold current will result. In general, great epitaxial homogeneity and control are needed to reproducibly fabricate a matching pass band and quantum dot ground state energy across a whole wafer. Otherwise the emission wavelength and threshold current will vary accordingly.

Optically pumped laser oscillation at $T = 77$ K has been obtained for a single sheet of quantum dots grown by MOVPE (Arakawa, 1996; Schur et al., 1997). The dots were fabricated by spinodal decomposition. The width of the photoluminescence spectrum of the dot ensemble in the vertical cavity is given by the width of the pass band (1.8 meV) and not by the inhomogeneous distribution of the dot ensemble. The

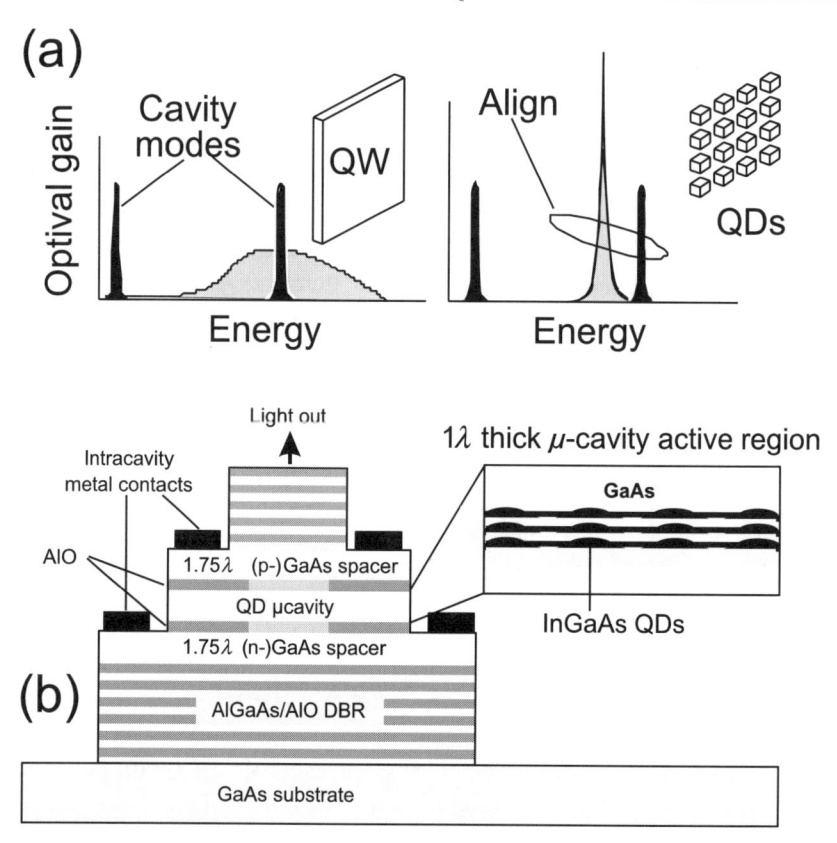

Fig. 8.16 (a) Matching of cavity mode and electronic density of states of the gain medium (QW and QDs). (b) Schematic structure of the quantum dot VCSEL with oxide aperture. (After Lott *et al.*, 1997a)

resonance wavelength was tuned to coincide with the ground state of the quantum dots. The light input–output characteristic indicates threshold behavior.

The first operation of an electrically driven VCSEL with self-organized quantum dots has been reported in (Saito *et al.* (1996)). The active medium is a tenfold stack of MOCVD-grown InGaAs/GaAs quantum dots with a density of 2×10^{10} cm^{-2} per layer. A schematic drawing of the structure is given in Fig. 8.17a. The room temperature quantum dot spectrum has its maximum at around 1000 nm, while the wetting layer is found at 920 nm. Devices of 25 μm × 25 μm were formed by wet etching. The threshold current for room temperature cw operation is 32 mA, corresponding to a threshold current density of about 5 kA/cm^2. The single-mode laser emission wavelength is 960.4 nm (Fig. 8.17b), determined by the cavity resonance peak. This wavelength lies in between the quantum dot peak at 1020 nm and the wetting layer at 940 nm for the device temperature at threshold (70 °C), indicating that mainly excited dot states contribute to the lasing.

PHOTONIC DEVICES

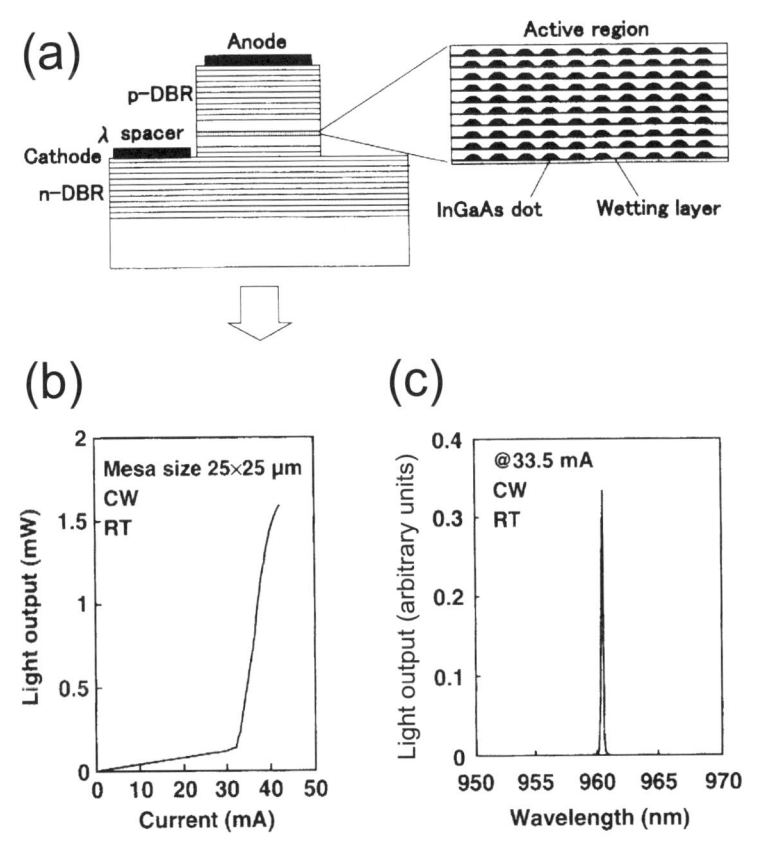

Fig. 8.17 (a) Schematic structure of the QD VCSEL. (b) Light output versus current. (c) Laser spectrum. (Reproduced by permission of American Institute of Physics from Saito *et al.*, 1996)

A room temperature pulsed threshold of 560 μA for a 7 μm diameter dielectric (oxide) aperture ($j_{th} = 1.5$ kA/cm^2) is reported in Huffaker *et al.* (1997).

In a VCSEL with GaAs/AlO mirrors and a dielectric aperture (Fig. 8.16b), a technique introduced by Dallesasse *et al.* (1990) vertically stacked and electronically coupled QDs as the active gain medium exhibit laser operation at room temperature. Emission on their ground state occurs with a threshold current of 200 μA (Lott *et al.*, 1997a) ($j_{th} = 400$ A/cm^2) for a mesa diameter of 8 μm. For smaller mesas with a 1 μm diameter, record values down to 68 μA were found (Fig. 8.18) (Lott *et al.*, 1997b), which compare to best values found for quantum well VCSEL (Huffaker and Deppe, 1997).

As discussed in Section 6.7.7, the quantum dot transition energy exhibits a blue shift upon thermal annealing. Upon annealing also a shift of the VCSEL emission wavelength (by 7 nm) was observed (see Fig. 8.19), although the dielectric mirrors

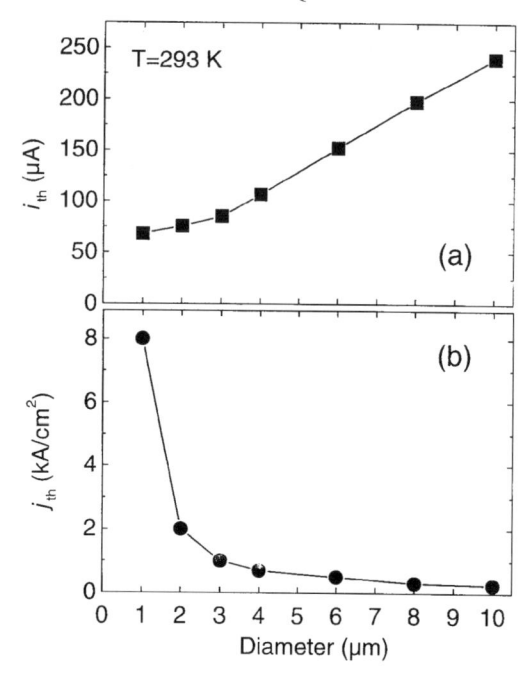

Fig. 8.18 (a) Threshold current and (b) threshold current density for the quantum dot VCSEL. (After Lott *et al.*, 1997b)

were not altered by the annealing process. This indicates an interaction of the (empty) resonator mode with the dispersion of the quantum dots. Such an effect is well known for atoms (Zhu *et al.*, 1990) and quantum wells (e.g. see Savona *et al.*, 1995).

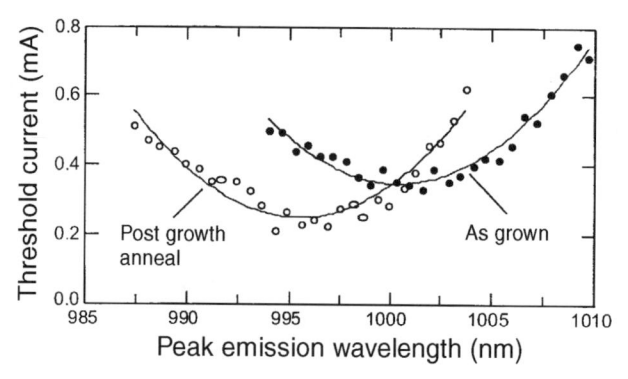

Fig. 8.19 Threshold current and peak emission wavelength of the quantum dot VCSEL along a stripe from a 2 inch wafer for as-grown and thermally annealed dots. (After Lott *et al.*, 1997a)

PHOTONIC DEVICES

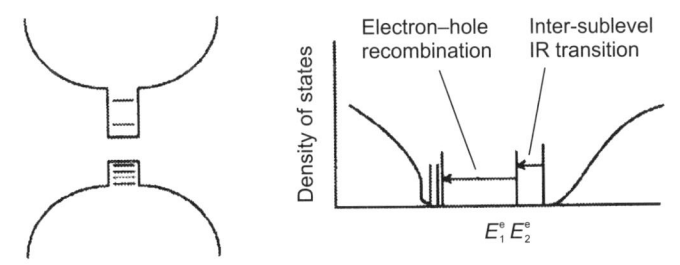

Fig. 8.20 Schematic band diagram of the quantum dot inter-sublevel laser. (After Singh, 1996)

If structural anisotropy of self-ordered dots is present, it allows emission of linearly polarized light from VCSELs (Saito et al., 1997); in the observed case the polarization is along [-110]. Emission at 1154 nm from a oxide-confined QD VCSEL has been reported by Huffaker et al. (1998). At 300 K a CW threshold of 502 µA was obtained for a device size of 10 µm.

8.2.5 INTER-SUBLEVEL IR LASER

Singh (1996) proposed the use of inter-sublevel optical transitions of quantum dots for building a laser in the infrared (IR). A schematic density of states is shown in Fig. 8.20. The quantum dot is assumed to have two electron levels and a number of hole levels. Besides the recombination from the lowest level E_1^e with ground state holes (E_1^h), the inter-sublevel transition $E_2^e \rightarrow E_1^e$ is considered which may have a typical energy of $\sim 50\text{--}100$ meV. For this spectral range semiconductor laser devices are hard to find, the most promising being the quantum cascade laser (Faist et al., 1994a, 1994b).

For the operation as an infrared laser, the following favorable situation is exploited. A small phonon bottleneck effect is present with an electron inter-level scattering time in the 10–100 ps range. The hole energy relaxation is fast, so that spontaneous recombination from E_2^e with excited hole states can be neglected. First stimulated emission on the $E_1^e \rightarrow E_1^h$ transition (a regular quantum dot laser) is started with current injection. This shortens the recombination time of the E_1^e level with an increasing number of photons n_ph in the cavity, going from spontaneous to stimulated emission:

$$\tau_\text{eh}^\text{stim} = \frac{\tau_\text{eh}^\text{spon}}{n_\text{ph}} \tag{8.8}$$

Solving a rate equation model, it was found (Singh, 1996) that the population of the electron ground state is pinned at $f(E_1^e) \sim 0.5$, regardless of injection current J. The population of the excited electron state is given by

$$f(E_2^e) \sim \frac{2J\tau_{21}}{e} \tag{8.9}$$

τ_{21} being the inter-sublevel relaxation time. The population of an excited level can thus be inverted with respect to the ground state, which will lead to inter-sublevel stimulated emission of mid-infrared photons. The longer τ_{21}, the faster inversion can be achieved. The maximum attainable gain, however, is smaller. In Fig. 8.21 gain spectra have been calculated for $\tau_{21} = 150\,\mathrm{ps}$ and an inhomogeneous broadening of $\sigma_E = 1.5\,\mathrm{meV}$. The material gain could reach values of $\sim 10^3\,\mathrm{cm}^{-1}$ for injection densities $\sim 10\,\mathrm{kA/cm^2}$. For different broadening and values of the inter-sublevel scattering time the gain g scales as

$$g \propto (\tau_{21}\sigma_E)^{-1} \tag{8.10}$$

A quantum dot version of the quantum well cascade laser (Faist *et al.*, 1994a, 1994b) has been proposed and theoretically analyzed by Wingreen and Stafford (1997). By eliminating single-phonon scattering processes in energy relaxation, the quantum dot device (necessary QD size $\sim 10\text{--}20\,\mathrm{nm}$) is predicted to lead to a reduction of

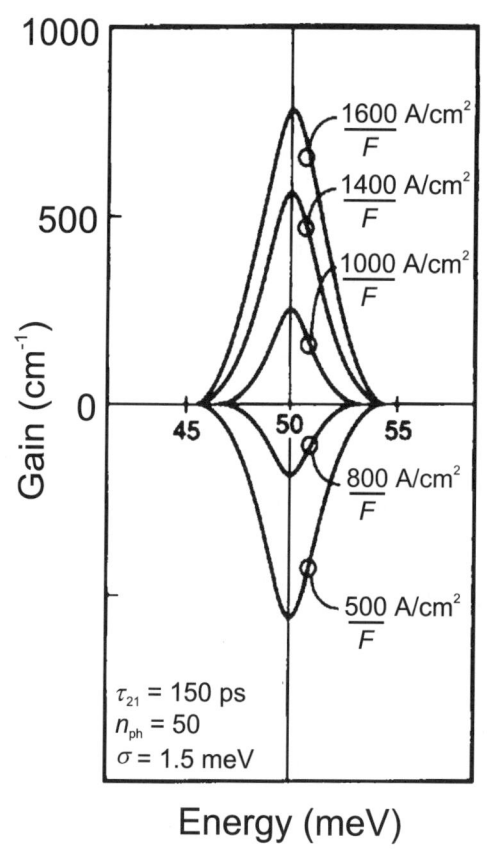

Fig. 8.21 Theoretical gain of the inter-sublevel laser based on quantum dots. (Reproduced by permission of American Institute of Physics from Singh, 1996)

PHOTONIC DEVICES

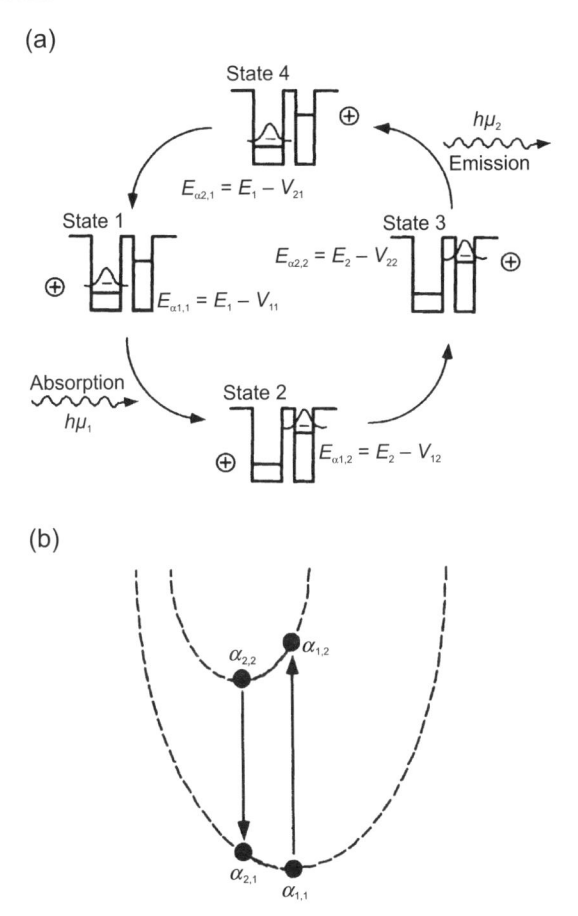

Fig. 8.22 (a) Absorption and emission processes in an asymmetric pair of coupled quantum dots. (b) Representation of electronic states in terms of the image charge distribution function. (After Lee and Khurgin, 1996)

threshold by several orders of magnitude compared to QW devices, even when nonuniformity typically found in current dot arrays is taken into account. However, the rate of multiphonon processes was treated as negligible (except in narrow energy bands; see Inoshita and Sakaki, 1992), an assumption not supported by experiments (Heitz et al., 1996a) (see Section 6.7.5.1).

Lee and Khurgin (1996) proposed the use of an asymmetrically coupled pair of quantum dots for an inter-sublevel laser. The laser operates between the $|1\rangle + |2\rangle$ and $|1\rangle - |2\rangle$ electron states of the coupled system, where $|1\rangle$ and $|2\rangle$ denote the uncoupled eigenstates of the two dots. The main idea is that a redistribution of image charges below and on top of the quantum dot pair leads to a relative shift of absorption and emission energies (Fig. 8.22). The purely dielectric effect for

AlAs/GaAs yields a shift of about 2 meV, which can be enhanced to 5 meV using high-doped or metallic layers close to the dots.

Another scheme to separate emission and absorption based on the quantum confined Stark effect in biased quantum dots is reported in Yamanishi *et al.* (1990) and Yamanishi and Yamamoto (1991).

The strong enhancement of spontaneous far infrared emission in a near infrared quantum dot laser ($\lambda = 940$ nm) as compared to a quantum cell laser has been reported by Vorob'ev *et al.* (1998).

References

Abstreiter, G., P. Schittenhelm, C. Engel, E. Silveira, A. Zrenner, D. Meertens, and W. Jäger (1996) *Semicond. Sci. Technol.* **11**, 1521.
Adler, F., M. Geiger, A. Bauknecht, F. Scholz, H. Schweizer, M. H. Pilkuhn, B. Ohnesorge, and A. Forchel (1996) *J. Appl. Phys.* **80**, 4019.
Adolph, B., S. Glutsch, and F. Bechstedt (1993) *Phys. Rev. B* **48**, 15077.
Agrawal, G. P. (1990) *Appl. Phys. Lett.* **57**, 1.
Aigouy, L., T. Holden, F. H. Pollak, N. N. Ledentsov, V. M. Ustinov, P. S. Kop'ev, and D. Bimberg (1997a) *Appl. Phys. Lett.* **70**, 3329.
Aigouy, L., T. Holden, F. H. Pollak, N. N. Ledentsov, V. M. Ustinov, P. S. Kop'ev, and D. Bimberg (1997b) In Proceedings of the 4th *Int. Symp. on Quantum Confinement*, The Electrochemical Society, Pennington, USA, Vol. 97–11, p. 146.
Akimoto, O and H. Hasegawa (1967) *J. Phys. Soc. Jap.* **22**, 181.
Alerhand, O. L., D. Vanderbilt, R. D. Meade and J. D. Joannopoulos (1988) *Phys. Rev. Lett.* **61**, 1973.
Alferov, Zh. I. (1967) *Fiz. Tekh. Poluprovodn.* **1**, 436 (*Sov. Phys. Semicond.* **1**, 358).
Alferov, Zh. I. (1998) *Semicond.* **32**, 1.
Alferov, Zh. I. and R. F. Kazarinov (1963) Semiconductor laser with electric pumping. Author's Certificate N181737.
Alferov, Zh. I., Yu. V. Zhiljaev, and Yu. V. Shmartsev (1971) *Fiz. Tekh. Polyprovodn.* **5**, 196 (*Sov. Phys. Semicond.* **5**, 174).
Alferov, Zh. I., A. Yu. Egorov, A. E. Zhukov, S. V. Ivanov, P. S. Kop'ev, N. N. Ledentsov, B. Ya. Mel'tser, and V. M. Ustinov (1992) *Fiz. Tekh. Poluprovodn.* **26**, 1715 (*Sov. Phys. Semicond.* **26**, 959).
Alferov, Zh. I., S. V. Ivanov, P. S. Kop'ev, A. V. Lebedev, N. N. Ledentsov, M. V. Maximov, I. V. Sedova, T. V. Shubina, and A. A. Toropov (1994) *Superlattices and Microstructs* **15**, 65.
Alferov, Zh. I., N. Yu. Gordeev, S. V. Zaitsev, P. S. Kop'ev, I. V. Kochnev, V. V. Komin, I. L. Krestnikov, N. N. Ledentsov, A. V. Lunev, M. V. Maximov, S. S. Ruvimov, A. V. Sakharov, A. F. Tsapul'nikov, Yu. M. Shernyakov, and D. Bimberg (1996) *Semicond.* **30**, 197.
Anand, S., N. Carlsson, M.-E. Pistol, L. Samuelson, and W. Seifert (1995) *Appl. Phys. Lett.* **67**, 3016.
Ando, T., A. B. Fowler, and F. Stern (1982) *Rev. Mod. Phys.* **54**, 437.
Andreani, L. C., F. Tassone, and F. Bassani (1991) *Solid State Commun.* **77**, 641.
Andreev, A. F. (1980) *Pis'ma Zh. Eksp. Teor. Fiz.* **32**, 654 (*JETP Lett.* **32**, 640).
Andreev, A. F. (1981) *Zh. Eksp. Teor. Fiz.* **80**, 2042 (*Sov. Phys. JETP* **53**, 1063).
Andreev, A. F. and Yu. A. Kosevich (1981) *Zh. Eksp. Teor. Fiz.* **81**, 1435 (*Sov. Phys. JETP* **54**, 761).
Andrews, S. R., H. Arnot, R. K. Rees, T. M. Kerr, and S. P. Beaumont (1990) *J. Appl. Phys.* **67**, 3472.
Androussi, Y., A. Lefebvre, B. Courboulès, N. Grandjean, J. Massies, T. Bouhacina, and J. P. Aimé (1994) *Appl. Phys. Lett.* **65**, 1162.
Aoki, K., E. Anastassakis, and M. Cardona (1984) *Phys. Rev. B* **30**, 681.
Arakawa, Y. (1996) In Proceedings of 23rd International Conference on *The Physics of Semiconductors*, Berlin, Germany, M. Scheffler and R. Zimmermann (eds.), World Scientific, Singapore, p. 1349.
Arakawa, Y. and H. Sakaki (1982) *Appl. Phys. Lett.* **40**, 939.
Arakawa, Y. and A. Yariv (1985) *IEEE J. Quantum Electron.* **QE-21**, 1666.

Arakawa, Y. and A. Yariv (1986) *IEEE J. Quantum Electron.* **QE-22**, 1887.
Arita, M., A. Avramescu, K. Uesugi, I. Suemune, T. Numai, H. Machida, and N. Shimoyama (1997) *Jpn J. Appl. Phys.* **36**, 4097.
Asada, M., A. Kameyama, and Y. Suematsu (1984) *IEEE J. Quantum Electron.* **QE-20**, 745.
Asada, M., Y. Miyamoto, and Y. Suematsu (1986) *IEEE J. Quantum Electron.* **QE-22**, 1915.
Asaro, R. J. and W. A. Tiller (1982) *Metall. Trans.* **3**, 1789.
Aspnes, D. E. (1985) *J. Vac. Sci. Technol. B* **3**, 1498.
Aspnes, D. E. and M. Cardona (1978) *Phys. Rev. B* **17**, 726.
Asryan, L. V. and R. A. Suris (1996) *Semicond. Sci. Technol.* **11**, 1.
Bahder, Th. B. (1990) *Phys. Rev.* **41**, 11992.
Baldereschi, A. and N. O. Lipari (1973) *Phys. Rev. B* **8**, 2697.
Bányai, L. (1989) *Phys. Rev. B* **39**, 8022.
Bányai, L. and S. W. Koch (1993) *Semiconductor Quantum Dots*, World Scientific Series on Atomic, Molecular and Optical Properties, Vol. 2, World Scientific, Singapore.
Bányai, L., P. Gilliot, Y. Z. Hu, and S. W. Koch (1992) *Phys. Rev. B* **45**, 14136.
Barabási, A.-L. (1997) *Appl. Phys. Lett.* **1997**, 2565.
Barenco, A. and M. A. Dupertuis (1995) *Phys. Rev. B* **52**, 2766.
Bartels, W. J. and W. Nijman (1978) *J. Cryst. Growth* **44**, 518.
Baski, A. A. and L. J. Whitman (1995) *Phys. Rev. Lett.* **74**, 956.
Bastard, G. (1988) *Wave Mechanics Applied to Semiconductor Heterostructures*, Les éditions de Physique, Les Ulis.
Bauer, E. (1958) *Z. Kristallogr.* **110**, 372.
Bawendi, M. G., W. L. Wilson, L. Rothberg, P. J. Carrol, T. M. Jedju, M. L. Steigerwald, and L. E. Brus (1990) *Phys. Rev. Lett.* **65**, 1623.
Bayer, M., T. Gutbrod, A. Forchel, V. D. Kulakovskii, A. Gorbunov, M. Michel, R. Steffen and K. H. Wang (1998) *Phys. Rev. B* **58**, 5740.
Bayer, M., V. B. Timofeev, T. Gutbrod, A. Forchel, R. Steffen, and J. Oshinowo (1995) *Phys. Rev. B* **52**, R11623.
Bayer, M., O. Schilling, A. Forchel, T. L. Reinecke, P. A. Knipp, Ph. Pagnod-Rossiaux, and L. Goldstein (1996) *Phys. Rev. B* **53**, 15810.
Beaumount, S. P. (1991) In *Low-Dimensional Structures in Semiconductors*, A. R. Peaker and H. G. Grimmeiss (eds.), Plenum Press, New York, p. 109.
Beenakker, C. W. J. (1997) *Rev. Mod. Phys.* **69**, 731.
Belousov, M. V., N. N. Ledentsov, M. V. Maximov, P. D. Wang, I. N. Yassievitch, N. N. Faleev, I. A. Kozin, V. M. Ustinov, P. S. Kop'ev, and C. M. Sotomayor Torres (1995) Phys. Rev. B **51**, 14346.
Benisty, H., C. M. Sotomayor-Torrès, and C. Weisbuch (1991) *Phys. Rev. B* **44**, 10945.
Benjamin, S. C. and N. F. Johnson (1995) *Phys. Rev. B* **51**, 14733.
Bennett, B. R., R. Magno, and B. V. Shanabrook (1996) *Appl. Phys. Lett.* **68**, 505.
Berendschet, T. T. J. M., H. A. J. M. Reinen, H. A. Bluyssen, C. Harder, and H. P. Meier (1989) *Appl. Phys. Lett.* **54**, 1827.
Bernard, J. and A. Zunger (1994) *Appl. Phys. Lett.* **65**, 165.
Berryman, K. W., S. A. Lyon, and M. Segev (1997) *Appl. Phys. Lett.* **70**, 1861.
Biegelsen, D. K., F. A. Ponce, A. J. Smith, and J. C. Tramontana (1987) *J. Appl. Phys.* **61**, 1856.
Bimberg, D. (1994) QUEST Workshop on *Quantum Structures*, Colloquium Abstracts, Santa Barbara, p. 15.
Bimberg, D. and H. J. Queisser (1972) In Proceedings of 11th International Conference on *The Physics of Semiconductors*, Polish Scientific Publishers, Warsaw, p. 157.
Bimberg, D., T. Wolf, and J. Böhrer (1991) In *Advances in Nonradiative Processes in Solids*, B. di Bartolo (ed.), Plenum Press, New York, p. 577.
Bimberg, D., F. Heinrichsdorff, R. K. Bauer, D. Gerthsen, D. Stenkamp, D. E. Mars, and J. N. Miller (1992) *J. Vac. Sci. Technol. B* **10**, 1793.

Bimberg, D., M. Grundmann, N. N. Ledentsov, S. S. Ruvimov, P. Werner, U. Richter, J. Heydenreich, V. M. Ustinov, P. S. Kop'ev, and Zh. I. Alferov (1995) *Thin Solid Films* **267**, 32.
Bimberg, D., N. N. Ledentsov, M. Grundmann, N. Kirstaedter, O. G. Schmidt, M.-H. Mao, V. M. Ustinov, A. Yu. Egorov, A. E. Zhukov, P. S. Kop'ev, Zh. I. Alferov, S. S. Ruvimov, U. Gösele, and J. Heydenreich (1996) *Phys. Stat. Sol. (b)* **194**, 159.
Bimberg, D., N. Kirstaedter, N. N. Ledentsov, Zh. I. Alferov, P. S. Kop'ev, and V. M. Ustinov (1997) *IEEE J. Selected Topics in Quantum Electronics* **3**, 1.
Bir, G. L. and G. E. Pikus (1974) *Symmetry and Strain-induced Effects in Semiconductors*, John Wiley, New York.
Bloch, F. (1928) *Z. Physik* **52**, 555.
Blood, P. and J. W. Orton (1992) *The Electrical Characterization of Semiconductors: Majority Carriers and Electron States*, Academic, London.
Bockelmann, U., and G. Bastard (1990) *Phys. Rev. B* **42**, 8947.
Bockelmann, U. (1993) *Phys. Rev.* **48**, 17637.
Bockelmann, U. and T. Egeler (1992) *Phys. Rev. B* **46**, 15574.
Bockelmann, U., Ph. Roussignol, A. Filoramo, W. Heller, G. Abstreiter, K. Brunner, G. Böhm, and G. Weimann (1996) *Phys. Rev. Lett.* **76**, 3623.
Bressler-Hill, V., A. Lorke, S. Varma, K. Pond, P. M. Petroff, and W. H. Weinberg (1994) *Phys. Rev. B* **50**, 8479.
Bressler-Hill, V., S. Varma, A. Lorke, B. Z. Nosho, P. M. Petroff, and W. H. Weinberg (1995) *Phys. Rev. Lett.*, **74**, 3209.
Bruinsma, R. and A. Zangwill (1986) *J. Phys.* **47**, 2055.
Brunkov, P. N., N. N. Faleev, Yu. G. Musikhin, A. A. Suvorova, V. M. Ustinov, A. E. Zhukov, A. Yu. Egorov, V. M. Maximov, A. F. Tsatsul'nikov, P. S. Kop'ev, and S. G. Konnikov (1996a) In Proceedings of 23rd International Conference on *The Physics of Semiconductors*, Berlin, Germany, M. Scheffler and R. Zimmermann (eds.), World Scientific, Singapore, p. 1361.
Brunkov, P. N., S. G. Konnikov, V. M. Ustinov, A. E. Zhukov, A. Yu. Egorov, M. V. Maximov, N. N. Ledentsov, and P. S. Kop'ev (1996b) *Semicond.* **30**, 492.
Brunner, K., J. Zhu, G. Abstreiter, O. Kienzle, and F. Ernst (1998) In *Proceedings 24th International Conference on the Physics of Semiconductors*, Jerusalem, Israel, D. Gershoni (ed.), World Scientific, Singapore.
Brunner, T. (1997) *Technical Digest of 1997 International Electron Devices Meeting*, Washington, IEEE Catalog 97CH36103, IEEE, Piscataway, p. 9.
Brunner, K., U. Bockelmann, G. Abstreiter, M. Walther, G. Böhm, G. Tränkle, and G. Weimann (1992) *Phys. Rev. Lett.* **69**, 3216.
Brunner, K., G. Abstreiter, G. Böhm, G. Tränkle, and G. Weimann (1994a) *Appl. Phys. Lett.* **64**, 3320.
Brunner, K., G. Abstreiter, G. Böhm, G. Tränkle, and G. Weimann (1994b) *Phys. Rev. Lett.* **73**, 1138.
Brus, L. E. (1984) *J. Chem. Phys.* **80**, 4403.
Bryant, G. W. (1988) *Phys. Rev. B* **37**, 8763.
Bryant, G. W. (1992) *Phys. Rev. B* **46**, 1893.
Bryant, G. W. (1993a) *Phys. Rev. B* **47**, 1683.
Bryant, G. W. (1993b) *Phys. Rev. B* **48**, 8024.
Bychkov, V. A. and E. I. Rashba (1984) *J. Phys. C*, **17**, 6039.
Campbell, J. C., D. L. Huffaker, H. Deng, D. G. Deppe (1997) *Electron. Lett.* **33**, 337.
Cardona, M. (1963) *J. Phys. Chem. Solids* **24**, 1543.
Caridi, E. A., T. Y. Chang, K. W. Goossen, and L. F. Eastman (1990) *Appl. Phys. Lett.* **56**, 659.
Carlsson, N., W. Seifert, A. Petersson, P. Castrillo, M. E. Pistol, and L. Samuelson (1994) *Appl. Phys. Lett.* **65**, 3093.
Casey, H. C. and M. B. Panish (1978) *Heterostructure Lasers*, Part A, Academic, New York.

Castrillo, P., D. Hessmann, M.-E. Pistol, S. Anand, N. Carlsson, W. Seifert, and L. Samuelson (1995) *Appl. Phys. Lett.* **67**, 1905.
Cerdeira, F., C. J. Buchenauer, F. H. Pollak, and M. Cardona (1972) *Phys. Rev. B* **5**, 580.
Cerva, H. and H. Oppolzer (1990) *Prog. Crystal Growth and Charact.* **20**, 231.
Chakraborty, T., V. Halonen, and P. Pietiläinen (1991) *Phys. Rev. B* **43**, 14289.
Chand, N., E. E. Becker, J. P. van der Ziel, S. N. G. Chu, and N. K. Dutta (1991) *Appl. Phys. Lett.* **1991**, 1704.
Chang, L. L., L. Esaki, and R. Tsu (1974) *Appl. Phys. Lett.* **24**, 593.
Chaudhari, P. (1969) *IBM J. Res. Develop.* **13**, 197.
Chen, M. and W. Porod (1995) *J. Appl. Phys.* **78**, 1050.
Chen, Y. and J. Washburn (1996) *Phys. Rev. Lett.* **77**, 4046.
Chen, A.-B., A. Sher, and W. T. Yost (1992) *Semicond. Semimetals* **37**, 1.
Chen, K. M., D. E. Jesson, S. J. Pennycook, T. Thundat, and R. J. Warmack (1995) *Proc. Mater. Res. Soc. Symp.* **399**, 271.
Chengguan, G., R. Wenying, and L. Youyang (1996) *Phys. Rev. B* **53**, 10820.
Chepic, D. I., Al. L. Efros, A. I. Ekimov, M. G. Ivanov, V. A. Kharchenko, I. A. Kudriavtsev, and T. V. Yazeva (1990) *J. Luminescence* **47**, 113.
Chernov, A. A. (1961) *Uspekhi Fiz. Nauk* **73**, 277 (*Sov. Phys. Uspekhi* **4**, 116).
Chernov, A. A. (1984) *Modern Crystallography*, Vol. III, Springer-Verlag, Berlin.
Chiba, Y. and S. Ohnishi (1988) *Phys. Rev. B* **38**, 12988.
Chiba, Y. and S. Ohnishi (1989) *Superlattices and Microstructs* **6**, 23.
Chidley, E. T. R., S. K. Haywood, R. E. Mallard, N. J. Mason, R. J. Nicholas, P. J. Walker, and R. J. Warburton (1989) *Appl. Phys. Lett.* **54**, 1241.
Cho, A. Y. and J. R. Arthur Jr (1975) *Prog. in Solid State Chem.* **10**, 157.
Christen, J. and D. Bimberg (1990) *Phys. Rev. B* **42**, 7213.
Christen, J., M. Krahl, and D. Bimberg (1990) *Superlattices and Microstructs* **7**, 1.
Christensen, N. E. (1988) *Phys. Rev. B* **38**, 12687.
Christiansen, S., M. Albrecht, H. P. Strunk, and H. J. Maier (1994) *Appl. Phys. Lett.* **64**, 3617.
Chuang, S. L. (1995) *Physics of Optoelectronic Devices* John Wiley, New York.
Chun, Y. J., S. Nakajima, and M. Kawabe (1996) *Jpn. J. Appl. Phys. 2, Lett.* **35**, L 1075.
Cibert, J., P. M. Petroff, G. J. Dolan, S. J. Pearton, A. C. Gossard, and J. H. English (1986) *Appl. Phys. Lett.* **49**, 1275.
Cirlin, G. E., G. M. Guryanov, A. O. Golubok, S. Ya. Tipissev, N. N. Ledentsov, P. S. Kop'ev, M. Grundmann, and D. Bimberg (1995) *Appl. Phys. Lett.* **67**, 97.
Cirlin, G. E., V. G. Dubrovskii, V. N. Petrov, N. K. Polyakov, N. P. Korneeva, V. N. Demidov, A. O. Golubok, S. A. Masalov, D. V. Kurochkin, O. M. Gorbenko, N. I. Komyak, V. M. Ustinov, A. F. Tsatsul'nikov, Zh. I. Alferov, N. N. Ledentsov, M. Grundmann, and D. Bimberg (1998) *Semicond. Sci. Technol.* **13**, October.
Citrin, D. S. (1993) *Phys. Rev. B* **47**, 3832.
Clausen, E. M., H. G. Craighead, J. M. Worlock, J. P. Harbison, L. M. Schiavone, L. Florez, and B. van der Gaag (1989) *Appl. Phys. Lett.* **55**, 1427.
Colvard, C., R. Merlin, M. V. Klein, and A. C. Gossard (1980) *Phys. Rev. Lett.* **43**, 298.
Craighead, H. G., R. E. Howard, L. D. Jackel, and P. M. Mankievich (1983) *Appl. Phys. Lett.* **42**, 38.
Cullis, A. G. (1996) *MRS Bull.* **21**, 31.
Cusack, M. A., P. R. Briddon, and M. Jaros (1996) *Phys. Rev. B* **54**, R2300.
Dabrowski, J., E. Pehlke, and M. Scheffler (1994) *Phys. Rev. B* **49**, 4790.
Dallesasse, J. M., N. Holonyak Jr, A. R. Sugg, T. A. Richard, and N. El-Zein (1990) *Appl. Phys. Lett.* **57**, 2844.
Darhuber, A. A., V. Holy, J. Stangl, G. Bauer, A. Krost, F. Heinrichsdorff, M. Grundmann, D. Bimberg, V. M. Ustinov, P. S. Kop'ev, A. O. Kosogov, and P. Werner (1997a) *Appl. Phys. Lett.* **70**, 955.

Darhuber, A. A., P. Schittenhelm, V. Holý, J. Stangl, G. Bauer, and G. Abstreiter (1997b) *Phys. Rev. B* **55**, 15652.
Darwin, C. G. (1930) *Proc. Cambridge Phil. Soc.* **27**, 86.
Daudin, B., F. Widmann, G. Feuillet, Y. Samson, M. Arlery, and J. L. Rouvière (1997) *Phys. Rev. B* **56**, R7069.
Davies, R. and H. Hosack (1963) *J. Appl. Phys.* **33**, 864.
Dekel, E., D. Gershoni, E. Ehrenfreund, D. Spektor, J. M. Garcia, and P. M. Petroff (1998) *Phys. Rev. Lett.* **80**, 4991.
de la Cruz, R. M., S. W. Teitsworth, and M. A. Stroscio (1995) *Phys. Rev. B* **52**, 1489.
Demel, T., D. Heitmann, P. Grambow, and K. Ploog (1990) *Phys. Rev. Lett.* **64**, 788.
de Oliveira, C. R. M., A. M. de Paula, F. O. Plentz Filho, J. A. Medeiros Neto, L. C. Barbosa, O. L. Alves, E. A. Menezes, J. M. M. Rios, H. L. Fragnito, C. H. Brito Cruz, and C. L. Cesar (1995) *Appl. Phys. Lett.* **66**, 439.
Deppe, D. G. and N. Holonyak Jr (1988) *J. Appl. Phys.* **64**, R93.
Deveaud, B., F. Clérot, N. Roy, K. Satzke, B. Sermage, and D. S. Katzer (1991) *Phys. Rev. Lett.* **67**, 2355.
Dingle, R. B. (1952) *Proc. Roy. Soc. Lond. A* **211**, 500.
Dingle, R., W. Wiegmann, and C. H. Henry, *Phys. Rev. Lett.* **33**, 827.
Dingle, R., H. L. Störmer, A. C. Gossard, and W. Wiegmann (1978) *Appl. Phys. Lett.* **33**, 665.
Dobbs, H. T., D. D. Vvedensky, A. Zangwill, J. Johansson, N. Carlsson, and W. Seifert (1997) *Phys. Rev. Lett.* **79**, 897.
Downes, J. R., D. A. Faux, and E. P. O'Reilly (1997) *J. Appl. Phys.* **82**, 3754.
Drexler, H., D. Leonard, W. Hansen, J. P. Kotthaus, and P. M. Petroff (1994) *Phys. Rev. Lett.* **73**, 2252.
Eaglesham, D. J. and M. Cerullo (1990) *Phys. Rev. Lett.* **64**, 1943.
Eaglesham, D. J., A. E. White, L. C. Feldman, N. Moriya, and D. C. Jacobson (1993) *Phys. Rev. Lett.* **70**, 1643.
Efros, Al. L. and A. L. Efros (1982) *Sov. Phys. Semicond.* **16**, 772.
Efros, Al. L. and A. V. Rodina (1993) *Phys. Rev. B* **47**, 10005.
Efros, Al. L., V. A. Kharchenko, and M. Rosen (1995) *Solid State Commun.* **93**, 281.
Ehrenreich, H. (1961) *J. Appl. Phys.* **32**, 2155.
Ejeckam, F. E., Y. H. Lo, S. Subramian, H. Q. Hou, and B. E. Hammons (1997) *Appl. Phys. Lett.* **70**, 1685.
Ekimov, A. I. and A. A. Onushenko (1984) *JETP Lett.*, **40**, 1137.
Ekimov, A. I., F. Hache, M. C. Schanne-Klein, D. Ricard, C. Flytzanis, I. A. Kudryatsev, T. V. Yazeva, A. Rodina, and Al. L. Efros (1993) *J. Opt. Soc. Am. B* **10**, 100.
Enders, P., A. Bärwolff, M. Woerner, and D. Suisky (1995) *Phys. Rev. B* **51**, 16695.
Esaki, L. and R. Tsu (1970) *IBM J. Res. Develop.* **14**, 61.
Esch, V., B. Fluegel, G. Khitrova, H. M. Gibbs, X. Jiajin, K. Kang, S. W. Koch, L. C. Liu, S. H. Risbud, and N. Peyghambarian (1990) *Phys. Rev. B* **42**, 7450.
Eshelby, J. D. (1957) *Proc. Roy. Soc. Lond. A* **241**, 376.
Fafard, S., D. Leonard, J. L. Merz, and P. M. Petroff (1994) *Appl. Phys. Lett.* **65**, 1388.
Fafard, S., R. Leon, D. Leonard, J. L. Merz, and P. M. Petroff (1995) *Phys. Rev. B* **52**, 5752.
Fafard, S., Z. Wasilewski, J. McCaffrey, S. Raymond, and S. Charbonneau (1996) *Appl. Phys. Lett.* **68**, 991.
Faist, J., F. Capasso, D. L. Sivco, C. Sirtori, A. L. Hutchinson, and A. Y. Cho (1994a) *Science* **264**, 553.
Faist, J., F. Capasso, D. L. Sivco, C. Sirtori, A. L. Hutchinson, and A. Y. Cho (1994b) *Electron. Lett.* **30**, 829, 865.
Feldmann, J., G. Peter, E. O. Göbel, P. Dawson, K. Moore, C. Foxon, and R. J. Elliott (1987) *Phys. Rev. Lett.* **59**, 2337.
Ferrer, J. C., F. Peiró, A. Cornet, J. R. Morante, T. Uztmeier, G. Armelles, and F. Briones (1996) *Appl. Phys. Lett.* **69**, 3887.

Feuillet, G., B. Daudin, F. Widmann, J. L. Rouvière, M. Arlery (1997) In Proceedings of 2nd International Conference on *Nitride Semiconductors (ICNS'97)*, Tokushima, Japan, p. 498.
Filatov, O. N. and I. A. Karpovich (1968) *Fiz. I Tekh. Poluprovodn. (Sov. Phys. Semicond.)* **10**, 9.
Finley, J. J., M. Skalitz, D. Heinrich, M. Arzberger, A. Zrenner, G. Böhm, and G. Abstreiter (1998a) In *Proceedings 24th International Conference on The Physics of Semiconductors*, Jerusalem, Israel, D. Gershoni (ed.), World Scientific, Singapore.
Fiorentini, V., M. Methfessel, and M. Scheffler (1993) *Phys. Rev. Lett.* **71**, 1051.
Flack. F., N. Samarth, V. Nikitin, P. A. Crowell, J. Shi, J. Levy, and D. D. Awschalom (1996) *Phys. Rev. B* **54**, R17312.
Fock, V. (1928) *Z. Phys.* **47**, 446.
Forchel, A., H. Leier, B. E. Maile, and R. Germann (1988) *Festkörperprobleme (Advances in Solid State Physics)* **28**, 99.
Forchel, A., R. Steffen, T. Koch, M. Michel, M. Albrecht, and T. L. Reinecke (1996a) *Semicond. Sci. Technol.* **11**, 1529.
Forchel, A., R. Steffen, M. Michel, A. Pecher, and T. L. Reinecke (1996b) In Proceedings of 23rd International Conference on *The Physics of Semiconductors*, Berlin, Germany, M. Scheffler and R. Zimmerman (eds.), World Scientific, Singapore, p. 1285.
Frank, F. C. and J. H. van der Merwe (1949) *Proc. Roy. Soc. Lond. A* **198**, 205.
Fricke, M., A. Lorke, J. P. Kotthaus, G. Medeiros-Ribeiro, and P. M. Petroff (1996) *Europhys. Lett.* **36**, 197.
Fu, H., L.-W. Wang, and A. Zunger (1997) *Appl. Phys. Lett.* **71**, 3433.
Fujita, S., S. Maruno, H. Watanabe, Y. Kusumi, and M. Ichikawa (1995) *Appl. Phys. Lett.* **66**, 2754.
Fukuda, M., K. Nakagawa, S. Miyazaki, and M. Hirose (1997) *Appl. Phys. Lett.* **70**, 2291.
Fukui, T., S. Ando, Y. Tokura, and T. Toriyama (1991) *Appl. Phys. Lett.* **58**, 2018.
Fukuyama, H. and T. Ando (1992) *Transport Phenomena in Mesoscopic Systems*, Springer, Berlin.
Gammon, D., E. S. Snow, and D. S. Katzer (1995) *Appl. Phys. Lett.* **67**, 2391.
Gammon, D., E. S. Snow, B. V. Shanabrook, D. S. Katzer, and D. Park (1996a) *Phys. Rev. Lett.* **76**, 3005.
Gammon, D., E. S. Snow, B. V. Shanabrook, D. S. Katzer, and D. Park (1996b) *Science* **273**, 87.
Garcia, A. and J. E. Northrup (1993) *Phys. Rev. B* **48**, 17350.
Gavrilovic, P., D. G. Deppe, K. Meehan, N. Holonyak Jr, J. J. Coleman, and R. D. Burnham (1985) *Appl. Phys. Lett.* **47**, 130.
Geels, R. S. and L. A. Coldren (1990) *Appl. Phys. Lett.* **57**, 1605.
Geiger, M., A. Bauknecht, F. Adler, H. Schweizer, and F. Scholz (1997) *J. Crystal Growth* **170**, 558.
Georgsson, K., N. Carlsson, L. Samuelson, W. Seifert, and L. R. Wallenberg (1995) *Appl. Phys. Lett.* **67**, 2981.
Gibbs, J. W. (1928) *Collected Works*, Vol. 1, *Thermodynamics*, Longmans, London.
Glaser, E. R., B. R. Bennett, B. V. Shanabrook, and R. Magno (1996) *Appl. Phys. Lett.* **68**, 3614.
Le Goff, S. and B. Stébé (1992) *Solid State Commun.* **83**, 555.
Goldstein, L., F. Glas, J. Y. Marzin, M. N. Charasse, and G. Le Roux (1985) *Appl. Phys. Lett.* **47**, 1099.
Golubok, A. O., G. M. Guryanov, V. N. Petrov, Yu. B. Samsonenko, S. Ya. Tipisev, G. E. Cirlin, and N. N. Ledentsov (1994) *Fiz. Tekh. Poluprovodn.* **28**, 516 (*Semiconductors* **28**, 317).
Gossard, A. C. (1981) *Treat. Mater. Sci. Technol.* **24**, 13.
Gotoh, H., H. Ando, and H. Kanbe (1996) *Appl. Phys. Lett.* **68**, 2132.

Gotoh, H., H. Ando, and T. Takagahara (1997) *J. Appl. Phys.* **81**, 1785.
Grabert, H. and M. H. Devoret (eds.) (1992) *Single Charge Tunneling—Coulomb Blockade Phenomena in Nanostructures*, Plenum Press, New York.
Grabert, H. and H. Horner (eds.) (1991) *Z. Phys. B* **85**, 317 (Special Issue on *Single Charge Tunneling*).
Grabmeier, A., A. Hangleiter, G. Fuchs, J. E. A. Whiteaway, and R. W. Glew (1991) *Appl. Phys. Lett.* **59**, 3024.
Grandjean, N., J. Massies, C. Delamarre, L. P. Wang, A. Dubon, and J. Y. Laval (1993) *Appl. Phys. Lett.* **63**, 66.
Griesinger, U. A., H. Schweizer, S.. Kronmüller, M. Geiger, D. Ottenwälder, F. Stolz, and M. Pilkuhn (1996) *IEEE Photon. Technol. Lett.* **8**, 587.
Grinfield, M. A. (1986) *Dokl. Akad. Nauk SSSR* **290**, 1358 (*Sov. Phys. Dokl.* **31**, 831).
Grosse, S., J. H. H. Sandmann, G. von Plessen, J. Feldmann, H. Lipsanen, M. Sopanen, J. Tulkki, and J. Ahopelto (1996) In Proceedings of 23rd International Conference on *The Physics of Semiconductors*, Berlin, Germany, M. Scheffler and R. Zimmermann (eds.), World Scientific, Singapore, p. 1401.
Grosse, S., J. H. H. Sandmann, G. von Plessen, J. Feldmann, H. Lipsanen, M. Sopanen, J. Tullki, and J. Ahopelto (1997) *Phys. Rev. B* **55**, 4473.
Grundmann, M. (1996) *Festkörperprobleme (Advances in Solid State Physics)* **35**, 123.
Grundmann, M., A. Krost, and D. Bimberg (1991a) *J. Vac. Sci. Technol. B* **9**, 2159.
Grundmann, M., A. Krost, and D. Bimberg (1991b) *Appl. Phys. Lett.* **58**, 284.
Grundmann, M., J. Christen, N. N. Ledentsov, J. Böhrer, D. Bimberg, S. S. Ruvimov, P. Werner, U. Richter, U. Gösele, J. Heydenreich, V. M. Ustinov, A. Yu. Egorov, A. E. Zhukov, P. S. Kop'ev, and Zh. I. Alferov (1995a) *Phys. Rev. Lett.* **74**, 4043.
Grundmann, M., N. N. Ledentsov, J. Christen, J. Böhrer, D. Bimberg, S. S. Ruvimov, P. Werner, U. Richter, U. Gösele, J. Heydenreich, V. M. Ustinov, A. Yu. Egorov, A. E. Zhukov, P. S. Kop'ev, and Zh. I. Alferov (1995b) *Phys. Stat. Sol. (b)* **188**, 249.
Grundmann, M., O. Stier, and D. Bimberg (1995c) *Phys. Rev. B* **52**, 11969.
Grundmann, M., R. Heitz, N. Ledentsov, O. Stier, D. Bimberg, V. M. Ustinov, P. S. Kop'ev, Zh. I. Alferov, S. S. Ruvimov, P. Werner, U. Gösele, and J. Heydenreich (1996a) *Superlattices and Microstructs* **19**, 81.
Grundmann, M., N. N. Ledentsov, R. Heitz, D. Bimberg, V. M. Ustinov, A. Yu. Egorov, M. V. Maximov, P. S. Kop'ev, Zh. I. Alferov, A. O. Kosogov, P. Werner, J. Heydenreich, and U. Gösele (1996b) In Proceedings of 8th International Conference on *Indium Phosphide and Related Materials (IPRM-8)*, IEEE Catalog 96CH35930, Library of Congress 96-75713, p. 738.
Grundmann, M., N. N. Ledentsov, R. Heitz, O. Stier, N. Kirstaedter, D. Bimberg, S. Ruvimov, A. O. Kosogov, P. Werner, J. Heydenreich, U. Gösele, V. M. Ustinov, M. Maximov, A. Yu. Egorov, P. S. Kop'ev, and Zh. I. Alferov (1996c) In Proceedings of 3rd International Symposium on *Quantum Confinement (ECS-188)*, Chicago, Illinois, PV 75-17, The Electrochemical Society, Pennington, p. 80.
Grundmann, M., N. N. Ledentsov, O. Stier, D. Bimberg, V. M. Ustinov, P. S. Kop'ev, and Zh. I. Alferov (1996d) *Appl. Phys. Lett.* **68**, 979.
Grundmann, M., N. N. Ledentsov, O. Stier, J. Böhrer, D. Bimberg, V. M. Ustinov, P. S. Kop'ev, and Zh. I. Alferov (1996e) *Phys. Rev. B* **53**, R10509.
Grundmann, M. and D. Bimberg (1997a) *Phys. Rev. B* **55**, 4054.
Grundmann, M. and D. Bimberg (1997b) *Phys. Rev. B* **55**, 9740.
Grundmann, M. and D. Bimberg (1997c) *Jpn. J. Appl. Phys.* **36**, 4181.
Grundmann, M., R. Heitz, D. Bimberg, J. H. H. Sandmann, and J. Feldman (1997) *Phys. Stat. Sol. (b)* **203**, 121.
Grundmann, M., O. Stier, and D. Bimberg (1998) *Phys. Rev. B* **58** (in press).
Guha, S., A. Madhukar, and K. C. Rajkumar (1990) *Appl. Phys. Lett.* **57**, 2110.

Guryanov, G. M., G. E. Cirlin, V. N. Petrov, N. K. Polyakov, A. O. Golubok, S. Ya. Tipissev, E. P. Misikhina, V. B. Gubanov, Yu. B. Samsonenko, and N. N. Ledentsov (1995a) *Surf. Sci.* **331-333**, 414.

Guryanov, G. M., G. E. Cirlin, V. N. Petrov, Yu. B. Samsonenko, V. B. Gubanov, N. K. Polyakov, A. O. Golubok, S. Ya. Tipisev, E. P. Musikhina, and N. N. Ledentsov (1995b) *Fiz. Tekh. Poluprovodn.* **29**, 1642 (*Semiconductors* **29**, 854).

Guryanov, G. M., G. E. Cirlin, A. O. Golubok, S. Ya. Tipisev, N. N. Ledentsov, V. A. Shchukin, M. Grundmann, and D. Bimberg (1996a) *Surf. Sci.* **352-354**, 646.

Guryanov, G. M., G. E. Cirlin, V. N. Petrov, N. K. Polyakov, A. O. Golubok, S. Ya. Tipisev, V. B. Gubanov, Yu. B. Samsonenko, N. N. Ledentsov, V. A. Shchukin, M. Grundmann, D. Bimberg, and Zh. I. Alferov (1996b) *Surf. Sci.* **352-354**, 651.

Gustafsson, A., M.-E. Pistol, L. Montelius, and L. Samuelson (1998) *J. Appl. Phys.* **84**, 1715.

Hanamura, E. (1988) *Phys. Rev. B* **37**, 1273.

Hansson, P. O., M. Albrecht, W. Dorsch, H. P. Strunk, and E. Bauser (1994) *Phys. Rev. Lett.* **73**, 444.

Harison, I., H. P. Hot, B. Tuck, M. Henini, and O. H. Hughes (1989) *Semicond. Sci. Technol.* **4**, 841.

Harris, L., D. J. Mowbray, M. S. Skolnick, M. Hopkinson, and G. Hill (1998) *Appl. Phys. Lett.* **73**, 969.

Harris Liao, M. C., Y. H. Chang, Y. F. Chen, J. W. Hsu, J. M. Linm, and W. C. Chou (1997) *Appl. Phys. Lett.* **70**, 2256.

Hartmann, A., L. Loubies, F. Reinhardt, and E. Kapon (1997) *Appl. Phys. Lett.* **71**, 1314.

Hatami, F., N. N. Ledentsov, M. Grundmann, J. Böhrer, F. Heinrichsdorff, M. Beer, D. Bimberg, S. S. Ruvimov, P. Werner, U. Gösele, J. Heydenreich, U. Richter, S. V. Ivanov, B. Ya. Meltser, P. S. Kop'ev, and Zh. I. Alferov (1995) *Appl. Phys. Lett.* **67**, 656.

Hatami, F., M. Grundmann, N. N. Ledentsov, R. Heitz, J. Böhrer, D. Bimberg, S. V. Ivanov, B. Ya. Meltser, V. M. Ustinov, P. S. Kop'ev, and Zh. I. Alferov (1998) *Phys. Rev. B* **57**, 4635.

Häusler, K., K. Eberl, F. Noll, and A. Trampert (1996) *Phys. Rev. B* **54**, 4913.

Hegarty, J. and M. D. Sturge (1985) *J. Opt. Soc. Am. B* **2**, 1143.

Heinrichsdorff, F. (1998) PhD thesis, TU Berlin.

Heinrichsdorff, A., A. Krost, M. Grundmann, D. Bimberg, A. O. Kosogov, and P. Werner (1996a) *Appl. Phys. Lett.* **68**, 3284.

Heinrichsdorff, F., A. Krost, M. Grundmann, R. Heitz, D. Bimberg, A. Kosogov, P. Werner, F. Bertram, and J. Christen (1996b) In Proceedings of 23rd International Conference on *The Physics of Semiconductors*, Berlin, Germany, M. Scheffler and R. Zimmermann (eds.), World Scientific, Singapore, p. 1321.

Heinrichsdorff, F., M. Grundmann, O. Stier, A. Krost, and D. Bimberg (1997a) (in press).

Heinrichsdorff, F., A. Krost, D. Bimberg, A. O. Kosogov, and P. Werner (1997b) *Appl. Surf. Sci.* (in press).

Heinrichsdorff, F., A. Krost, M. Grundmann, D. Bimberg, F. Bertram, J. Christen, A. Kosogov, and P. Werner (1997c) *J. Cryst. Growth* **170**, 569.

Heinrichsdorff, F., M.-H. Mao, N. Kirstaedter, A. Krost, D. Bimberg, A. O. Kosogov, and P. Werner (1997d) *Appl. Phys. Lett.* **71**, 22.

Heitz, R., M. Grundmann, N. N. Ledentsov, L. Eckey, M. Veit, D. Bimberg, V. M. Ustinov, A. Yu. Egorov, A. E. Zhukov, P. S. Kop'ev, and Zh. I. Alferov (1996a) *Appl. Phys. Lett.* **68**, 361.

Heitz, R., A. Kalburge, Q. Xie, M. Veit, M. Grundmann, P. Chen, A. Madhukar, and D. Bimberg (1996b) In Proceedings of 23rd International Conference on *The Physics of Semiconductors*, Berlin, Germany, M. Scheffler and R. Zimmermann (eds.), World Scientific, Singapore, p. 1425.

Heitz, R., T. R. Ramachandran, A. Kalburge, Q. Xie, I. Mukhametzhanov, P. Chen, and A. Madhukar (1997a) *Phys. Rev. Lett.* **1997**, 4071.

REFERENCES

Heitz, R., M. Veit, N. N. Ledentsov, A. Hoffmann, D. Bimberg, V. M. Ustinov, P. S. Kop'ev, and Zh. I. Alferov (1997b) *Phys. Rev. B* **56**, 10435.
Heitz, R., A. Kalburge, Q. Xie, M. Grundmann, P. Chen, A. Hoffmann, A. Madhukar, and D. Bimberg (1998) *Phys. Rev. B*, **57**, 9050.
Heller, O., Ph. Lelong, and G. Bastard (1997) *Phys. Rev. B* **56**, 4702.
Heller, W. and U. Bockelmann (1997) *Phys. Rev. B* **55**, R4871.
Heller, W., U. Bockelmann, and G. Abstreiter (1998) *Phys. Rev. B* **57**, 6270.
Henneberger, F., J. Puls, A. Schülzgen, V. Jungnickel, and Ch. Spiegelberg (1992) *Festkörperprobleme* (*Advances in Solid State Physics*) **32**, 279.
Herman, M. A. and H. Sitter (1989) *Molecular Beam Epitaxy: Fundamentals and Current Status*, Springer, Berlin.
Herring, C. (1951a) In *The Physics of the Powder Metallurgy*, W. E. Kingston (ed.), McGraw-Hill, New York.
Herring, C. (1951b) *Phys. Rev.* **82**, 87.
Hersee, S. D., M. Baldy, P. Assenat, B. de Cremoux, and J. P. Duchemin (1982) *Electron. Lett.* **18**, 870.
Hess, H. F., E. Betzig, T. D. Harris, L. N. Pfeiffer, and K. W. West (1994) *Science* **264**, 1740.
Hessmann, D., P. Castrillo, M.-E. Pistol, C. Pryor, and L. Samuelson (1996) *Appl. Phys. Lett.* **69**, 749.
Hibino, H., T. Fukuda, M. Suzuki, Y. Hommo, T. Sato, M. Iwatsuki, K. Miki, and H. Tokumoto (1993) *Phys. Rev. B* **47**, 13027.
Hirayama, Y., Y. Suzuki, S. Tarucha, and K. Okamoto (1985) *Jpn. J. Appl. Phys.* **24**, 516.
Hirayama, H., K. Matsunaga, M. Asada, and Y. Suematsu (1994) *Electron. Lett.* **30**, 142.
Hirayama, H., S. Tanaka, P. Ramvall, and Y. Aoyagi (1997) In Proceedings of 2nd International Conference on *Nitride Semiconductors (ICNS'97)*, Tokushima, Japan, p. 472.
Hogg, R. A., K. Suzuki, K. Tachibana, L. Finger, K. Hirakawa, and Y. Arakawa (1998) *Appl. Phys. Lett.* **72**, 2856.
Hommel, D., K. Leonardi, H. Heineke, H. Selke, K. Ohkawa, F. Gindele, and U. Woggon (1997) *Phys. Stat. Sol. (b)* **202**, 835.
Hopfield, J. J. (1958) *Phys. Rev.* **112**, 1555.
Horiguchi, N., T. Futatsugi, Y. Nakata, and N. Yokoyama (1997) *Appl. Phys. Lett.* **70**, 2294.
Howard, R. E., L. D. Jackel, and W. J. Skocpol (1985) *Microelectronic Engineering* **3**, 3.
Hu, Y. Z., M. Lindberg, and S. W. Koch (1990) *Phys. Rev. B* **42**, 1713.
Hu, S. Y., B. Young, S. W. Corzine, A. C. Gossard, and L. A. Coldren (1994) *J. Appl. Phys.* **76**, 3932.
Hu, Y. Z., H. Gieen, N. Peyghambarian, and S. W. Koch (1996) *Phys. Rev. B* **53**, 4814.
Huffaker, D. L. and D. G. Deppe (1997) *Appl. Phys. Lett.* **70**, 1781.
Huffaker, D., H. Deng, and D. G. Deppe (1998) *IEEE Photon. Technol. Lett.* **10**, 185.
Huffaker, D. L., O. Baklenov, L. A. Graham, B. G. Sreetman, and D. G. Deppe (1997) *Appl. Phys. Lett.* **70**, 2356.
Hull, R. and A. Fischer-Colbrie (1987) *Appl. Phys. Lett.* **50**, 853.
Hurle, D. T. J. (1979) *J. Phys. Chem. Solids* **40**, 613.
Ibach, H. and H. Lüth (1991) *Solid-State Physics* Springer, Berlin.
IBM (1993) *IBM J. Res. Develop.* **37**, 288 (Special Issue on *X-ray Lithography*).
Ide, T., A. Yamashida, and T. Mizutani (1992) *Phys. Rev. B* **46**, 1950.
Iga, K., F. Koyama, and S. Kinoshita (1988) *IEEE J. Quantum Electron.* **QE-24**, 1845.
Imamura, K., Y. Sugiyama, Y. Nakata, Sh. Muto, and N. Yokoyama (1995) *Jpn. J. Appl. Phys.* **34**, L1445.
Inoshita, T. and H. Sakaki (1992) *Phys. Rev. B* **46**, 7260.
Inoue, K., K. Kimura, K. Maehashi, S. Hasegawa, H. Nakashima, O. Matsuda, and K. Murase (1993) *J. Crystal Growth* **127**, 1041.

Iogansen, L. V. (1963) *Zh. Exp. i Teor. Fiziki* (*Sov. Phys. JETP*) **45**, 207.
Iogansen, L. V. (1964) *Zh. Exp. i Teor. Fiziki* (*Sov. Phys. JETP*) **47**, 270.
Ipatova, I. P., V. G. Malyshkin, and V. A. Shchukin (1993) *J. Appl. Phys.* **74**, 7198.
Ipatova, I. P., V. G. Malyshkin, and V. A. Shchukin (1994) *Phil. Mag. B* **70**, 557.
Ipatova, I. P., V. G. Malyshkin, A. A. Maradudin, V. A. Shchukin, and R. F. Wallis (1996) in Proceedings of International Symposium on *Compound Semiconductors*, St Petersburg (in press).
Ishida, S., Y. Arakawa, and K. Wada (1998) *Appl. Phys. Lett.* **72**, 800.
Itskevich, I. E., T. Ihn, A. Thornton, M. Henini, T. J. Foster, P. Moriarty, A. Nogaret, P. H. Beton, L. Eaves, and P. C. Main (1996) *Phys. Rev. B* **54**, 16401.
Itskevich, I. E., T. Ihn, A. Thornton, M. Henini, H. de Andrade Carmona, L. Eaves, P. C. Main, D. K. Maude, and J.-C. Portal (1997a) *Jpn. J. Appl. Phys.* **36**, 4037.
Itskevich, I.E., M. Henini, H. A. Carmona, L. Eaves, P. C. Main, D. K. Maude, and J. C. Portal (1997b) *Appl. Phys. Lett.* **70**, 505.
Ivanov, S. V., P. S. Kop'ev, and N. N. Ledentsov (1990) *J. Cryst. Growth* **104**, 345.
Ivchenko, E. L. and A. V. Kavokin (1992) *Sov. Phys. Solid State* **34**, 968.
Ivchenko, E. L., A. V. Kavokin, V. P. Kochereshko, P. S. Kop'ev, and N. N. Ledentsov (1992) *Superlattices and Microstructures* **12**, 317.
Jaziri, S. (1994) *Solid State Commun.* **91**, 171.
Jeppesen, S., M. S. Miller, D. Hessman, B. Kowalski, I. Maximov, and L. Samuelson (1996) *Appl. Phys. Lett.* **68**, 2228.
Jesson, D. E., S. J. Pennycook, J.-M. Baribeau, and D. C. Houghton (1993) *Phys. Rev. Lett.* **71**, 1744.
Jesson, D. E., K. M. Chen, and S. J. Pennycook (1996a) *MRS Bull.* **21**, 31.
Jesson, D. E., K. M. Chen, S. J. Pennycook, T. Thundat, and R. J. Warmack (1996b) *Phys. Rev. Lett.* **77**, 1330.
Jiang, H., and J. Singh (1997a) *Phys. Rev. B* **56**, 4696.
Jiang, H., and J. Singh (1998) *IEEE J. Quantum Electronics* **34**, 1188.
Jimenez, J. L., L. R. C. Fonseca, D. J. Brady, J. P. Leburton, D. E. Wohlert, and K. Y. Cheng (1997) *Appl. Phys. Lett.* **71**, 3558.
Johnson, N. F. (1995) *J. Phys.:Condens. Matter* **7**, 965.
Joós, B., T. L. Einstein, and N. C. Bartelt (1991) *Phys. Rev. B* **43**, 8153.
Joyce, B. A., J. H. Neave, P. J. Dobson, and P. K. Larsen (1984) *Phys. Rev. B* **29**, 814.
Jusserand, B. and M. Cardona (1989) *Light Scattering in Solids V: Superlattices and Other Microstructure*, Topics in Applied Physics, Vol. 66, Springer, Berlin, p. 49.
Jusserand, B., D. Paquet, and A. Regreny (1984) *Phys. Rev. B* **30**, 6245.
Kalosha, V. P., G. Ya. Slepyan, S. A. Maksimenko, O. Stier, M. Grundmann, N. N. Ledentsov, and D. Bimberg (1998) In *Proceedings 24th International Conference on The Physics of Semiconductors*, Jerusalem, Israel, D. Gershoni (ed.), World Scientific, Singapore.
Kamada, H., J. Temmyo, M. Notomi, T. Furuta, and T. Tamamura (1996b) *Jpn. J. Appl. Phys.* **36**, 4194.
Kamath, K., P. Bhattacharya, T. Sosnowski, T. Norris, and J. Phillips (1996a) *Electron. Lett.* **32**, 1374.
Kamath, K., N. Chervela, K. K. Linder, T. Sosnowski, H.-T. Jiang, T. Norris, J. Singh, and P. Battacharya (1997a) *Appl. Phys. Lett.* **71**, 927.
Kamath, K., J. Philips, H. Jiang, J. Singh, and P. Bhattacharya (1997b) *Appl. Phys. Lett.* **70**, 2952.
Kaminishi, K. (1987) *Solid State Technol.* **9**, 91.
Kamins, T. I. and R. S. Williams (1997) *Appl. Phys. Lett.* **71**, 1201.
Kamiya, I., D. E. Aspnes, H. Tanaka, L. T. Florez, J. P. Harbison, and R. Bhat (1992) *Phys. Rev. Lett.* **68**, 627.
Kane, E. O. (1957) *Phys. Chem. Solids* **1**, 249.
Kane, E. O. (1985) *Phys. Rev. B* **31**, 7865.

REFERENCES

Kapon, E., M. C. Tamargo, and D. M. Hwang (1987) *Appl. Phys. Lett.* **50**, 347.
Kapon, E., D. M. Hwang, and R. Bhat (1989) *Phys. Rev. Lett.* **63**, 430.
Kapteyn, C. M. A., F. Heinrichsdorff, O. Stier, M. Grundmann, and D. Bimberg (1998) In *Proceedings 24th International Conference on The Physics of Semiconductors*, Jerusalem, Israel, D. Gershoni (ed.), World Scientific, Singapore.
Kash, K., J. M. Worlock, M. D. Sturge, P. Grabbe, J. P. Harbison, A. Scherer, and P. S. D. Lin (1988) *Appl. Phys. Lett.* **53**, 782.
Kash, K., B. P. Van der Gaag, D. D. Mahoney, A. S. Gozdz, L. T. Florez, J. P. Harbison, M. D. Sturge (1991) *Phys. Rev. Lett.* **67**, 1326.
Kastner, M. A. (1993) *Physics Today* **1**, 25.
Kasu, M. and N. Kobayashi (1993) *Appl. Phys. Lett.* **62**, 1262.
Kavokin, A. V. (1994) *Phys. Rev. B* **50**, 8000.
Kayanuma, Y. (1988) *Phys. Rev. B* **38**, 9797.
Kayanuma, Y. (1993) *J. Phys. Soc. Jap.* **62**, 346.
Kazarinov, R. F. and R. A. Suris (1971) *Fiz. I Tekh. Poluprovodn.* (*Sov. Phys. Semicond.*) **5** 797.
Kazarinov, R. F. and R. A. Suris (1972) *Fiz. I Tekh. Poluprovodn.* (*Sov. Phys. Semicond.*) **6**, 148.
Keating, P. N. (1966) *Phys. Rev.* **145**, 637.
Keldysh, L. V. (1962) *Fizika Tverdogo Tela* (*Sov. Phys. Solid State*) **4**, 2265.
Kern, K., H. Niehus, A. Schatz, P. Zeppenfeld, J. George, and G. Comsa (1991) *Phys. Rev. Lett.* **67**, 855.
Khachaturyan, A. G. (1974) *Theory of Phase Transformations and the Structure of Solid Solution* (in Russian), Moscow, Nauka.
Khachaturyan, A. G. (1983) *Theory of Structural Transformations in Solids*, John Wiley, New York.
Kim, J., L.-W. Wang, and A. Zunger (1998a) *Phys. Rev. B* **57**, R9408.
Kim, S., H. Mohseni, M. Erdtmann, E. Michel, C. Jelen, and M. Razeghi (1998b) *Appl. Phys. Lett.* **73**, 963.
Kirstaedter, N., N. N. Ledentsov, M. Grundmann, D. Bimberg, V. M. Ustinov, S. S. Ruvimov, M. V. Maximov, P. S. Kop'ev, Zh. I. Alferov, U. Richter, P. Werner, U. Gösele, and J. Heydenreich (1994) *Electron. Lett.* **30**, 1416.
Kirstaedter, N., O. Schmidt, N. N. Ledentsov, D. Bimberg, V. M. Ustinov, A. Yu. Egorov, A. E. Zhukov, M. V. Maximov, P. S. Kop'ev, and Zh. I Alferov (1996) *Appl. Phys. Lett.* **69**, 1226.
Kitamura, M., M. Nishioka, J. Oshinowo, and Y. Arakawa (1995) *Appl. Phys. Lett.* **66**, 3663.
Kittel, C. (1959) *Introduction to Solid State Physics*, John Wiley, New York.
Kittel, C. (1963) *Quantum Theory of Solids*, John Wiley, New York.
Klitzing, K. v., G. Dorda, and M. Pepper (1980) *Phys. Rev. Lett.* **45**, 494.
Klotzin, D., K. Kamath, K. Vineberg, P. Bhattacharya, R. Murty, and J. Laskar (1998) *IEEE Photon. Technol. Lett.* **10**, 932.
Knipp, P. A. and T. L. Reinecke (1992) *Phys. Rev. B* **46**, 10310.
Knipp, P. A. and T. L. Reinecke (1996) *Solid State Electron.* **40**, 343.
Knox, W. H., C. Hirlimann, D. A. Miller, J. Shah, D. S. Chemla, and C. V. Shank (1986) *Phys. Rev. Lett.* **56**, 1191.
Ko, H.-C., D.-C. Park, Y. Kawakami, S. Fujita, and S. Fujita (1997) *Appl Phys. Lett.* **70**, 3278.
Kobayashi, N. P., T. R. Ramachandran, P. Chen, and A. Madhukar (1996) *Appl. Phys. Lett.* **68**, 3299.
Koch, R., M. Borbonus, O. Haase, and K. H. Rieder (1991) *Phys. Rev. Lett.* **67**, 3416.
Kohn, M., D. Heitman, P. Grambow and K. Ploog (1989) *Phys. Rev. Lett.* **63**, 2124.
Kohn, W. (1961) *Phys. Rev.* **123**, 1242.
Komuro, M., H. Hiroshima, H. Tanoue, and T. Kanayama (1983) *J. Vac. Sci. Technol. B* **4**, 985.
Konkar, A., K. C. Rajkumar, Q. Xie, P. Chen, A. Madhukar, H. T. Lin, and D. H. Rich (1995) *J. Cryst. Growth* **150**, 311.

Konkar, A., A. Madhukar, and P. Chen (1998) *Appl. Phys. Lett.* **72**, 220.
Kop'ev, P. S., N. N. Ledentsov, B. Ya. Meltser, I. N. Uraltsev, Al. L. Efros, and D. R. Yakovlev (1986) in Proceedings of 18th International Conference on *The Physics of Semiconductors*, O. Engström (ed.), World Scientific, Singapore, p. 219.
Kop'ev, P. S. and N. N. Ledentsov (1988) *Soviet Physics—Semiconductors* **22**, 1093.
Kosogov, A. O., P. Werner, U. Gösele, N. N. Ledentsov, D. Bimberg, V. M. Ustinov, A. Yu. Egorov, A. E. Zhukov, P. S. Kop'ev, and Zh. I. Alferov (1996) *Appl. Phys. Lett.* **69**, 3072.
Kouwenhoven, L. P., Ch. M. Marcus, P. L. McEuen, S. Tarucha, R. M. Westervelt, and N. S. Wingreen (1997) In Proceedings of the Advanced Study Institute on *Mesoscopic Electron Transport*, L. L. Sohn and L. P. Kouwenhoven (eds.), Kluwer, Dordrecht.
Král, K., and Z. Khás (1998a) *Phys. Rev. B* **57**, R2061.
Král, K., and Z. Khás (1998b) *Phys. Stat. Sol. (b)* **208**, R5.
Krestnikov, I. L., M. V. Maximov, A. V. Sakharov., P. S. Kop'ev, *et al.* (1998) *J. Crystal Growth* 184.
Krauss, P. R. and S. Y. Chou (1997) *Appl. Phys. Lett.* **71**, 3174.
Kroemer, H. (1963) *Proc. IEEE* **51**, 1782.
Kröger, F. A. (1964) *The Chemistry of Imperfect Crystals*, North-Holland, Amsterdam.
Krost, A., G. Bauer, and J. Woitok (1996a) In *Optical Characterization of Epitaxial Semiconductor Layers*, G. Bauer and W. Richter (eds.), Springer, Berlin, p. 287.
Krost, A., F. Heinrichsdorff, D. Bimberg, A. Darhuber, and G. Bauer (1996b) *Appl. Phys. Lett.* **68**, 785.
Kuhl, J., A. Honold, L. Schultheis, and C. W. Tu (1989) *Festkörperprobleme (Advances in Solid State Physics)* **29**, 157.
Kurtenbach, A., K. Eberl, and T. Shitara (1995) *Appl. Phys. Lett.* **66**, 361.
Kuttler, M., M. Strassburg, V. Türck, R. Heitz, U. W. Pohl, D. Bimberg, E. Kurtz, G. Landwehr, and D. Hommel (1996) *Appl. Phys. Lett.* **69**, 2647.
Kuttler, M., M. Strassburg, O. Stier, U. W. Pohl, D. Bimberg, E. Kurtz, J. Nürnberger, G. Landwehr, M. Behringer, and D. Hommel (1997) *Appl. Phys. Lett.* **71**, 243.
Kwok On, N. and D. Vanderbilt (1995) *Phys. Rev. B* **52**, 2177.
Lacombe, D., A. Ponchet, J.-M. Gérard, and O. Cabrol (1997) *Appl. Phys. Lett.* **70**, 2398.
Laheld, U. E. H., F. B. Pedersen, and P. C. Hemmer (1993) *Phys. Rev. B* **48**, 4659.
Laheld, U. E. H., F. B. Pedersen, and P. C. Hemmer (1995) *Phys. Rev. B* **52**, 2697.
Lampert, M. A. and P. Mark (1970) *Current injection in Solids*, Academic Press, New York.
Landau, L. D. and E. M. Lifshits (1959) *Theory of Elasticity*, Pergamon, New York.
Landau, L. D. and E. M. Lifshits (1960) *Electrodynamics of Continuous Media*, Pergamon, New York.
Lang, D. V. (1974) *J. Appl. Phys.* **45**, 3023.
Larsen, P. K. and P. J. Dobson (eds.) (1988) *Reflection High Energy Electron Diffraction and Reflection Electron Imaging of Surfaces*, NATO ASI Series B: Physics Vol. 188, Plenum Press, New York.
Lebens, J. A., Ch. S. Tsai, K. J. Vahala, and T. F. Kuech (1990) *Appl. Phys. Lett.* **56**, 2642.
Ledentsov, N. N., R. Nötzel, P. S. Kop'ev, and K. Ploog (1992) *Appl. Phys. A* **55**, 533.
Ledentsov, N. N., G. M. Guryanov, G. E. Cirlin, V. N. Petrov, Yu. B. Samsonenko, A. O. Golubok, and S. Ya. Tipisev (1994a) *Fiz. Tekh. Poluprovodn.* **28**, 903 (*Semicond.* **28**, 526).
Ledentsov, N. N., P. D. Wang, C. M. Sotomayor Torres, A. Yu. Egorov, M. V. Maximov, V. M. Ustinov, A. E. Zhukov, and P. S. Kop'ev (1994b) *Phys. Rev. B* **50**, 12171.
Ledentsov, N. N., J. Böhrer, M. Beer, F. Heinrichsdorff, M. Grundmann, D. Bimberg, S. V. Ivanov, B. Ya. Meltser, S. V. Shaposhnikov, I. N. Yassievich, N. N. Faleev, P. S. Kop'ev, and Zh. I. Alferov (1995a) *Phys. Rev. B* **52**, 14058.
Ledentsov, N. N., M. Grundmann, N. Kirstaedter, J. Christen, R. Heitz, J. Böhrer, F. Heinrichsdorff, D. Bimberg, S. S. Ruvimov, P. Werner, U. Richter, U. Gösele, J. Heydenreich, V. M. Ustinov, A. Yu. Egorov, M. V. Maximov, P. S. Kop'ev, and Zh. I. Alferov (1995b) In Proceedings of 22nd International Conference on *The Physics of Semiconductors*, Vancouver, Canada, D. J. Lockwood (ed.), Vol. 3, World Scientific, Singapore, p. 1855.

REFERENCES

Ledentsov, N. N., M. V. Maximov, P. S. Kop'ev, V. M. Ustinov, M. V. Belousov, B. Ya. Meltser, S. V. Ivanov, V. A. Shchukin, Zh. I. Alferov, M. Grundmann, D. Bimberg, S. S. Ruvimov, W. Richter, P. Werner, U. Gösele, J. Heydenreich, P. D. Wang, and C. M. Sotomayor Torres (1995c) *Microelectron. J.* **26**, 871.

Ledentsov, N. N., J. Böhrer, D. Bimberg, I. V. Kochnev, M. V. Maximov, P. S. Kop'ev, Zh. I. Alferov, A. O. Kosogov, S. S. Ruvimov, P. Werner, and U. Gösele (1996a) *Appl. Phys. Lett.* **69**, 1095.

Ledentsov, N. N., M. Grundmann, N. Kirstaedter, O. Schmidt, R. Heitz, J. Böhrer, D. Bimberg, V. M. Ustinov, V. A. Shchukin, P. S. Kop'ev, Zh. I. Alferov, S. S. Ruvimov, A. O. Kosogov, P. Werner, U. Gösele, J. Heydenreich (1996b) *Solid State Electron.* **40**, 875.

Ledentsov, N. N., I. L. Krestnikov, M. V. Maximov, S. V. Ivanov, S. L. Sorokin, P. S. Kop'ev, Zh. I. Alferov, D. Bimberg, and C. M. Sotomayor Torres (1996c) *Appl. Phys. Lett.* **69**, 1343.

Ledentsov, N. N., V. A. Shchukin, M. Grundmann, N. Kirstaedter, J. Böhrer, O. Schmidt, D. Bimberg, V. M. Ustinov, A. Yu. Egorov, A. E. Zhukov, P. S. Kop'ev, S. V. Zaitsev, N. Yu. Gordeev, Zh. I. Alferov, A. I. Borovkov, A. O. Kosogov, S. S. Ruvimov, P. Werner, U. Gösele, and J. Heydenreich (1996d) *Phys. Rev. B* **54**, 8743.

Lee, H., R. Lowe-Webb, W. Yang, and P. C. Sercel (1998) *Appl. Phys. Lett.* **72**, 812.

Lee, S. J. and J. B. Khurgin (1996) *Appl. Phys. Lett.* **69**, 1038.

Legrand, B., B. Grandidier, J. P. Nys, D. Stiévenard, J. M. Gérard, and V. Thierry-Mieg (1998) *Appl. Phys. Lett.* **73**, 96.

Lelong, Ph. and G. Bastard (1996a) *Solid State Commun.* **98**, 819.

Lelong, Ph. and G. Bastard (1996b) In Proceedings of 23rd International Conference on *The Physics of Semiconductors*, Berlin, Germany, M. Scheffler and R. Zimmermann (eds.), World Scientific, Singapore, p. 1377.

Lent, C. S. and P. D. Tougaw (1993) *J. Appl. Phys.* **74**, 6227.

Lent, C. S., P. D. Tougaw, and W. Porod (1993a) *Appl. Phys. Lett.* **62**, 714.

Lent, C. S., P. D. Tougaw, W. Porod, and G. H. Bernstein (1993b) *Nanotechnol.* **4**, 49.

Leon, R., S. Fafard, D. Leonard, J. L. Merz, and P. M. Petroff (1995) *Appl. Phys. Lett.* **67**, 521.

Leon, R., Y. Kim, C. Jagadish, M. Gal, J. Zou, and D. J. H. Cockayne (1996) *Appl. Phys. Lett.* **69**, 1888.

Leon, R., T. J. Senden, Y. Kim, C. Jagadish, and A. Clark (1997) *Phys. Rev. Lett.*, 4942.

Leon, R., C. Lobo, T. P. Chin, J. M. Woodall, S. Fafard, S. Ruvimov, Z. Liliental-Weber, and M. A. Stevens Kalceff (1998) *Appl. Phys. Lett.* **72**, 1356.

Leonard, D., M. Krishnamurthy, C. M. Reaves, S. P. DenBaars, and P. M. Petroff (1993) *Appl. Phys. Lett.* **63**, 3203.

Leonard, D., K. Pond, and P. M. Petroff (1994) *Phys. Rev. B* **50**, 11687.

Li, H., J. Wu, B. Xu, J. Liang, and Z. Wang (1998) *Appl. Phys. Lett.* **72**, 2123.

Li, S.-S., and J.-B. Xia (1998) *Phys. Rev. B* **58**, 3561.

Li, T. S. and K. J. Kuhn (1994) *J. Comp. Phys.* **110**, 292.

Li, Sh.-Sh., J.-B. Xia, Z. L. Yuan, Z. Y. Xu, W. Ge, X. R. Wang, Y. Wang, J. Wang, and L. L. Chang (1996) *Phys. Rev. B* **54**, 11575.

Liao, X. Z., J. Zou, X. F. Duan, D. J. H. Cockayne, R. Leon, and C. Lobo (1998) *Phys. Rev. B* **58**, R4235.

Lippens, P. E. and M. Lannoo (1990) *Phys. Rev. B* **41**, 6079.

Lipsanen, H., M. Sopanen, and J. Ahopelto (1995) *Phys. Rev. B* **51**, 13868.

Lobo, C., R. Leon, S. Fafard, and P. G. Piva (1998) *Appl. Phys. Lett.* **72**, 2850.

Lott, J. A., N. N. Ledentsov, V. M. Ustinov, A. Yu. Egorov, A. E. Zhukov, P. S. Kop'ev, Zh. I. Alferov, and D. Bimberg (1997a) *Electron. Lett.* **33**, 1150.

Lott, J. A., *et al.* (1997b) (in press).

Lowisch, M., M. Rabe, B. Stegemann, F. Henneberger, M. Grundmann, V. Türck, and D. Bimberg (1996) *Phys. Rev. B* **54**, R11074.

Luttinger, J. M. (1956) *Phys. Rev.*, **102**, 1030.

Lutskii, V. N. and L. A. Kulik (1968) *Pis'ma v Zh. Eksp. And Teor. Fiz (JETP Lett.)* **8**, 3.

Lutskii, V. N. (1970) *Phys. Stat. Sol. (a)* **1**, 199.
MacLeod, R. W., C. M. Sotomayor Torres, Y.-S. Tang, and A. Kohl (1993) *J. de Physique IV, Colloque 5* **3**, 335.
McMurry, H. L., A. W. Solbrig Jr, J. K. Boyter, and C. Noble (1967) *J. Phys. Chem. Solids* **28**, 2359.
McSkimin, H. J. and P. Andreatch (1964) *J. Appl. Phys.* **35**, 3312.
McSkimin, H. J. and P. Andreatch (1967) *J. Appl. Phys.* **38**, 2610.
Madhukar, A. (1993) *Thin Solid Films* **231**, 8.
Madhukar, A., K. C. Rajkumar, and P. Chen (1993) *Appl. Phys. Lett.* **62**, 1547.
Madhukar, A., Q. Xie, P. Chen, and A. Konkar (1994) *Appl. Phys. Lett.* **64**, 2727.
Maile, B. E., A. Forchel, R. Germann, and D. Grützmacher (1989) *Appl. Phys. Lett.* **54**, 1552.
Mailhiot, C. and D. L. Smith (1987) *Phys. Rev. B* **35**, 1242.
Maksym, P. A. (1996) *Phys. Rev. B* **53**, 10871.
Maksym, P. A. and T. Chakraborty (1990) *Phys. Rev. Lett.* **65**, 108.
Maksym, P. A. and T. Chakraborty (1992) *Phys. Rev. B* **45**, 1947.
Makino, T. (1996) *IEEE J. Quantum Electron.* **QE-32**, 493.
Malik, S., Ch. Roberts, R. Murray, and M. Pate (1997) *Appl. Phys. Lett.*, **71**, 1987.
Malyshkin, V. G. and V. A. Shchukin (1993) *Fiz. Tekh. Poluprovodn.* **27**, 1932 (*Semicond.* **27**, 1062).
Mansfield, M. and R. J. Needs (1990) *J. Phys. Condens. Matter* **2**, 2361.
Mao, M.-H., F. Heinrichsdorff, A. Krost, and D. Bimberg (1997) *Electron. Lett.* **33**, 1641.
Maradudin, A. A. and R. F. Wallis (1980) *Surf. Sci.* **91**, 423.
Marchand, H., P. Desjardins, S. Guillon, J.-E. Paultre, Z. Bougrioua, Y.-F. Yip, and R. A. Masut ((1997) *Appl. Phys. Lett.* **71**, 527.
Marchenko, V. I. (1981a) *Zh. Eksp. Teor. Fiz.* **81**, 1141 (*Sov. Phys. JETP* **54**, 605).
Marchenko, V. I. (1981b) *Pis'ma Zh. Eksp. Teor. Fiz.* **33**, 397 (*JETP Lett.* **33**, 381).
Marchenko, V. I. and A. Ya. Parshin (1980) *Zh. Eksp. Teor. Fiz.* **79**, 257 (*Sov. Phys. JETP* **52**, 129).
Martin, R. M. (1970) *Phys. Rev. B* **1**, 4005.
Marzin, J.-Y. and G. Bastard (1994) *Solid State Commun.* **92**, 437.
Marzin, J.-Y., J.-M. Gerard, A. Izraël, D. Barrier, and G. Bastard (1994) *Phys. Rev. Lett.* **73**, 716.
Mateeva, E., P. Sutter, J. C. Bean, and M. G. Lagally (1997) *Appl. Phys. Lett.* **71**, 3233.
Matthews, J. W. and A. E. Blakeslee (1976) *J. Cryst. Growth* **32**, 265.
Maximov, M. V., N. Yu. Gordeev, S. V. Zaitsev, P. S. Kop'ev, I. V. Kochnev, N. N. Ledentsov, A. V. Lunev, S. S. Ruvimov, A. V. Sakharov, A. F. Tsatsul'nikov, Yu. M. Shernyakov, Zh. I. Alferov, and D. Bimberg (1997a) *Semicond.* **31**, 124.
Maximov, M. V., I. V. Kochnev, Yu. M. Shernakov, S. V. Zaitsev, N. Yu. Gordeev, A. F. Tsatsulnikov, A. V. Sakharov, I. L. Krestnikov, P. S. Kop'ev, Zh. I. Alferov, N. N. Ledentsov, D. Bimberg, A. O. Kosogov, P. Werner, and U. Gösele (1997b) *Jpn. J. Appl. Phys.* **36**, 4221.
Medeiros-Ribeiro, G., D. Leonard, and P. M. Petroff (1995) *Appl. Phys. Lett.* **66**, 1767.
Medeiros-Ribeiro, G., F. G. Pikus, P. M. Petroff, and A. L. Efros (1997) *Phys. Rev. B* **55**, 1568.
Men, F. K., W. E. Packard, and M. B. Webb (1988) *Phys. Rev. Lett.* **61**, 2469.
Michel, M., A. Forchel, and F. Faller (1997) *Appl. Phys. Lett.* **70**, 393.
Miller, D. A. B. (1990) *Optics and Photonics News* **2**, 7.
Miller, D. A. B., D. S. Chemla, T. C. Damen, A. C. Gossard, W. Wiegmann, T. Wood, and C. A. Burus (1984) *Appl. Phys. Lett.* **45**, 13.
Miller, B. T., W. Hansen, S. Manus, A. Lorke, J. P. Kotthaus, G. Medeiros-Ribeiro, and P. M. Petroff (1997) *Phys. Rev. B* **56**, 6764.
Ming, Z. H., Y. L. Soo, S. Huang, Y. H. Kao, K. Stair, G. Devane, and C. Choi-Feng (1995) *Appl. Phys. Lett.* **66**, 165.
Mirin, R., J. P. Ibbetson, K. Nishi, A. C. Gossard, and J. E. Bowers (1995) *Appl. Phys. Lett.* **67**, 3795.

REFERENCES

Mirin, R., A. Gossard, and J. Bowers (1996) *Electron. Lett.* **32**, 1732.
Miyamoto, Y., M. Cao, Y. Shinagi, K. Furuya, Y. Suematsu, K. G. Ravikumar, and S. Arai (1987) *Jpn. J. Appl. Phys.* **26**, L225.
Miyamoto, Y., Y. Miyake, M. Asada, and Y. Suematsu (1989) *IEEE J. Quantum Electron.* **QE-25**, 2001.
Mo, Y.-W., B. S. Swartzentruber, R. Kariotis, M. B. Webb, and M. G. Lagally (1989) *Phys. Rev. Lett.* **63**, 2393.
Mo, Y.-W., D. E. Savage, B. S. Swartzentruber, and M. G. Lagally (1990) *Phys. Rev. Lett.* **65**, 1020.
Moison, J. M., F. Houzay, F. Barthe, L. Leprince, E. André, and O. Vatel (1994) *Appl. Phys. Lett.* **64**, 196.
Moll, N. (1998a) PhD Thesis, TU Berlin, Wissenschaft und Technik Verlag, Berlin.
Moll, N., M. Scheffler, and E. Pehlke (1998b) *Phys. Rev. B* **58**, 4566.
Moritz, A., R. Wirth, A. Hangleiter, A. Kurtenbach, and K. Eberl (1996) *Appl. Phys. Lett.* **69**, 212.
Mui, D. S., D. Leonard, L. A. Coldren, and P. M. Petroff (1995) *Appl. Phys. Lett.* **66**, 1620.
Mukai, K., N. Ohtsuka, M. Sugawara, and S. Yamazaki (1994) *Jpn. J. Appl. Phys.* **33**, L1710.
Mukai, K., N. Ohtsuka, H. Shoji, and M. Sugawara (1996a) *Appl. Phys. Lett.* **68**, 3013.
Mukai, K., N. Ohtsuka, and M. Sugawara (1996b) *Jpn. J. Appl. Phys.* **35**, L262.
Mukherjee, S., E. Pehlke, and J. Tersoff (1994) *Phys. Rev. B* **49**, 1919.
Mullins, W. W. (1963) In *Metal Surfaces: Structure, Energetics and Kinetics*, American Society for Metals, Metals Park, Ohio, p. 17.
Murray, C. W., S. C. Racine, and E. R. Davidson (1992) *J. Comp. Phys.* **103**, 382.
Musgrave, M. J. P. and J. A. Pople (1962) *Proc. Roy. Soc. Lond. A* **268**, 474.
Muto, S. (1995) *Jpn. J. Appl. Phys.* **34**, L210.
Nabetani, Y., T. Ishikawa, S. Noda, and A. Sasaki (1994) *J. Appl. Phys.* **76**, 347.
Nagarajan, R., I. Ishikawa, T. Fukushima, R. Geels, and J. E. Bowers (1992) *IEEE J. Quantum Electron.* **28**, 1990.
Nagamune, Y., M. Nishioka, S. Tsukamoto, and Y. Arakawa (1994) *Appl. Phys. Lett.* **64**, 2495.
Nagamune, Y., H. Watabe, M. Nishioka, and Y. Arakawa (1995) *Appl. Phys. Lett.* **67**, 3257.
Nakayama, H. and Y. Arakawa (1996) In Proceedings of 15th IEEE International Conference on *Semiconductor Lasers*, Haifa, IEEE Catalog 96CH35896, p. 41.
Nambu, Y. and K. Asakawa (1995) *Appl. Phys. Lett.* **67**, 1509.
Narasimhan, S. and D. Vanderbilt (1992) *Phys. Rev. Lett.* **69**, 1564.
Narihiro, M., G. Yusa, Y. Nakamura, T. Noda, and H. Sakaki (1997) *Appl. Phys. Lett.* **70**, 105.
Narukawa, Y., Y. Kawakami, M. Funato, Sh. Fujita, Sh. Fujita, and Sh. Nakamura (1997) *Appl. Phys. Lett.* **70**, 981.
Needs, R. J. (1987) *Phys. Rev. Lett.* **58**, 53.
Neumann, W., H. Hofmeister, D. Conrad, K. Scheerschmidt, and S. Ruvimov (1996) *Z. Kristallogr.* **211**, 147.
Ngo, T. T., P. M. Petroff, H. Sakaki, and J. L. Merz (1996) *Phys. Rev. B* **53**, 9618.
Nie, H., O. Baklenov, P. Yuan, C. Lenox, B. G. Streetman, and J. C. Campbell (1998) *IEEE Photon. Technol. Lett.* **10**, 1009.
Nikitin, V., P. A. Crowell, J. A. Gupta, D. D. Awschalom, F. Flack, and N. Samarth (1997) *Appl. Phys. Lett.* **71**, 1213.
Nirmal, M., D. J. Norris, M. Kuno, M. G. Bawendi, Al. L. Efros, and M. Rosen (1995) *Phys. Rev. Lett.* **75**, 3728.
Nishi, K., T. Anan, A. Gomyo, S. Kohmoto, and S. Sugou (1997) *Appl. Phys. Lett.* **70**, 3579.
Nomura, S. and T. Kobayashi (1990a) *Solid State Commun.* **73**, 425.
Nomura, S. and T. Kobayashi (1990b) *Solid State Commun.* **74**, 1153.
Nötzel, R. (1996) *Semicond. Sci. Technol.* **11**, 1365.

Nötzel, R., T. Fukui, H. Hasegawa, J. Temmyo, and T. Tamamura (1994c) *Appl. Phys. Lett.* **65**, 2854.
Nötzel, R., N. N. Ledentsov, L. Däweritz, M. Hohenstein, and K. Ploog (1991) *Phys. Rev. Lett.* **67**, 3812.
Nötzel, R., J. Temmyo, H. Kamada, T. Furuta, and T. Tamamura (1994a) *Appl. Phys. Lett.* **65**, 457.
Nötzel, R., J. Temmyo, and T. Tamamura (1994b) *Nature* **369**, 131.
Nötzel, R., J. Temmyo, A. Kozen, T. Tamamura, T. Fukui, and H. Hasegawa (1995) *Appl. Phys. Lett.* **66**, 2525.
Nozières, P. and D. E. Wolf (1988) *Z. Phys. B* **70**, 399, 507.
Ohnesorge, B., M. Albrecht, J. Oshinowo, A. Forchel, and Y. Arakawa, *Phys. Rev. B* **54**, 11532.
Olson, J. M. and A. Kibbler (1986) *J. Cryst. Growth* **77**, 182.
Onodera, Y., and Y. Toyozawa (1967) *J. Phys. Soc. Jpn.* **22**, 833.
Osbourn, G. C. (1982) *J. Appl. Phys.* **53**, 1586.
Oshinowo, J., M. Nishioka, S. Ishida, and Y. Arakawa (1994) *Appl. Phys. Lett.* **65**, 1421.
Ostwald, W. (1900) *Z. Phys. Chem.* **34**, 495.
Ourmazd, A., F. H. Baumann, M. Bode, and Y. Kim (1990) *Ultramicrosc.* **34**, 237.
Ozawa, K., Y. Aoyagi, Y. J. Park, and L. Samuelson (1997) *Appl. Phys. Lett.* **71**, 797.
Pacheco, M., and Z. Barticevic (1997) *Phys. Rev. B* **55**, 10688.
Pan, D., Y. P. Zeng, M. Y. Kong, J. Wu, Y. Q. Zhu, C. H. Zhang, J. M. Li, and C. Y. Wang (1996) *Electron. Lett.* **32**, 1726.
Pan, J. L. (1994) *Phys. Rev. B* **49**, 2536.
Paz, J. P. and G. Mahler (1993) *Phys. Rev. Lett.* **71**, 3235.
Pedersen, F. B., and Y.-C. Chang (1996) *Phys. Rev. B* **53**, 1507.
Pedersen, F. B., and Y.-C. Chang (1997) *Phys. Rev. B* **55**, 4580.
Pehlke, E., N. Moll, and M. Scheffler (1996) In Proceedings of 23rd International Conference on *The Physics of Semiconductors*, Berlin, Germany, M. Scheffler and R. Zimmerman (eds.), World Scientific, Singapore, p. 1301.
Pehlke, E., N. Moll, A. Kley, and M. Scheffler (1997) *Appl. Phys. A* **65**, 525.
Petroff, P. M. and G. Medeiros-Ribeiro (1996) *MRS Bull.* **21**, 50.
Petroff, P. M., A. C. Gossard, R. A. Logan, and W. Wiegmann (1982) *Appl. Phys. Lett.* **41**, 635.
Pettersson, H., S. Anand, H. G. Grimmeiss, and L. Samuelson (1996) *Phys. Rev. B* **53**, R10497.
Pfeffer, P. and W. Zawadzki (1995) *Phys. Rev. B* **52**, R14332.
Phillips, J., K. Kamath, and P. Bhattacharya (1997a) *Electron. Lett.*
Phillips, J., K. Kamath, X. Zhou, N. Chervela, and P. Bhattacharya (1997b) *Appl. Phys. Lett.* **71**, 2079.
Phillips, J., K. Kamath, and P. Bhattacharya (1998a) *Appl. Phys. Lett.* **72**, 2020.
Phillips, J., K. Kamath, T. Brock, and P. Bhattacharya (1998b) *Appl. Phys. Lett.* **72**, 3509.
Pino, R. (1998) *Phys. Rev.B* **58**, 4644.
Ploog, K. (1988) *Angew. Chem. Int. Ed. Engng* **27**, 593.
Ploog, K. H. and R. Nötzel (1998) *Physica E* (in press).
Podgorny, M., M. T. Czyzyk, A. Balzarotti, P. Letardi, N. Motta, A. Kisiel, and M. Zimnal-Starnawska (1985) *Solid State Commun.* **55**, 413.
Polimeni, A., A. Patanè, M. Capizzi, F. Martelli, L. Nasi, and G. Salviati (1996) *Phys. Rev. B* **53**, R4213.
Pollak, F. H. (1990) *Semicond. Semimetals* **32**, 17.
Ponchet, A., A. Le Corre, H. L'Haridon, B. Lambert, and S. Salaün (1995) *Appl. Phys. Lett.* **67**, 1850.
Priester, C. and M. Lannoo (1995) *Phys. Rev. Lett.* **75**, 93.
Prins, F. E., G. Lehr, H. Schweizer, M. H. Pilkuhn, and G. W. Smith (1993) *Appl. Phys. Lett.* **62**, 1365.

REFERENCES

Pryor, C. (1998) *Phys. Rev. B* **57**, 7190.
Pryor, C., M.-E. Pistol, and L. Samuelson (1997) *Phys. Rev. B* **56**, 10404.
Pryor, C., J. Kim, L. W. Wang, A. Williamson, and A. Zunger (1998) *J. Appl. Phys.* **83**, 2548.
Qian, G.-X., R. M. Martin, and D. J. Chadi (1988) *Phys. Rev. B* **38**, 7649.
Que, W. (1992) *Solid State Commun.* **81**, 721.
Raghuraman, R., N. Yu, R. Engelmann, H. Lee and C. L. Shieh (1993) *IEEE J. Quantum Electron.* **29**, 69.
Rahmati, B., W. Jäger, H. Trinkaus, R. Loo, L. Vescan, and H. Lüth (1996) *Appl. Phys. A* **62**, 575.
Rajkumar, K. C., K. Kaviani, J. Chen, P. Chen, A. Madhukar, and D. Rich (1992) *Mater. Res. Soc. Symp. Proc.* **263**, 163.
Ralston, J., A. L. Moretti, R. K. Jain, and F. A. Chambers (1987) *Appl. Phys. Lett.* **50**, 1817.
Rashba, E. I. (1975) *Sov. Phys. Semicond.* **8**, 807.
Ratsch, C. and A. Zangwill (1993) *Surf. Sci.* **293**, 123.
Razeghi, M. (1989) *The MOCVD Challenge*, Vol. 1, Adam Hilger, Bristol.
Razeghi, M. (1995) *The MOCVD Challenge*, Vol. 2, Adam Hilger, Bristol.
Razeghi, M., F. Omnes, M. Defour, and P. Maurel (1988) *Appl. Phys. Lett.* **52**, 209.
Reaves, C. M., R. I. Pelzel, G. C. Hsueh, W. H. Weinberg, and S. P. DenBaars (1996) *Appl. Phys. Lett.* **69**, 3878.
Reed, M., R. T. Bate, K. Bradshaw, W. M. Duncan, W. R. Frensley, J. W. Lee, and H. D. Smith (1986) *J. Vac. Sci. Technol. B* **4**, 358.
Reimer, L. (1984) *Transmission Electron Microscopy*, Springer Series in Optical Sciences, Vol. 36, Springer, Berlin.
Reinhardt, F., W. Richter, A. B. Müller, D. Gutsche, P. Kurpas, K. Ploska, K. C. Rose, and M. Zorn (1993) *J. Vac. Sci. Technol. B* **11**, 1427.
Richter, W. and D. R. T. Zahn (1996) In *Optical Characterization of Epitaxial Semiconductor Layers*, G. Bauer and W. Richter (eds.), Springer, Berlin, p. 12.
Rinaldi, R., P. V. Giugno, R. Cingolani, H. Lipsanen, M. Sopanen, J. Tulkki, and J. Ahopelto (1996) *Phys. Rev. Lett.* **77**, 342.
Rockenberger, J., A. Rogach, L. Tröger, M. Tischer, M. Grundmann, A. Eychmüller, and H. Weller (1998) *J. Chem. Phys.* **108**, 7807.
Rocksby, H. P. (1932) *J. Soc. Glass Technol.* **16**, 171.
Roitburd, A. L. (1976) *Phys. Stat. Sol. (a)* **37**, 329.
Rorison, J. M. (1993) *Phys. Rev. B* **48**, 4643.
Rottman, C. and M. Wortis (1984) *Phys. Rep.* **103**, 59.
Roussignol, P., D. Ricard, C. Flytzanis, and N. Neuroth (1989) *Phys. Rev. Lett.* **62**, 312.
Ruan, W. Y., Y. Y. Liu, C. G. Bao, and Z. Q. Zhang (1995) *Phys. Rev. B* **51**, 7942.
Ruggerone, P., C. Ratsch, and M. Scheffler (1997) In *Growth and Properties of Ultrathin Epitaxial Layers*, D. A. King and D. P. Woodruff (eds.), The Chemical Physics of Solid Surfaces, Vol. 8, Elsevier Science, Amsterdam, p. 490.
Ruvimov, S. and K. Scheerschmidt (1995) *Phys. Stat. Sol. (a)* **150**, 471.
Ruvimov, S., P. Werner, K. Scheerschmidt, J. Heydenreich, U. Richter, N. N. Ledentsov, M. Grundmann, D. Bimberg, V. M. Ustinov, A. Yu. Egorov, P. S. Kop'ev, and Zh. I. Alferov (1995) *Phys. Rev. B* **51**, 14766.
Saada, A. S. (1974) *Elasticity Theory and Applications*, Pergamon, New York.
Sahara, R., M. Matsuda, H. Shoji, K. Morito, and H. Soda (1996) *IEEE Phot. Technol. Lett.* **8**, 1477.
Saito, H., K. Nishi, I. Ogura, S. Sugou, and Y. Sugimoto (1996) *Appl. Phys. Lett.* **69**, 3140.
Saito, H., K. Nishi, Sh. Sugou, and Y. Sugimoto (1997) *Appl. Phys. Lett.* **71**, 590.
Sakaki, H., G. Yusa, T. Someya, Y. Ohno, T. Noda, H. Akiyama, Y. Kadoya, and H. Noge (1995) *Appl. Phys. Lett.* **67**, 3444.
Sauvage, S., P. Boucaud, F. H. Julien, J.-M. Gerard, and J.-Y. Marzin (1997) *J. Appl. Phys.* **82**, 3396.

Savona, V., L. C. Andreani, P. Schwendimann, and A. Quattropani (1995) *Solid State Commun.* **93**, 733.
Schedelbeck, G., W. Wegscheider, M. Bichler, and G. Abstreiter (1998) *Science* **278**, 1792.
Scheffler, M., J. P. Vigneron, and G. B. Bachelet (1985) *Phys. Rev. Lett.* **49**, 1765.
Scherer, A. and H. G. Craighead (1986) *Appl. Phys. Lett.* **49**, 1284.
Schittenhelm, P., M. Gail, J. Brunner, J. F. Nützel, and G. Abstreiter (1995) *Appl. Phys. Lett.* **67**, 1292.
Schlesinger, T. E. and T. Kuech (1986) *Appl. Phys. Lett.* **49**, 519.
Schmidt, K. H., G. Medeiros-Ribeiro, M. Oesterreich, P. M. Petroff, and G. H. Döhler (1996a) *Phys. Rev. B* **54**, 11346.
Schmidt, O. G., N. Kirstaedter, N. N. Ledentsov, M.-H. Mao, D. Bimberg, V. M. Ustinov, A. Yu. Egorov, A. E. Zhukov, M. V. Maximov, P. S. Kop'ev, and Zh. I. Alferov (1996b) *Electron. Lett.* **32**, 1302.
Schmitt-Rink, S., D. A. B. Miller, and D. S. Chemla (1987) *Phys. Rev. B* **35**, 8113.
Schultheis, L. and C. W. Tu (1985) *Phys. Rev. B* **32**, 6978.
Schur, R., F. Sogawa, M. Nishioka, S. Ishida, Y. Arakawa (1997) *Jpn. J. Appl. Phys.* **36**, L357.
Segmüller, A., I. C. Noyan, and V. S. Speriosu (1989) *Prog. Cryst. Growth and Charact.* **18**, 21.
Seifert, W., N. Carlsson, M. Miller, M.-E. Pistol, L. Samuelson, and L. R. Wallenberg (1996a) *Proc. Cryst. Growth Charact. Mater.* **33**, 423.
Seifert, W., N. Carlsson, A. Petersson, L.-E. Wernersson, and L. Samuelson (1996b) *Appl. Phys. Lett.* **68**, 1684.
Sercel, P. C. (1996) *Phys. Rev. B* **53**, 14532.
Sercel, P. C. and K. J. Vahala (1990a) *Phys. Rev. B* **42**, 3690.
Sercel, P. C. and K. J. Vahala (1990b) *Appl. Phys. Lett.* **57**, 1569.
Shaklee, K. L., R. E. Nahory, and R. F. Laheny (1985) *J. Lumin.* **7**, 284.
Shchukin, V. A. and D. Bimberg (1998) *Rev. Modern Physics* (in press).
Shchukin, V. A., A. I. Borovkov, N. N. Ledentsov, and D. Bimberg (1995a) *Phys. Rev. B* **51**, 10104.
Shchukin, V. A., A. I. Borovkov, N. N. Ledentsov, and P. S. Kop'ev (1995b) *Phys. Rev. B* **51**, 17767.
Shchukin, V. A., N. N. Ledentsov, P. S. Kop'ev, and D. Bimberg (1995c) *Phys. Rev. Lett.* **75**, 2968.
Shchukin, V. A., N. N. Ledentsov, M. Grundmann, P. S. Kop'ev, and D. Bimberg (1996) *Surf. Sci.* **352–354**, 117.
Shernyakov, Yu. M., A. Yu. Egorov, A. E. Zhukov, A. V. Zaitsev, A. R. Kovsh, I. L. Krestnikov, A. V. Lunev, N. N. Ledentsov, M. V. Maximov, A. V. Sakharov, V. M. Ustinov, Zhao Zhen, P. S. Kop'ev, Zh. I. Alferov, and D. Bimberg (1997) *Pis'ma v Zh. Tekhn. Fiz.* **23** (1), 51 (*Tech. Phys. Lett.* **23**).
Shinada, M. and S. Sugano (1966) *J. Phys. Soc. Jpn.* **21**, 1936.
Shiraki, Y., Ota, K. and Usami, N. 1998) In *Proceedings 24th International Conference on The Physics of Semiconductors*, Jerusalem, Israel, D. Gershoni (ed.), World Scientific, Singapore.
Shoji, H., Y. Nakata, M. Mukai, Y. Sugiyama, M. Sugawara, N. Yokoyama, and H. Ishikawa (1996) *Electron. Lett.* **32**, 2032.
Shoji, H., Y. Nakata, K. Mukai, Y. Sugiyama, M. Sugawara, N. Yokoyama, and H. Ishikawa (1997) *Appl. Phys. Lett.* **71**, 193.
Shuttleworth, R. (1950) *Proc. Phys. Soc. Lond. A* **63**, 444.
Sikorski, Ch. and U. Merkt (1989) *Phys. Rev. Lett.* **62**, 2164.
Singh, J. (1996) *IEEE Photon. Technol. Lett.* **8**, 488.
Sleight, J. W., R. E. Welser, L. J. Guido, M. Amman, and M. A. Reed (1995) *Appl. Phys. Lett.* **66**, 1343.
Smith, D. L. (1986) *Solid State Commun.* **57**, 919.

REFERENCES

Snyder, C. W., B. G. Orr, D. Kessler, and L. M. Sander (1991) *Phys. Rev. Lett.* **66**, 3032.
Snow, E. S. and P. M. Campbell (1994) *Appl. Phys. Lett.* **64**, 1932.
Snow, E. S., P. M. Campbell, and P. J. McMarr (1993) *Appl. Phys. Lett.* **63**, 749.
Soga, T., T. George, T. Suzuki, T. Jimbo, M. Umeno, and E. R. Weber (1991) *Appl. Phys. Lett.* **58**, 2108.
Solbrig Jr, A. W. (1971) *J. Phys. Chem. Solids* **32**, 1761.
Solomon, G. S., J. A. Trezza, A. F. Marshall, and J. S. Harris Jr (1996) *Phys. Rev. Lett.* **76**, 952.
Someya, T., H. Akiyama, and H. Sakaki (1995) *Phys. Rev. Lett.* **74**, 3664.
Sood, A. K., J. Menéndez, M. Cardona, and K. Ploog (1985) *Phys. Rev. Lett.* **54**, 2115.
Sopanen, M., H. Lipsanen, and J. Ahopelto (1995) *Appl. Phys. Lett.* **65**, 1662.
Sotomayor Torres, C. M., A. P. Smart, M. A. Foad, and C. D. W. Wilkinson (1992) *Festkörperprobleme* (*Advances in Solid State Physics*) **32**, 265.
Sotomayor Torres, C. M., P. D. Wang, N. N. Ledentsov, and Y.-S. Tang (1994) In *Spectroscopic Characterization Techniques for Semiconductor Technology V*, O. Glembocki (ed.), SPIE Vol. 2141, p. 2.
Spencer, B. J., P. W. Voorhees, and S. H. Davis (1991) *Phys. Rev. Lett.* **67**, 3696.
Srolovitz, D. (1989) *Acta Metall.* **37**, 621.
Stafford, C. A. and S. Das Sarma (1994) *Phys. Rev. Lett.* **72**, 3590.
Steer, M. J., D. J. Mowbray, W. R. Tribe, M. S. Skolnick, M. D. Sturge, M. Hopkinson, A. G. Cullis, C. R. Whitehoue, and R. Murray (1996) *Phys. Rev. B* **54**, 17738.
Steffen, R., F. Faller, and A. Forchel (1994) *J. Vac. Sci. Technol. B* **12**, 3653.
Steffen, R., A. Forchel, T. Reinecke, T. Koch, M. Albrecht, J. Oshinowo, and F. Faller (1996a) *Phys. Rev. B* **54**, 1510.
Steffen, R., Th. Koch, J. Oshinowo, F. Faller, and A. Forchel (1996b) *Surf. Sci.* **361/362**, 805.
Steffen, R., Th. Koch, J. Oshinowo, and A. Forchel (1996c) *Appl. Phys. Lett.* **68**, 225.
Steimetz, E., J.-T. Zettler, F. Schienle, T. Trepk, T. Wethkamp, W. Richter, and I. Sieber (1996) *Appl. Surf. Sci.* **107**, 203.
Stern, F. (1963) *Solid State Physics* **15**, 299.
Stern, M. B., H. G. Craighead, P. F. Liao, and P. M. Mankievich (1984) *Appl. Phys. Lett.* **45**, 410.
Stier, O., M. Grundmann, and D. Bimberg (1996) In Proceedings of 23rd International Conference on *The Physics of Semiconductors*, Berlin, Germany, M. Scheffler, and R. Zimmermann (eds.), World Scientific, Singapore, p. 1177.
Stier, O., M. Grundmann, and D. Bimberg (1998) In *Proceedings 24th International Conference on The Physics of Semiconductors*, Jerusalem, Israel, D. Gershoni (ed.), World Scientific, Singapore.
† Stranski, I. N. and L. Krastanow (1937) Sitzungsberichte d. Akad. d. Wissenschaften in Wien, Abt. IIb, Band 146, p. 797.
Strassburg, M., V. Kutzer, U. W. Pohl, A. Hoffmann, I. Broser, N. N. Ledenstov, D. Bimberg, A. Rosenauer, U. Fischer, D. Gerthsen, I. L. Krestnikov, M. V. Maximov, P. S. Kop'ev, and Zh. I. Alferov (1998) *Appl. Phys. Lett.* **72**, 942.
Stringfellow, G. B. (1989) *Organometallic Vapor Phase Epitaxy: Theory and Practice*, Academic Press, San Diego.
Sugawara, M. (1995) *Phys. Rev. B* **51**, 10743.
Sugiyama, Y., Y. Sakuma, S. Muto, and N. Yokoyama (1995) *Appl. Phys. Lett.* **67**, 256.
Sugiyama, Y., Y. Nakata, K. Imamura, Sh. Muto, and N. Yokoyama (1996) *Jpn. J. Appl. Phys.* **35**, 1320.
Sugiyama, Y., Y. Nakata, T. Futatsugi, M. Sugawara, Y. Awano, and N. Yokoyama (1997) *Jpn. J. Appl. Phys.* **36**, L158.

---

† Note that the frequently used similar reference involving the Akademie in Mainz is incorrect.

Susa, N. (1996) *IEEE J. Quantum Electron.* **QE-32**, 1760.
Tabuchi, M., S. Noda, and A. Sasaki (1992) In *Science and Technology of Mesoscopic Structures*, S. Namba, C. Hamaguchi, and T. Ando (eds.), Springer, Tokyo, p. 379.
Takagahara, T. (1987) *Phys. Rev. B* **36**, 9293.
Takagahara, T. (1993) *Phys. Rev. B* **47**, 4569.
Takeuchi, M., K. Shiba, K. Sato, H. K. Hung, K. Inoue, and H. Nakashima (1995) *J. Appl. Phys.* **34**, 4411.
Tan, I.-H., Y.-L. Chang, R. Mirin, E. Hu, J. Merz, T. Yasuda, and Y. Segawa (1993) *Appl. Phys. Lett.* **62**, 1376.
Tanaka, S., S. Iwai, and Y. Aoyagi (1996) *Appl. Phys. Lett.* **69**, 4096.
Tapfer, L. and K. Ploog (1986) *Phys. Rev. B***33**, 5565.
Tarucha, S., D. G. Austing, T. Honda, R. J. van der Hage, and L. P. Kouwenhoven (1996) *Phys. Rev. Lett.* **77**, 3613.
Taskinen, M., M. Sopanen, H. Lipsanen, J. Tulkki, T. Tuomi, J. Ahopelto (1997) *J. Cryst. Growth* **170**, 60.
Temmyo, J., A. Kozen, T. Tamamura, R. Nötzel, T. Fukui, and H. Hasegawa (1996) *J. Electron. Mater.* **25**, 431.
Tersoff, J. and R. M. Tromp (1993) *Phys. Rev. Lett.* **70**, 2782.
Tersoff, J., C. Teichert, and M. G. Lagally (1996) *Phys. Rev. Lett.* **76**, 1675.
Thibado, P. M., B. R. Bennett, M. E. Twigg, B. V. Shanabrook, and L. J. Whitman (1996) *J. Vac. Sci. Technol. A* **14**, 885.
Thornton, A. S. G., T. Ihn, P. C. Main, L. Eaves, and M. Henini (1998) *Appl. Phys. Lett.* **73**, 354.
Tournié, E. and K. Ploog (1993) *Appl. Phys. Lett.* **62**, 858.
Tournié, E., N. Grandjean, A. Trampert, J. Massies, and K. H. Ploog (1995) *J. Cryst. Growth* **150**, 460.
Toyozawa, Y. (1959) *Prog. Theor. Phys.*, Suppl. 12, 111.
Trampert, A., E. Tournié, and K. H. Ploog (1994) *Phys. Stat. Sol. (b)* **145**, 481.
Tran Thoai, D. B., Y. Z. Hu, and S. W. Koch (1990a) *Phys. Rev. B* **42**, 11261.
Tran Thoai, D. B., R. Zimmermann, M. Grundmann, and D. Bimberg (1990b) *Phys. Rev. B* **42**, 5906.
Tsai, F.-Y., and C. P. Lee (1998) *J. Appl. Phys.* **84**, 2624.
Tsang, W. T. (1981) *Appl. Phys. Lett.* **38**, 835.
Tsatsul'nikov, A. F., G. E. Cirlin, A. Yu. Egorov, A. O. Golubok, P. S. Kop'ev, A. R. Kovsh, N. N. Ledentsov, S. A. Masalov, M. V. Maximov, V. N. Petrov, V. M. Ustinov, B. V. Volovik, A. E. Zhukov, R. Heitz, P. Werner, M. Grundmann, D. Bimberg, and Zh. I. Alferov (1998) In *Proceedings 24th International Conference on The Physics of Semiconductors*, Jerusalem, Israel, D. Gershoni (ed.), World Scientific, Singapore.
Tsiper, E. V. (1996) *Phys. Rev. B* **54**, 1959.
Tsui, D. C., H. L. Störmer, and A. C. Gossard (1982) *Phys. Rev. Lett.* **48**, 1559.
Tsui, R., R. Zhang, K. Shiralagi, and H. Goronkin (1997) *Appl. Phys. Lett.* **71**, 3254.
Tuck, B. (1985) *J. Phys. D: Appl. Phys.* **18**, 557.
Tulkki, J. and A. Heinämäki (1995) *Phys. Rev. B* **52**, 8239.
Ugajin, R. (1995) *Phys. Rev. B* **51**, 10714.
Ugajin, R. (1996) *Phys. Rev. B* **53**, 6963.
Uskov, A. V., J. McInerney, F. Adler, H. Schweizer, and M. H. Pilkuhn (1998) *Appl. Phys. Lett.* **72**, 58.
US Patent 5,260,957, 9 November 1993.
Uztmeier, T., P. A. Postigo, J. Tamayo, R. García, and F. Briones (1996) *Appl. Phys. Lett.* **69**, 2674.
Vahala, K. J. (1988) *IEEE J. Quantum Electron.* **QE-24**, 523.
Vahala, K. J. and P. C. Sercel (1990) *Phys. Rev. Lett.* **65**, 239.
Vahala, K. J., Y. Arakawa, and A. Yariv (1987) *Appl. Phys. Lett.* **51**, 365.

REFERENCES

Vanderbilt, D. (1992) *Surf. Sci.* **268**, L300.
Vanderbilt, D. and L. K. Wickham (1991) *Proc. Mater. Res. Soc. Symp.* **202**, 555.
van Roosbroeck, W. and W. Shockley (1954) *Phys. Rev.* **94**, 1558.
Venkatesan, T., S. A. Schwarz, D. M. Hwang, R. Bhat, M. Koza, H. W. Yoon, P. Mei, Y. Arakawa, and A. Yariv (1986) *Appl. Phys. Lett.* **49**, 701.
Volmer, M. and A. Weber (1926) *Z. Phys. Chem.* **119**, 277.
Vollmer, M., E. J. Mayer, W. W. Rühle, A. Kurtenbach, and K. Eberl (1996) *Phys. Rev. B* **54**, R17292.
Vorob'ev, L. E., D. A. Firsov, V. A. Shalygin, V. N. Tulupenko et al. (1998) *Technical Phys. Lett.* **24**, 590.
Vurgaftman, I. and J. Singh (1994) *Appl. Phys. Lett.* **64**, 232.
Wang, P. D. and C. M. Sotomayor Torres (1993) *J. Appl. Phys.* **74**, 5047.
Wang, L.-W. and A. Zunger (1994) *J. Phys. Chem.* **98**, 2158.
Wang, L.-W. and A. Zunger (1996a) *Phys. Rev. B* **53**, 9579.
Wang, L.-W. and A. Zunger (1996b) *Phys. Rev. B* **54**, 11417.
Wang, P. D., N. N. Ledentsov, C. M. Sotomayor Torres, P. S. Kop'ev, and V. M. Ustinov (1994) *Appl. Phys. Lett.* **64**, 1526.
Wang, J., U. A. Griesinger, M. Geiger, D. Ottenwaelder, F. Scholz, and H. Schweizer (1996a) *IEEE Photon. Technol. Lett.* **8**, 1585.
Warburton, R. J., C. S. Dürr, K. Karrai, J. P. Kotthaus, G. Medeiros-Ribeiro, and P. M. Petroff (1997) *Phys. Rev. Lett.* **79**, 5282.
Warren, A. C., I. Plotnik, E. H. Anderson, M. L. Schattenburg, D. A. Antoniadis, and H. I. Smith (1986) *J. Vac. Sci. Technol. B* **4**, 365.
Watson, G. M., D. Gibbs, D. M. Zehner, M. Yoon, and S. G. J. Mochrie (1993) *Phys. Rev. Lett.* **71**, 3166.
Wegscheider, W., G. Schedelbeck, G. Abstreiter, M. Rother, and M. Bichler (1997) *Phys. Rev. Lett.* **79**, 1917.
Wen, G., J. Y. Lin, H. X. Jiang, and Z. Chen (1995) *Phys. Rev. B* **52**, 5913.
Werner, J., E. Kapon, N. G. Stoffel, E. Colas, S. A. Schwarz, and N. Andreadakis (1989) *Appl. Phys. Lett.* **55**, 540.
Wiesendanger, R. (1994) *Scanning Probe Microscopy and Spectroscopy*, Cambridge University Press, Cambridge.
Wiesner, P. and U. Heim (1975) *Phys. Rev. B* **11**, 3071.
Willatzen, M., A. Uskov, J. Mork, H. Olesen, B. Tromborg, and A. P. Jauho (1991) *Photon. Technol. Lett.* **3**, 606.
Williams, E. D., R. J. Phaneuf, J. Wei, N. C. Bartelt, and T. L. Einstein (1993) *Surf. Sci.* **294**, 219.
Wingreen, N. S. and Ch. A. Stafford (1997) *IEEE J. Quantum Electron.* **QE-33**, 1170.
Woggon, U. (1997) *Optical Properties of Semiconductor Quantum Dots*, Springer Tracts in Modern Physics 136, Springer, Berlin.
Woggon, U., W. Langbein, J. M. Hvam, A. Rosenauer, T. Remmele, and D. Gerthsen (1997) *Appl. Phys. Lett.* **71**, 377.
Wojs, A. and P. Hawrylak (1996) *Phys. Rev. B* **53**, 10841.
Wolf, D. (1993) *Phys. Rev. Lett.*, **70**, 627.
Wu, W.-Y., J. N. Schulman, T. Y. Hsu, and U. Efron (1987) *Appl. Phys. Lett.* **51**, 710.
Wu, W., J. R. Tucker, G. S. Solomon, and J. S. Harris Jr, (1997) *Appl. Phys. Lett.* **71**, 1083.
Wulff, G. (1901) *Kristallogr. Mineral.* **34**, 449.
Xia, J.-B. (1989) *Phys. Rev. B* **40**, 8500.
Xie, Q., P. Chen, and A. Madhukar (1994) *Appl. Phys. Lett.* **65**, 2051.
Xie, Q., A. Madhukar, P. Chen, and N. Kobayashi (1995) *Phys. Rev. Lett.* **75**, 2542.
Xie, Q., A. Kalburge, P. Chen, and A. Madhukar (1996) *IEEE Photon. Technol. Lett.* **8**, 965.
Xie, Y. H., S. B. Samavedam, M. Bulsara, T. A. Langdo, and E. A. Fitzgerald (1997) *Appl. Phys. Lett.* **71**, 3567.
Xu, Z. and P. M. Petroff (1991) *J. Appl. Phys.* **69**, 6564.

Xu, Z. Y., Z. D. Lu, X. P. Yang, Z. L. Yuan, B. Z. Zheng, J. Z. Xu, W. K. Ge, Y. Wang, J. Wang, and L. L. Chang (1996) *Phys. Rev. B* **54**, 11528.
Yamada, T., M. Tachikawa, T. Sasaki, H. Mori, and Y. Kadota (1997) *Appl. Phys. Lett.* **70**, 1614.
Yamaguchi, H., M. R. Fahy, and B. A. Joyce (1996) *Appl. Phys. Lett.* **69**, 776.
Yamanishi, M. and I. Suemune (1984) *Jpn. J. Appl. Phys.* **23**, L35.
Yamanishi, M. and Y. Yamamoto (1991) *Jpn. J. Appl. Phys.* **30**, L60.
Yamanishi, M., Y. Osaka, and M. Kurosaki (1990) *Jpn. J. Appl. Phys.* **29**, L308.
Yan, R. H., S. W. Corzine, L. A. Coldren, and I. Suemune (1990) *IEEE J. Quantum Electron.* **QE-26**, 213.
Yang, W. H. and D. J. Srolovitz (1993) *Phys. Rev. Lett.* **71**, 1593.
Yao, J. Y., T. G. Andersson, and G. L. Dunlop (1991) *J. Appl. Phys.* **69**, 2224.
Yusa, G. and H. Sakaki (1996) *Electron. Lett.* **32**, 491.
Yusa, G. and H. Sakaki (1997) *Appl. Phys. Lett.* **70**, 345.
Zahari, M. D. and B. Tuck (1985) *J. Phys. D: Appl. Phys.* **18**, 1585.
Zaitsev, S., N. Yu. Gordeev, V. I. Kopchatov, V. M. Ustinov, A. E. Zhukov, A. Yu. Egorov, N. N. Ledentsov, M. V. Maximov, P. S. Kop'ev, A. O. Kosogov, and Zh. I. Alferov (1997) *Jpn. J. Appl. Phys.* **36**, 4219.
Zeppenfeld, P., M. Krzyzowski, C. Romainczuk, G. Comsa, and M. G. Lagally (1994) *Phys. Rev. Lett.* **72**, 2737.
Zhang, B. P., T. Yasuda, Y. Segawa, H. Yaguchi, K. Onabe, E. Edamatsu, and T. Itoh (1997a) *Appl. Phys. Lett.* **70**, 2413.
Zhang, B. P., W. X. Wang, T. Yasuda, Y. Segawa, K. Edamatsu, and T. Itho (1997b) *Appl. Phys. Lett.* **71**, 3370.
Zhang, G. (1994a) *Electron. Lett.* **21**, 1230.
Zhang, Y. (1994b) *Phys. Rev. B* **49**, 14352.
Zhang, Y., M. D. Sturge, K. Kash, B. P. Van der Gaag, A. S. Gozdz, L. T. Florez, and J. P. Harbison (1995) *Phys. Rev. B* **51**, 1303.
Zhu, Y., D. J. Gauthier, S. E. Morin, Q. Wu, H. J. Carmichael, and T. W. Mossberg (1990) *Phys. Rev. Lett.* **64**, 2499.
Zhukov, A. E., V. M. Ustino, A. Yu. Egorov, A. R. Kovsh, A. F. Tsatsulnikov, N. N. Ledentsov, S. V. Zaitsev, N. Yu. Gordeev, P. S. Kop'ev, and Zh. I. Alferov (1997) *Jpn. J. Appl. Phys.* **36**, 4216.
Zimmermann, R. and D. Bimberg (1993) *J. de Physique IV* **3**, C261.
Zimmermann, R. and E. Runge (1994) *J. Luminescence* **60, 61**, 320.
Zimmermann, R. and E. Runge (1997) *Phys. Stat. Sol. (a)* **164**, 511.
Zrenner, A., L. V. Butov, M. Hagn, G. Abstreiter, G. Böhm, and G. Weimann (1994) *Phys. Rev. Lett.* **72**, 3382.
Zunger, A. (1998) *MRS Bull.* **23**(2), 35.
Zuo, J.-K., R. J. Warmack, D. M. Zehner, and J. F. Wendelken (1993) *Phys. Rev.* **47**, 10743.

Index

2D-3D transition, 63ff, 76ff

absorption, 223, 239ff, 243
   coefficient, 151
   two-photon, 217ff
acoustic phonons, 174ff, 229
alloy fluctuation, 154
angular momentum, 116ff
annealing, 255ff, 297ff
antimony compounds, 82
asymmetric barrier, 121
   dense, 48ff
   dilute, 41ff
artificial alignment, 92ff
atom, 2
atomic force microscopy (AFM), 59ff
Auger process, 165

band structure, 115ff
band lineup, 261, 143
bandwidth, 291ff
biexciton, 141ff, 149
biexciton binding energy, 141ff
Bose distribution, 247

calorimetric absorption spectroscopy (CAS), 223, 239ff
capacitance voltage spectroscopy (CV), 265ff
capacitance, 147
carrier dynamics, 246ff, 163ff
carrier, capture, 246ff
   relaxation, 246ff
cascade laser, 3, 300
cathodoluminescence (CL), 201, 208, 216ff, 226ff, 262
characteristic temperature (T_0), 180, 287ff
chirp, 294ff
cleaved edge overgrowth (CEO), 19, 124ff, 220ff
coherent islands, 20
compliance coefficients, 96, 109, 126
confinement regime, intermediate, 130ff
   strong, 129ff
   weak, 131ff

contact pressure, 97
correlation, 182
corrugated substrate, 33ff, 92ff
Coulomb blockade, 146ff, 266
Coulomb interaction, 128ff
coupled quantum dots, 145ff, 168ff, 173ff, 252ff
critical coverage, 63ff, 76ff
cutoff frequency, 292
cylinder, 97ff

de Broglie wavelength, 1, 4
deep level transient spectroscopy (DLTS), 269ff
deformation potential, 111ff
diamagnetic shift, 172, 235ff
dielectric constant, 128
diffraction, reflection high electron energy (RHEED), 59ff
   X-ray (XRD), 59ff
dipole approximation, 131, 135ff

edge emitting laser, 178ff, 281ff
eight-band $k \cdot p$ theory, 125ff
electric field, 167ff
electron beam lithography, 10ff
electron energy loss spectroscopy (EELS), 60
electron–phonon interaction, 174ff
electroreflectance, 241ff
energy states, excitonic, 137ff
   single particle, 115ff
epitaxy, metal–organic vapor phase (MOVPE), 2, 59, 76ff
   molecular beam (MBE), 2, 59, 61ff
equilibrium crystal shape (ECS), 24ff
etching, 12
exchange interaction, 132
exciton, binding energy, 128ff, 139
   charged, 140ff
   energy levels, 137ff, 167ff, 171ff
   oscillator strength, 130ff
excitonic waveguiding, 196ff

faceting, 24ff
far infrared, absorption, 243ff, 278ff

far infrared (*continued*)
  emission, 246, 302
  quantum dot devices, 278ff, 299ff
Fermi distribution, 157, 190
Fermi energy, 2, 265, 271
few electron systems, 172ff
field, electric, 167ff
  magnetic, 170ff
  strain, 95ff
fine structure, 132
focused ion beam lithography, 11ff
force monopole, 28ff
Fourier spectroscopy, 240, 243ff
Frank–van der Merwe growth mode, 37ff

g factor, 238
gain, 179, 184ff, 281ff
  compression, 292
  material, 179, 181, 281
  mechanism, 193ff
  modal, 179, 281
  saturation, 185, 282
Ge on Si, 82ff
glass, 4
graded index separate confinement heterostructure (GRINSCH), 279
growth mode, Frank–van der Merwe, 37ff
  Stranski–Krastanow, 20, 37ff
  Volmer–Weber, 37ff

harmonic oscillator, 115ff, 134ff, 232
Hartree approximation, 147
high index substrates, 17, 75, 79ff

III–V on Si, 84
II–VI compounds, 84ff, 129, 219ff
image charge, 128, 147
interaction, Coulomb, 128
  exchange, 132
  island-island, 48ff
interband transitions, 148ff
interface condition, 104
interface fluctuations, 19, 212ff
intermixing, 15ff
inter-sublevel scattering lime, 159ff, 165, 249ff
inter-sublevel transitions, 154ff, 243ff
intrinsic surface stress, 26ff
inversion, 180

kinetic theories of ordering, 51ff
Kramers–Kronig relations, 179

Landau level, 273
Landau regime, 171
laser, 177ff, 279ff
  cascade, 3, 300
  chirp, 294ff
  dynamic properties, 291ff
  edge emitting, 178ff, 281ff
  high power, 291
  inter-sublevel IR, 299ff
  quantum dot, 4, 177ff, 279ff
  quantum well, 178, 279
  static properties, 281ff
  vertical cavity surface emitting (VCSEL), 179, 295ff
lasing threshold, 179, 184
lifetime, 132ff
line width, homogeneous, 228ff
  inhomogeneous, 152, 205, 252
lithography, 10ff
LO phonons, 174ff, 232ff, 246
longitudinal modes, 290
loss, 179, 284, 286
Luttinger Hamiltonian, 117ff, 171

magnetic field, 170ff
magnetoluminescence, 209ff, 235ff
master equations, 160ff
matrix diagonalization technique, 142
metal-organic vapor phase epitaxy (MOVPE), 2, 59, 76ff
microstates, 160ff
minizone, 2, 3
misfit dislocation, 20
modulation bandwidth, 291ff
molecular beam epitaxy (MBE), 2, 59, 61ff

nano-imprint lithography, 10
negative differential resistance, 2, 3
nitrides, 85ff
non-thermal distribution, 155ff, 192, 284

optical confinement factor, 178, 183ff, 282
optical transitions, excitonic, 128ff
  far infrared, 154ff, 243ff
  interband, 148ff, 198ff
  inter-sublevel, 243ff
  quantum dot ensemble, 150ff
  single dot, 199, 201ff, 215ff, 225ff, 228ff
ordering, lateral, 48ff
  shape, 43ff
  size, 44ff
  vertical, 86ff

INDEX

oscillator strength, 130ff, 251
Ostwald ripening, 44ff

patterned substrate, 17, 93ff
perturbation theory, 129, 138, 141
phonon bottleneck, 174ff, 202, 224
phonons, acoustic, 174ff, 229, 246
phonons, LO, 174ff, 232ff
photoconductivity, 243
photocurrent, 278
photodetector, 278ff
photoluminescence (PL)
photoluminescence excitation spectroscopy (PLE), 215ff, 226ff, 232ff
piezoelectricity, 114ff, 121
Poisson ratio, 96
polarization, 216ff, 219, 299
population statistics, 158ff, 187ff
pseudopotential theory, 125ff

quantum confinement, 115ff
quantum dot, array, 37ff, 48ff, 87
  buried, 53
  conical, 118ff, 139
  coupling, 145ff, 168ff, 173ff, 253
  ensemble, 150ff
  etched, 12ff, 198ff
  formation, 64ff, 76ff
  harmonic oscillator, 115ff, 134ff
  in glass, 4
  intermixed, 15ff, 201ff
  laser, 177ff, 279ff
  lens-shape, 122
  natural, 19ff, 212ff
  on patterned substrate, 17, 93ff, 208ff
  photodetector, 278ff
  pyramidal, 100ff, 104ff, 119ff
  self-organized, 19ff, 59ff, 222ff
  spherical, 97ff, 116ff, 137ff
  stress-induced, 16ff, 123, 202ff
  two-fold CEO, 19, 124ff, 220ff
  type-II, 143ff, 260ff
quantum efficiency, 286ff
quantum well, 9, 97ff, 137
quantum well laser, 178ff, 279
quantum wire, 3

rate equations, 158ff
recombination, 249ff
recombination current, 162ff
reflectance anisotropy spectroscopy (RAS), 59ff

reflection high electron energy diffraction (RHEED), 59ff
relaxation kinetics, 246, 163ff
relaxation oscillation, 291ff
resonant tunneling, 2, 270ff
ripening, 44ff, 51
Rydberg energy, 144

scanning near-field optical microscopy (SNOM), 213, 238
scanning tunneling microscopy (STM), 59ff
selection rules, 205, 213
selective growth, 17ff, 208ff
self-assembly, 5
self-ordering, 5
self-organization, 5, 22ff, 69ff
sequential tunneling, 3
shape fluctuation, 152ff
single electron transistor, 4
size distribution, 150ff, 183, 224, 232
size quantization, 3, 4, 6, 115ff
spherical quantum dot, 97ff, 116ff, 137ff
spin splitting, 172, 238ff
Stark shift, 167ff, 169, 220
state filling, 159ff, 211, 231ff
step bunches, 32ff
stiffness coefficients, 96
Stokes shift, 241
strain, atomistic, 107ff
strain, biaxial, 100
  compressive, 100
  continuum mechanical, 95ff
  distribution, 95ff
  energy, 98
  hydrostatic, 98, 111
  impact on bandstructure, 111ff
  impact on phonons, 113ff, 174, 234
  tensile, 100
  tetragonal, 100
Stranski–Krastanow growth mode, 20, 37ff
stressor, 16ff, 202ff
stress-strain relations, 96
sub-monolayer coverage, 61ff
substrate bending, 96
superlattice, 3, 222
surface, domain, 28, 35ff
  strained, 53ff
surface photovoltage spectroscopy, 243
surface strain, 106ff

theory, 8-band $k \cdot p$, 125ff
theory, pseudopotential, 125ff

thermal distribution, 155ff
thermal evaporation, 6, 268
threshold current, 281ff
time-resolved luminescence, 202, 206, 220, 233, 246ff, 262ff
transients, 163ff
transmission electron microscopy (TEM), 59ff, 70
transparency current, 179
transport, lateral, 274ff
    vertical, 270ff
tunneling, 270ff
type-I structure, 143, 222ff
type-II structure, 143ff, 260ff

uniformity, 6, 7

valence force field model, 107ff
variational calculation, 131, 139

vector potential, 170
vertical cavity surface emitting laser (VCSEL), 179, 295ff
vertically correlated growth, 53ff, 88ff, 106ff
vertically coupled quantum dots (VCQD), 56ff, 86ff, 252ff
Volmer–Weber growth mode, 37ff
volume filling factor, 6, 7

waveguide, 180

X-ray diffraction (XRD), 59ff, 89ff
X-ray lithography, 9

Young's modulus, 37, 96, 104

Zeeman splitting, 235ff